ROYAL HISTORICAL SOCIETY

STUDIES IN HISTORY

New Series

THE BIRTH OF MILITARY AVIATION

THE
BIRTH OF MILITARY AVIATION

BRITAIN, 1903–1914

Hugh Driver

THE ROYAL HISTORICAL SOCIETY
THE BOYDELL PRESS

First published 1997

A Royal Historical Society publication
Published by The Boydell Press
an imprint of Boydell & Brewer Ltd
PO Box 9   Woodbridge   Suffolk IP12 3DF, UK
and of Boydell & Brewer Inc.
PO Box 41026   Rochester   NY 14604–4126, USA

ISBN 0 86193 234 X

A catalogue record for this book is available
from the British Library

Library of Congress Cataloging-in-Publication Data
Driver, Hugh, 1961–
    The birth of military aviation : Britain, 1903–1914 / Hugh Driver.
        p.   cm. – (Royal Historical Society studies in history,
    0269–2244. New series)
    Includes bibliographical references and index.
    ISBN 0–86193–234–X (hardback : alk. paper)
    1. Aeronautics, Military – Great Britain – History. 2. Military-
industrial complex – Great Britain – History. I. Title.
II. Series.
UG635.G7D75 1997
358.4'00941–dc21   97–4456

This book is printed on acid-free paper

Printed in Great Britain by
St Edmundsbury Press Ltd, Bury St Edmunds, Suffolk

List of Illustrations

1

Technological and entrepreneurial background

The new age

The opening decade of the twentieth century, the Edwardian period, tends to be viewed in retrospect as the end of an era; yet the actual evidence of the period seems to suggest that the opposite was true, that contemporaries actually viewed 'their' era (so far as they were conscious of living in a definable era) as being the beginning of a new age, the almost discernible dawn of a new epoch. Despite the fact that the nineteenth century is popularly felt to have reached its conclusion in 1914, rather than at the turn of the century,[1] the period can now, as it was then, be perceived as the spring of a new current of civilisation: the contrast between the nineteenth and twentieth century was to be that dramatic. The world itself was to be changed as a direct result of the mass adoption of new technology.

In the mind of contemporaries, particularly in Britain, the great psychological turning point was indeed the turning of the century, coinciding conveniently with the death of Queen Victoria. The twentieth century was ushered in like the sweeping of a new broom, and with it returned a feeling of more youthful vigour, of anticipation and progress; almost, it might be said, of relief. As Peter Clare has written, 'The death of Victoria . . . like the death of Churchill more recently, brought a sense of psychological release from the burden of a past that no longer seemed wholly consonant with the needs of the present.'[2] Edward VII was outward-looking and gregarious; he exuded an air of permissiveness. At the age of sixty he became the constitutional head of state of a nation which had already changed profoundly from that which his mother had inherited some sixty-four years before. By 1901 77 per cent of the population, more than in any other country, was urbanised, and the proportion was rising.[3] The trend was probably long discernible, but the effect was no less disconcerting to many contemporaries. One of the foremost social commentators of the time, and a junior member of Asquith's Cabinet, Charles Masterman, wrote in *The condition of England* in 1909 that 'the multitude . . . all unnoticed ['unnoticed', presumably, to those who considered themselves above the multitude: it can hardly have been unnoticed to the multitude themselves]

---

[1] For instance, Barbara Tuchman begins her narrative, *The proud tower: a portrait of the world before the war, 1890–1914*, London 1966, on this note.
[2] P. Clare, 'The Edwardians and the constitution', in D. Read (ed.), *Edwardian England*, London 1972, 40.
[3] D. Fraser, 'The Edwardian city', ibid. 56.

1

and without clamour or protest had passed through the largest secular change of a thousand years; from the life of the fields to the life of the city'.[4]

Mass urbanisation was the foundation of that prosperity which the Victorian age bequeathed upon its passing; and underpinning everyday life in these growing urban centres was the employment of new technology. Residential suburbia, of all classes, was the child of improved communications and new systems of transport, conveying the assembled masses to their places of work by tram, train, bicycle and, ultimately, omnibus and motor car. Although what may be described as Victorian Whig-derived aspirations of progressive constitutional government had largely been realised, it was – and was seen to be – a time of great economic, social and technological upheaval, and it is only in relation to what came afterwards that this period can be portrayed as the last lingering 'Indian summer' of the nineteenth century.

The virtual realisation of democratic government, inseparable from the increase in population, primary education and varying degrees of literacy, and the development of a mass market economy, had created a climate of consensus politics and popular culture. These trends found their most immediate expression in the expansion of the tabloid press, which improved communications had so transformed that London daily papers became national daily papers. Through this medium proprietors and politicians could at once inform the mass electorate of the various practical manifestations of the age's latest technological advances and popularise their continued development.[5] This was evident in the Northcliffe Press's fervent adoption of the cause of automobiling, which had become firmly established in western society by the advent of the First World War and which perhaps most obviously symbolises the technological growth of this period. What began as a trickle of custom made designs for the wealthy and sporting was revolutionised between 1907 and 1913 with Henry Ford's refinement of the assembly line method of production in his Detroit factory.

One result of this particular technological development was the increasing significance of oil as a national commodity. This was to have far-reaching repercussions – and not merely in domestic affairs. In seeking to modernise the Royal Navy Lord Fisher superintended the conversion of a large part of it from coal to oil.[6] By 1914 supplies had become a matter of considerable strategic importance. In the event, Britain was to be dragged into a costly Mesopotamian campaign merely in an effort to maintain them.

On the domestic front, every city corporation invested in vast public works in order to modernise their municipalities through, for example, the

4   C. Masterman, *The condition of England*, London 1909, 96.
5   Fleet Street was at this time at the height of its power, Northcliffe taunting that 'every extension of the franchise renders more powerful the newspaper and less powerful the politician': S. Koss, *The rise and fall of the political press in Britain*, ii, London 1984, introduction.
6   R. F. MacKay, *Fisher of Kilverstone*, Oxford 1971, 436ff. See also 'A new Navy' (1912), Fisher papers, FISR 6/2. This memo was widely distributed to colleagues.

installation of tram-tracks and electrical supply. The latter was soon evident on the landscape in the form of cables and pylons. This electrical supply could provide the power to run virtually any machinery (in contrast to huge and cumbersome steam generators), as well as to both heat and light houses. The appliance of modern technology had, by the war, come to affect all of society.

Within the international community the cumulative effect of rapid technological innovation in industry and elsewhere soon acquired a political dimension as such activity became the pre-eminent gauge of a nation's social and economic vitality. In this respect the technological age saw Britain, the former workshop of the world, in relative decline, particularly in relation to Germany – and, perhaps more significantly, in relation to the huge resources that the United States was now harnessing for its economic and technological growth. Following the nation's political unification German industry expanded at a significant and steady rate. By the turn of the century both her population and national income had doubled, and her industrial production had already surpassed Britain's. There was, furthermore, little evidence in the German schools and universities of the anti-technological bias that had for some time existed in Britain. This partly accounts for the early German advances in, for example, the chemical and electrical industries. Meanwhile, by 1890 American industrial production was 75 per cent of all Europe's (not just Britain's); and by 1910 it was 86 per cent and rising.[7]

The effect of this was felt particularly keenly in Britain, where there was a widespread reluctance to adopt new manufacturing techniques, and where improved transatlantic communications and a stubborn commitment to Free Trade had led to the market being flooded with United States agricultural surplus. Growing reliance on these imports had made British trade – on which depended her very survival, not simply her prosperity – increasingly vulnerable to enemy intervention in time of conflict. The historian C. R. L. Fletcher, in his *History of England*, published for school use in 1911, was not the first to identify what he believed to be a disturbing causality. By depopulating the countryside, urbanisation and industrialisation had, it was argued, 'undermined agriculture and placed the nation's security in the hands of foreign farmers and fleets'.[8] Such sentiments, clearly echoing the earlier fears of industrialism

---

[7] Figures taken from P. Kennedy, 'British and German reactions to the rise of American power', in R. J. Bullen, H. Pogge von Strandmann and A. B. Polonsky (eds), *Ideas into politics: aspects of European history 1880–1950 (essays in honour of James Joll)*, London 1984. See also S. B. Saul, 'The American impact on British industry', *Business History* iii (1960), 19ff, and D. H. Aldcroft, 'The entrepreneur and the British economy, 1870–1914', *Economic History Review* xvii (1964), 124. Aldcroft notes that, in absolute terms, the British economy was still growing. It was only in relation to the growth of these other major economies that the deterioration of Britain's leading industrial position became evident. Export markets, in particular, were becoming increasingly competitive.

[8] C. R. L. Fletcher and R. Kipling, *A history of England*, London 1911, 235, quoted in M. J. Wiener, *English culture and the decline of the industrial spirit 1850–1980*, Cambridge 1981, 84.

articulated by John Ruskin and the socialist craftsman William Morris, and perpetuated by writers like G. K. Chesterton, were increasingly focused on particular political circumstances, taking the debate beyond the confines of Victorian pastoral revisionism. E. E. Williams, for example, had sparked controversy in the 1890s with such provocatively entitled polemics as *Made in Germany* (1896) and *The foreigner in the farmyard* (1897). The concept of diplomatic isolation was worthless, he affirmed, when 'we are humbly dependent on other countries for much of our daily bread'.[9] A series of *Daily Mail* articles by F. A. McKenzie, reprinted in a 1901 pamphlet entitled *The American invaders*, was equally alarmist and even more popularly received. This was all symptomatic of the often ill-defined sense of vulnerability which developed in the wake of the perceived technological and economic growth of Britain's main competitors.

International competition was often viewed through postulated teleological concepts of social evolution. Deterministic prediction was well established in the Victorian era, for example within the Whig tradition of constitutional progress, or, in philosophy, through the Hegelian concept of a rational historical process. Yet coinciding with the technological revolution of the late nineteenth and early twentieth centuries there developed a spate of socioeconomic, geopolitical and biological (Social Darwinistic) teleological ideas relating to scientific rationales of change and progression. Nowhere was this inclination stronger than in the industrial context. The huge natural and economic resources of certain continents were seen by some, notably Halford MacKinder and J. R. Seeley, to portend a progress with far-reaching political ramifications. In the Darwinian parlance of the day, nations had either to adapt to the changing, progressive environment, or face political extinction. Deterministic thinking is equally evident in Benjamin Kidd's influential 1893 study, *Social evolution*. Here he postulates the need for the nation state to cultivate 'social efficiency'.[10] This meant, as much as anything, technological efficiency and adaptability, as therein lay the current of progress. Through technological advancement in, for example, communications, Britain might retain its dominance over a far-flung empire, and hence its dominant global position. Ideas of national survival and prestige became increasingly associated with technological advancement.

However, it is questionable just how deeply into the national consciousness concepts of technological change and adaption initially permeated. Because the rise of the new industrial forces was carried (as were political advances) within a largely harmonious process of change – thereby not tearing the social fabric of British society – the nation's old traditional values were to a significant

9   E. E. Williams, *The foreigner in the farmyard*, London 1897, 1.
10   B. Kidd, *Social evolution*, London 1893, 186–8, quoted in M. Howard, 'Empire, race and war in pre-1914 Britain', in idem, *The lessons of history*, Oxford 1991, 71. See also B. Kidd, *The principles of western civilisation*, London 1902: 'It is not into the end but into the beginning of an era that we have been born', he comments (p. 340).

extent retained even in a rapidly changing environment. Thus a pre-techno-logical culture coexisted with a modern, industrial society, helping to legitimise and prolong anti-industrial sentiments.[11] It has been argued that these were particularly nurtured in the schools and universities.

The nature and extent of any technological backwardness in British educa-tion has been disputed.[12] It is widely accepted that technical and scientific education was, broadly speaking, further advanced in Germany in the period immediately preceding the First World War, but a good deal of technological research was undertaken in Britain at, for example, the University of London and the Manchester College of Technology.[13] Much retrospective criticism was to be levelled at the universities generally for failing to adapt sufficiently to the needs of a modern industrial society. But the truth was that even such science graduates as there were, were for the most part ignored by the indigenous industrial community.[14] Industry was more responsive to the emergent tech-nical colleges of Britain's urban centres. Given the vocational nature of the training these institutions provided, however – often to working apprentices – they have too often been ignored by educational historians, just as they were disregarded by conventional society at the time.

Regarding secondary education in the schools there is less doubt. The problem was most acute in the public schools, to which the sons of socially-ambitious industrialists were invariably sent.[15] These institutions reflected rather than deflected the mores and aspirations of the upper middle classes, and it was against their ingrained, and often unconscious, social conservatism that many of the young men who were to utilise the internal combustion engine were effectively to rebel. In that respect, the emergence of the automo-bile and aviation pioneers of the first decade of the twentieth century was to reveal a significant generation gap.

It was therefore primarily within industry itself that the new technology, in its varied and inter-related forms, was to be forged. In the new 'locomotive industries', indeed, it is possible to discern something of an evolutionary line of progression in technological and industrial development. This led from the

---

[11] This thesis has been convincingly argued by Wiener, *English culture*. See also D. C. Coleman, 'Gentlemen and players', *Economic History Review* xxvi (1973), 93ff.

[12] See Aldcroft, 'The entrepreneur', 119–21, and, from a broader perspective, D. Edgerton, *England and the aeroplane: an essay on a militant and technological nation*, Basingstoke 1991, passim. So far as this revisionist survey addresses pre-1914 developments, however, it extrapolates widely (and polemically) from slender foundations and it is tainted by its uncritical acceptance of Haldane's own evaluation of himself.

[13] Notably King's College London's pioneering research into electrical engineering, which led to the development of the national grid system of transmission: M. Sanderson, 'The University of London and industrial progress, 1880–1914', *Journal of Contemporary History* vii (1972), 246–9. See also idem, *The universities and British industry 1850–1970*, London 1972.

[14] Sanderson, 'University of London', 259–60.

[15] Coleman, 'Gentlemen and players', 101–7; D. Ward, 'The public schools and industry in Britain after 1870', *Journal of Contemporary History* ii (1967), 37–52.

late Victorian modification of the bicycle – which first awoke the technological interest of many later pioneers – to the development of the automobile – a progression through which many aviationary pioneers passed – to, ultimately, the achievement of heavier-than-air powered flight. Earlier technological developments paved the way for the foundation of an aviation industry in both the technological and the intellectual environment that they created: technologically in the establishment of a mechanical manufacturing base and intellectually by providing the sociological framework within which a new breed of mechanical engineer could be inspired and sustained. In the ensuing development the role of such individuals was to be a vital factor, for it was above all progressive individuals who not only pioneered the invention and patenting of the new locomotive devices, but also first sought their practical utilisation and exploitation. In other words, whereas economic historians often refer to the importance of the commercial entrepreneur in industry, in the embryonic locomotive industries, such as the bicycle and, more particularly, the automobile and aviation industries, it was initially left to what may be called the 'technological entrepreneur' not only to develop, but also to exploit his product on the open market.[16] There was, at first, little in the way of support from either established business or government. Consequently any examination of the development of aviation needs to begin by tracing the technological foundations of the industry and the roots within it of those who would later come to figure so largely in its expansion.

The roots of progress: the bicycle

The bicycle has quite a long history. The introduction of the first so-called 'hobby horse' can be traced back to the opening decades of the nineteenth century, but at that time there was no mechanical means of propulsion.[17] It was 1839–40 before Kirkpatrick MacMillan, a Scottish blacksmith from Courthill, Dumfriesshire, devised the first crude pedal-propelled bicycle, incorporating a rear-wheel crank-axle treadle device. It made little immediate impact. The conveyance lacked suspension and, as with later models, the colloquial title 'bone-shaker', which properly belongs to the 1860s, would have been appropriate. The first commercially successful cycle – or 'velocipede', as such vehicles were known – was that devised by the ironsmith-derived Michaux family, in Paris, in 1861. This construction was again two-wheeled (most earlier velocipedes had been multi-wheeled), but incorporated front-wheel pedals. The type was displayed as part of the fringe of the 1867 Paris Exhibition, and

---

[16] On the question of the importance of the entrepreneur in industrial development see Neil McKendrick's introduction to the *Europa Library of Business Biography*, in R. J. Overy, *William Morris: Viscount Nuffield*, London 1976, p. xxxvi. See also Aldcroft, 'The entrepreneur', 113.

[17] D. Roberts, *Cycling history: myths and queries*, Birmingham 1991, 5–13.

was thereafter rapidly imitated.[18] It would take a number of further modifications to transform these early machines into the progenitor of the modern bicycle, among them the incorporation of the wire-spoked tension or suspension wheel, patented by E. A. Cowper in 1868 and simultaneously developed by Meyer of Paris, and the emergence of the pneumatic tyre, patented by J. B. Dunlop in 1888. However, this was effectively the start of the industry.

By adapting to the rapid development of machine technology the city of Coventry was to establish itself as the centre of this new industry. Many of its indigenous employers had for some years been depressed by foreign competition, particularly those working in the ribbon weaving and the American-hit clock, watch and sewing-machine industries. In acute need of opportunities for diversification, they discovered that they could adapt their manufacturing capabilities to the needs of the embryonic bicycle industry. This began in the late 1860s, when the Paris agent of the Coventry Sewing Machine Company, Rowley Turner, discerned a growing French demand for velocipedes and realised that his employers could, without too great a disruption, adapt to meet it.

Rowley Turner had originally gone to Paris as a student. The French capital at the height of the Second Empire held many attractions and interests for an active youth and Turner evidently immersed himself in them, but from the beginning he also maintained informal contacts with his uncle, Josiah Turner, who was the managing director of the Coventry Sewing Machine Company. Thus his official appointment as the firm's continental agent was probably not so much nepotism as recognition of a *de facto* situation. His activities on behalf of the company naturally brought him into contact with representatives of several related businesses, prominent among which were the early Parisian velocipede constructors. In fact the Coventry Sewing Machine Company had been represented at the 1867 Paris Exhibition that had first brought the Michaux velocipede to national prominence. It was doubtless on this occasion that Turner had first perceived the potential for diversification. By the following year he was able to suggest to one of the new firms that they sub-contract surplus orders to the Coventry Sewing Machine Company. This resulted in a contract for 400 machines. Responsibility for the order was delegated to the firm's works manager, James Starley.

James Starley originated from Lewisham, where he had begun his working life as a market gardener. (His family were of Sussex farming stock.) Of innate mechanical bent, he had early on developed watch and clock repairing skills and, in contemporary jargon, become a spare-time 'clock jobber'. From this base he diversified into full-time sewing machine repair work,[19] before moving to Coventry in 1861 with the more entrepreneurially-minded Josiah Turner, who was at that time working for the same sewing machine manufacturers,

---

[18] J. Althuser, *Pierre Michaux et ses fils*, trans. D. Roberts, Kenilworth 1988, passim. For the 1867 Paris Exhibition see Roberts, *Cycling history*, 72.

[19] W. Starley, *The life and inventions of James Starley*, Coventry 1902, 13–14.

Newton Wilson & Co. of Holborn. Drawing on the area's skilled labour they aimed to manufacture improved machines. This was the inception of the Coventry Sewing Machine Company, which was officially registered in 1863.[20]

The company proceeded to develop a succession of Starley patents, and among the future bicycle manufacturers brought to Coventry by Turner and Starley at this time were William Hillman, Thomas Bayliss and George Singer. Starley, indeed, ventured to construct his own four-wheel suspension veloci-pede in 1865: but it was only after the French sub-contract in 1868 that cycle (specifically bicycle) manufacture was adopted in earnest. Starley immediately drafted improvements in design and construction, and early the following year the firm's Articles of Association were amended to officially incorporate the manufacture of these upgraded machines.

The advent of the Franco-Prussian War in 1870 compelled the 'Coventry Machinists', as the firm had by then been redesignated, actively to cultivate the home market. This was soon to be dominated by the 'high bicycle' – the huge front-wheeled 'penny-farthing' as it was later dubbed – constructed in an effort to increase both speed and range potential. In an increasingly competi-tive market, however, the Coventry Machinists quickly lost their leading edge. In fact the most immediately successful of these new designs was the 7 ft 'Ariel', patented by James Starley and William Hillman in 1870. James Starley had left the Coventry Machinists that year and established his own manufacturing company at St Agnes Lane in Coventry, where he continued to produce both sewing machines and bicycles.

The 'Ariel' was the first indigenous all-metal bicycle to be assembly-pro-duced and incorporated wire-spoked tension wheels, a speed-gear and solid rubber tyres. It retailed at about £8 (£12 with the speed-gear), and was subsequently manufactured under licence by the firm of Haynes & Jefferies, of Spon Street, Coventry – their premises becoming the city's first workshop devoted entirely to bicycle manufacture. Despite its success in sporting circles, however, the 'high's' uncomfortable configuration and difficult mounting prevented it from being adopted for everyday use, so over the period 1876–9 the controversial transport entrepreneur, Henry Lawson, redeveloped the low-balanced bicycle, with rear-wheel chain drive. This presently acquired the sobriquet 'safety bicycle' to distinguish it from the 'high' or 'ordinary', as the penny-farthing came to be called, and its superior practicability was soon established, although it was J. K. Starley, James Starley's nephew, who success-fully initiated the commercial exploitation of the type with his production of the diamond-framed 'Rover', first manufactured in Coventry in 1885. There-after, in the words of an early historian of the cycle industry, 'the use of the bicycle expanded from Coventry and the midlands like a ripple on a pond spreads from the spot where a stone is thrown in'.[21] Improved production

---

[20] Ibid. 16.
[21] W. F. Grew, *The cycle industry*, London 1921, 76. For Lawson's early innovations see D. Roberts, *The invention of the safety bicycle*, Mitcham 1990, passim, and *Cycling history*, 58–63.

methods, the development of a versatile components industry (including, by the 1890s, the widespread adoption of the pneumatic tyre), and increased competition soon brought prices down. Thus the reason these new locomotive industries were initially based in the Midlands was above all because therein lay the original manufacturing base.

The next stage of technological development was largely fuelled by the competitive nature of the market. The popularity of the modified bicycle continued to increase, peaking in the boom years of 1895–6. At that point demand was perceived to outstrip supply and many more manufacturing concerns, including some armament manufacturers, such as the Birmingham Small Arms Company, adapted their production and vied for a share of the market.[22] Unsatisfied demand also left the way clear for an American entry into the European market, with the result that, yet again, those companies left behind in what rapidly became an over-subscribed industry had to identify new product-markets to exploit. Fortunately, by this time the motor cycle and automobile had begun to emerge.[23]

It was during this period that an interest in technological development was first aroused in many of those pioneers who were to play a central part in the continued evolution of the mechanical locomotive industries. The Wright brothers' careers, for instance, sprang directly from the extended season of bicycle supremacy. As early as 1892 the brothers had acquired one of the new pneumatic tyre 'safety' bicycles, and within a matter of months they had decided to launch their own bicycle retail business. This soon involved not only supplying and repairing established models, but also designing and manufacturing their own. Indeed, it was as the Wrights' Cycle Company that the brothers continued to conduct their affairs even after aeronautics began to dominate their lives.[24] Their principal rival, Glenn H. Curtiss, followed a similar path. He also established a cycle repair and manufacturing shop, and then passed via motor cycle production to the development of powered flight. The man who was to procure the first English Wright Flyer (the Short-Wright Flyer No.1), Charles Stewart Rolls, had corresponding, if more rarefied, antecedents. Educated at Eton and Cambridge, the young Rolls eschewed the usual activities associated with his class and developed instead a passionate interest in the new technology. While at Cambridge he chose to read for an Ordinary Applied Science degree and in his spare time nurtured his enthusiasm

[22] A. E. Harrison, 'The competitiveness of the British cycle industry, 1890–1914', *Economic History Review* xxii (1969), 288. One can discern a further example of diversification by arms manufacturers in the Austrian Small Arms Factory's acquisition of the Coventry Machinists' 'Swift' cycle patent in 1896. The former gained distribution rights throughout the Austro-Hungarian empire and Italy: *Coventry up-to-date* (trade review), Coventry 1896, 26–7.

[23] S. B. Saul, 'The motor industry in Britain to 1914', *Business History* v (1962), 22ff; D. Thoms and T. Donnelly, *The motor car industry in Coventry since the 1890s*, London 1985, 26–7.

[24] F. C. Kelly, *The Wright brothers*, London 1944, 35–6.

further by embracing bicycle racing – ultimately winning a half-blue and reputedly captaining the university team in 1897.[25] The early career of the future automobile manufacturer, William Morris (Lord Nuffield), could also be cited. He began with a bicycle repair workshop at his family's home in Cowley, Oxford, after having discovered his engineering vocation through reassembling his own cross-framed 'safety' bicycle when still a boy. He was soon not just repairing, but designing and manufacturing bicycles.[26]

The automobile and aero-engine manufacturer, John Siddeley (Lord Kenilworth), owed his industrial ascendancy entirely to the openings provided for him by the late Victorian bicycling environment in which he grew up. Born at Chorlton near Manchester in 1866, he joined the Anfield Bicycle Club (one of the period's best known provincial cycling associations) and, on a machine of his own design, became the first person to cycle from Lands End to John O'Groats. Thereafter he became a draughtsman in the Humber bicycle works at Beeston, in Nottinghamshire, and in 1895 manager of the Rover Cycle Company's racing team, based in Coventry. Diversification into automobile manufacture followed his recruitment by the Dunlop Tyre Company of Coventry, whose managing director, Arthur du Cros, was himself a noted cycling champion whose name would reappear in connection with early aviation. At Dunlop, Siddeley's work involved the adaptation of pneumatic tyres for motor vehicles. By 1900 he was managing director of the Clipper Tyre Company and the following year he became an agent for Peugeot cars. This induced him to devise his own automobile designs, which he marketed under the 'Siddeley Autocar Company' banner. His patents were presently acquired by the Vickers corporation, for manufacture by their subsidiary, Wolseley. Siddeley himself became general manager of Wolseley in 1905, leaving four years later to manage the Deasy (subsequently Siddeley-Deasy) Motor Company of Parkside, Coventry. The war then introduced this firm to aviation work and, through the recruitment of experienced former Royal Aircraft Factory employees such as F. M. Green and S. D. Heron, it eventually developed and mass-produced the 240 h.p. six-cylinder Siddeley-Puma engine.

The Siddeley-Deasy Motor Company was acquired by Armstrong, Whitworth & Co. in the post-war slump, becoming part of the latter's own aircraft subsidiary: but Siddeley was himself in a position to buy out Armstrong's aircraft and motor sub-divisions in 1926. Two years later Armstrong-Siddeley acquired A. V. Roe & Co., which before the war had been one of Britain's foremost pioneering aviation firms. In 1935 Siddeley then sold his controlling

---

[25] Curiously, Lord Montagu of Beaulieu makes light of Rolls's bicycling activities. His use of evidence is, however, somewhat selective: *Rolls of Rolls-Royce*, London 1966, 16–17.

[26] P. Andrews and E. Brunner, *The life of Lord Nuffield: a study in enterprise and benevolence*, Oxford 1955, 37–42. For further examples of the same progression see J. B. Rae, 'The engineer-entrepreneur in the American automobile industry', *Explorations in Entrepreneurial History* viii (1955), 1–11.

interest in this conglomeration to Thomas Sopwith's Hawker Engineering Company, and from this further amalgamation Hawker-Siddeley was formed.

Alliot Verdon-Roe, perhaps the most distinguished pre-war pioneer of British aviation, himself acknowledged an early interest in bicycles as a seminal influence. When still an apprentice with the Lancashire and Yorkshire Railway Locomotive Works he would devote as much of his spare time as possible to them, winning in the course of his activities several hundred pounds worth of prizes from cycling events. The money gleaned in this way helped finance his early aeronautical experiments.[27] The controversial aviation journalist, C. G. Grey, and the early aero-engine designer and manufacturer, Gustavus Green, were likewise inspired. Grey began his working life as a draughtsman with the Swift Cycle Company of Coventry (formerly the Coventry Machinists), while Green started in the trade with a bicycle repair shop at Bexhill in Sussex.

The bicycle's very availability created an environment which influenced a whole generation of those with any degree of engineering aptitude towards furthering the development of mechanical propulsion technology. By 1897 the shipping industrialist-cum-sociologist, Charles Booth, was recording in his study of the *Life and labour of the people in London* how young unmarried men, earning 18–20s. per week, were in a position to purchase their own bicycles by regular instalments.[28] This was the source of the widening world described in the early novels of H. G. Wells, notably *The wheels of chance*, published in 1896. It was an environment which also helped shape the nation's first mogul of mass-communication, Lord Northcliffe (Alfred Harmsworth: created Baron Northcliffe in 1905).

Northcliffe's interest in cycling in fact predated the introduction of the 'safety' bicycle.[29] When the device's popularity really began to gather momentum, however, it brought with it the first editorship of his career, that of the paper *Bicycling News*. This title had been published since 1876. The young Alfred Harmsworth, who had for some time been contributing cycling-related articles to contemporary journals, acted as the paper's editor or, some report, sub-editor from 1886 to 1888, following its acquisition by the Coventry firm of Iliffe & Son. During this period he took all the major decisions, and was instrumental in promoting the new pneumatic tyre.[30] Thus the considerable

[27] A. V. Roe, *The world of wings and things*, London 1939, 22.

[28] D. Rubinstein, 'Cycling in the 1890s', *Victorian Studies* xxi (1977), 51.

[29] Northcliffe to Lord Montagu of Beaulieu, 12 May 1921, Northcliffe papers, Add. 62165; R. Pound and G. Harmsworth, *Northcliffe*, London 1959, 34–5.

[30] Ibid. 55–60. Iliffes acquired the paper in October 1885. George Lacy Hillier acted as their negotiator and became ostensible editor, but although writing leaders he remained in London. Harmsworth took over all editorial duties – sometimes, as with the promotion of the pneumatic tyre, opposing Hillier: H. W. Bartleet, *Bartleet's bicycle book*, London 1931, 147–8.

impetus Northcliffe was later to give to all forms of technological innovation, particularly the automobile and aeroplane, can again be traced back to an early enthusiasm for bicycling.

Development of the motor cycle and automobile

The bicycle itself was soon to be not so much superseded as further developed by the advent of the internal combustion engine. An early historian of the cycle industry, W. F. Grew, commented in 1921 – with regard to the bicycle and what he termed the 'petrol propelled type of cycle' – that 'the two industries are . . . so closely allied that one hardly knows where one begins and the other leaves off'.[31] In other words, sections of the 'old' industry again adapted to an increasingly dynamic industrial environment. This development came rather speedily on the heels of the bicycle's boom years, which were followed almost immediately by a slump, when there was a fall in demand for the products of an already over-subscribed industry.[32] The earliest motorised cycles were those produced by Daimler and Benz in the 1880s. A variety of British machines were then tentatively marketed throughout the following decade as the commercial situation necessitated further diversification. The Coventry engineering firm of Perks & Birch, for example, constructed a cheap and practical self-propelled cycle, with the engine positioned inside the rear wheel of the machine, in the late 1890s. Commercial production then resulted from the patent's prompt acquisition by Singer & Co., a firm hitherto renowned as bicycle manufacturers. Thus here can be discerned a broad-based mechanical manufacturing concern adapting to meet new needs.[33] The prominent Coventry-based engineering firm of Triumph (consisting of the William Andrews Manufacturing Company, revamped by the German-Jewish expatriate, Siegfried Bettmann – formerly of the White Sewing Machine Company of Cleveland, Ohio), which, like Singer, had diversified into bicycle manufacture, also initiated motor cycle production, as did Lea Francis. Then, just as many of the manufacturing techniques employed in the production of bicycles could be adapted to the production of motor cycles, so with motor cycles to automobiles. Such a development can be discerned, for example, in William Morris's fledgling manufacturing organisation.

At about the turn of the century Morris and his mechanics had become intensely interested in the practicalities of petrol propulsion. In 1900 they

---

31 Grew, *Cycle industry*, 105.

32 'The cycle industry: trade and prices: a note of alarm', *Coventry Herald*, 16 July 1897 (Arthur Lowe collection, 1897. 81ff).

33 The Perks & Birch 'motor bicycle' (officially designated the Singer motor-wheel) was patented in 1899. The rear wheel contained, within its spokes, a single-cylinder engine unit: C. F. Caunter, *Motor cycles: an historical survey*, London 1982, 11; Grew, *Cycle industry*, 106; K. Atkinson, *The Singer story*, Godmanston 1996, 31–40.

they were invariably too heavy, and vice versa. When A. V. Roe embarked upon his first attempted flight trials at Brooklands in 1907 he was forced to employ a lightweight 9 h.p. JAP motor-cycle engine. This was only ever an expedient.[49] It was some time before suitable aero-engines – notably the Gnôme rotary – were perfected, and even then automobile manufacturers continued to target the market.[50]

All these factors are evident in Lanchester's early aviationary development. His interest in the entire subject in fact pre-dated that of the Wrights, his first aeronautical investigations dating back as far as 1892. To a greater extent than the Wrights, or any of the early pioneers, however, Lanchester developed into an aerodynamicist of the highest quality. From his early experiments with model gliders he formulated the 'vortex' or 'circulation' theory of sustentation in flight, which, in the words of the engineering historian, L. T. C. Rolt, was to become the 'foundation of the modern science of aerodynamics'.[51] Yet from the first, Lanchester realised that the actual achievement of heavier-than-air powered flight ultimately depended upon the power-to-weight ratio of any potential engine. Indeed, he had planned to develop a suitably adapted engine as early as 1894, only to be dissuaded by sceptical colleagues.[52]

This was indicative of the problems faced by the early pioneers. If there was little investment in or backing for the incipient automobile industry, there was to be even less for aviation. Figures of influence, such as Rolls, Montagu of Beaulieu and Northcliffe, who did accept the possibility of powered flight, and who used their positions to advance the cause of aviation, were motoring enthusiasts already, whose interests were then carried forward with the development of technology. Thus the development of aviation in Britain can in retrospect be seen as not simply the culmination of a much wider technological progression, but more particularly as having resulted from the efforts of a relatively small group of individuals who had come of age in a new technological-cum-mechanical engineering environment. It was pioneers such as A. V. Roe, the Short brothers, de Havilland, Handley Page and so forth, who may perhaps best be described as 'technological entrepreneurs', rather than merely engineers, who not only brought the aeroplane to some degree of practicability, but also first sought its commercial exploitation. It was they who developed aviation as an industry.

---

[49] A. V. Roe, World of wings, 40. A. V. evidently secured the loan of this engine through the Automotor Journal: H. Penrose, British aviation: the pioneer years 1903–1914, London 1967, 110.

[50] Witness the makes of aero-engines displayed at the last Olympia Aero Show prior to the war: The Aeroplane, 19 Mar. 1914, 328–34. The list includes such established automobiling names as the Austin Motor Co. of Birmingham, Milnes-Daimler, Mercedes Ltd, Renault, and the Sunbeam Motor Co. of Wolverhampton. See also A. Nahum, The rotary aero engine, London 1987.

[51] Rolt, Victorian engineering, 277; P. W. Kingsford, F. W. Lanchester: a life of an engineer, London 1960, passim.

[52] Claim for recognition, Lanchester papers, 1/12/2; Kingsford, Lanchester, 30–1.

2

The establishment of an aeronautical environment

The early pioneers

So much has been written of aviation's first pioneers that little is to be gained from retelling the story of the invention of the aeroplane *per se*, through the nineteenth century and up to the successes of the Wright brothers.[1] The more immediate concern is instead with following the thread of industrial progression through to the beginnings of a military application for heavier-than-air machines. This involves examining what might be called the second wave of pioneers, comprising those figures who were to begin to develop aviation as an industry.

The most influential individuals in this group in Great Britain were probably A. V. Roe and J. T. C. Moore-Brabazon. They were in many ways the precursors of what was to follow. Brabazon, from a privileged and cosmopolitan background, was to travel to France, where he became the confidant of the early French pioneers, Gabriel Voisin and Henri Farman. In collaboration with C. S. Rolls, he was to set the Short brothers on the road to aeroplane manufacturing. An account of his early years effectively unites many of the otherwise disparate elements involved in the emergence of British aviation in these years. Roe, on the other hand, emerged from a staid, provincial background, and was to struggle in near isolation and, at times, against open hostility, to establish what was to become by the advent of war the most influential design of aircraft: the tractor-biplane. Through these contrasting yet complementary lives can be traced the beginning of the hesitant and gradual re-emergence of Britain as a leading aviationary nation.

A. V. Roe

Alliott Verdon-Roe was born in the Patricroft district of Manchester on 26 April 1877. Patricroft was a prosperous but rapidly industrialising suburb, situated between Eccles and Irlam. The area would shortly by transformed by the building of the Manchester Ship Canal. It has often been assumed that

---

1 The best introduction to the subject can be found in the writings of Charles Gibbs-Smith, notably in *The invention of the aeroplane, 1809–1909*, London 1966, and *Aviation: an historical survey from its origins to the end of World War II*, London 1970.

Roe's father was himself an engineer or industrialist, but he was, in fact, a local GP. Alliott was thus born into a decidedly middle-class home.[2]

The family seem to have been thrifty and hard-working, but with perhaps unusually progressive views on many social issues. Alliott's mother, Sofia Verdon-Roe, felt obliged to found and run what has been called an urban orphanage, although the precise object, as Alliott's brother, Humphrey Verdon-Roe, later attested, seems to have been to provide a day care centre, or crèche, for the children of working mothers (women whose circumstances necessitated their working).[3] In this one can perhaps discern the source of Humphrey's later agitation for the municipal adoption of family planning clinics. He was subsequently to marry the birth-control pioneer, Marie Stopes, whom he had met in the course of his work in this field.[4]

The family presently moved to Earl's Court, in west London, and later to Clapham Common, and A. V. and H. V. Roe both attended St Paul's School. Neither was happy there, and A. V., the elder by a year, was the first to leave at the age of fourteen.[5] In 1892 he sailed for British Columbia to train as a civil engineer. This scheme, however, fell through and by the following year he was back in England, apprenticed to the Lancashire and Yorkshire Railway Locomotive Works at Horwich, near Manchester. He lived in local digs for the next five years, devoting nearly all his spare time to riding and amending his newfangled pneumatic-tyred safety bicycle. This interest soon devoured the greater part of his energies, and within a short time he had established a local reputation as both a skilled cyclist and mechanic. Indeed, through winning a variety of the then numerous cycling competitions held up and down the country he was able to accumulate the small nest-egg with which he would later finance his earliest aviationary experiments.[6]

At the conclusion of his apprenticeship, however, A. V.'s immediate desire was to become an engineering officer in the Royal Navy. He therefore spent some time as a fitter at Portsmouth dockyard and studied marine engineering at King's College, London. When he was subsequently turned down he became a merchant seaman instead, securing a position as 5th Engineer on ships of the British & South African Royal Mail Company. It was during this period at sea, Roe later recalled, that his all-consuming passion for aviation flowered.[7] Seeing the albatross glide in the wake of the company's ships induced him to build

---

[2] H. V. Roe to C. G. Grey, 18 Jan. 1916, AVRO papers, file 47. He had been christened Edwin Alliott Verdon Roe (Verdon was his mother's maiden name). As his father was also Edwin, the Christian name Alliott was generally adopted. He officially assumed the surname Verdon-Roe in 1933, but had long been known by that title. He was the fourth of seven children.

[3] H. V. Roe, 'A couple of pioneers', AVRO papers, file 110.

[4] A. Maude, *The authorised life of Marie C. Stopes*, London 1924, 115–16.

[5] A. V. Roe, *World of wings*, 19; H. V. Roe, 'A. V. and H. V.', AVRO papers, file 110.

[6] A. V. Roe, *World of wings*, 22; Avro reminiscences, PRO, AIR 1 2310/218/1.

[7] A. V. Roe, *World of wings*, 23; Avro reminiscences; *Times*, 6 Jan. 1958.

imitatory glider-models, of various wing and tail dispositions. Before long he was completely captivated by the experience. Resolving to continue this experimental work he left the Merchant Navy in 1902, and became a draughts-man in the nascent automobile industry, joining the firm of Brotherton (sometimes reported as Brotherhood)-Crockers Ltd. By this means he hoped to acquire such practical mechanical engineering experience as he already knew would be essential to the furthering of his incipient ambitions. Few of his colleagues took more than a cursory interest in his ideas, although he did succeed in designing an improved gear-change mechanism, for which he was to receive royalties. Indeed, by contrast to Brabazon, confidant of the leisured ballooning set, Roe was to begin his serious aviationary endeavours in some penury and professional isolation. It was his family, particularly his younger brother, Humphrey, who first took an active interest in his efforts.

This is of more than incidental significance. H. V. Roe was to play an increasingly important part in the financial development of Alliott's schemes. The pair had initially kept in only intermittent communication since their schooldays. Humphrey had become a regular soldier with the Manchester Regiment and been posted to South Africa to serve in the Boer War. Indeed, he was with the regiment throughout the siege of Ladysmith, and was therefore probably a witness to the use of captive balloons by the British army. Using the £200–£300 he had accumulated during his service, he then resigned his commission and in 1902 assumed control of a Manchester firm of webbing manufacturers, Everard & Co. As a professional and increasingly adroit busi-nessman he was soon in a position to offer his brother progressively useful advice as to how he should realistically pursue his aviationary aims. In this almost inadvertent manner he was to become one of the first financial speculators to appreciate both the innovative quality and commercial poten-tial of A. V.'s work in particular – and of aviation in general.[8]

The Roe family seem to have considered A. V. rather naive in financial matters and, in fact, encouraged H. V. to keep a fraternal eye on him. This aspect of the brothers' relationship was to give rise to a good deal of tension and ultimately lead to complete estrangement – resulting in the loss of the family's controlling interest in A. V. Roe & Co. Ltd. Perhaps already aware of the difficulty of infringing on A. V.'s independence, Humphrey was initially unwilling to offer anything more than advice;[9] he became financially entan-gled in the new undertaking almost in spite of himself.

For his part, A. V. soon tired of being patronised by people who clearly regarded him as a self-indulgent fantasist. The insecurity of his position made him increasingly susceptible to slights, imagined or otherwise. So, in an effort to elicit some measure of moral support, he wrote to the Wright brothers in December 1903. This was the month of their first heavier-than-air powered

---

8 H. V. Roe to C. G. Grey (draft), 11 May 1928, AVRO papers, file 64; H. V. Roe, 'I financed an aircraft pioneer', ibid. file 110.
9 Ibid.

flight and he may have been prompted by the incomplete reports of the event published in such prominent newspapers as the *Daily Mail* and the recently launched and as yet far from established *Daily Mirror* within two days. Roe continued to write periodically to the Wright brothers over the next decade. Through most of the early years of his aviationary development, however, he remained an isolated figure. Undeterred, he persisted in maintaining that not only had the Wrights achieved powered flight but that he could achieve it also. Rashly, he attempted to air this view in the letters column of the London *Times*. Writing at the close of 1905 he confidently dismissed the long-term value of lighter-than-air flight and declared that the future lay unquestionably with what he called 'the aeroplane system'. If immediate steps were taken, he affirmed, after describing his own experiments, there was no reason why a motor-driven machine could not be 'gliding over England' by the following summer.

The response can hardly have been as Roe had envisaged. The letter appeared tucked away in the specialist columns of the *Times Engineering Supplement* on 24 January 1906, coupled with an editorial disclaimer. 'It is not to be supposed', a footnote ran, 'that we in any way adopt the writer's estimate of his undertaking, being of the opinion, indeed, that all attempts at artificial aviation [*sic*] on the basis he describes are not only dangerous to human life, but foredoomed to failure from the engineering standpoint.'

With characteristic understatement A. V. was later to admit that 'this was certainly not encouraging', but he refused to be deflected.[10] Indeed, rebuffs made him the more determined. By this time he was searching in all directions for some form of employment concerned exclusively with aeronautics. On 3 February 1906 he wrote to the Wrights again, vainly offering himself as their agent in Britain, and when that failed (he was informed that they were planning to sell their invention directly to national governments as a piece of military hardware)[11] he applied for the recently advertised position of secretary to the Aero Club. This proved more successful. He was called for interview, where he discovered that his principal interviewer was to be none other than Brabazon's best friend, C. S. Rolls – by then already a well-known personality on account of his automobiling exploits. Also on the panel was another Aero Club committee member, Stanley Spooner, who went on to edit the journal *Flight*. Roe's aeronautical enthusiasm was patent (he presumably suppressed his dismissive views on ballooning) and Rolls offered him the appointment. The position, although poorly paid, would entail only two or three hours' work each day, leaving Roe sufficient time to develop his own interests.[12] But no sooner

[10] A. V. Roe, *World of wings*, 36–7.
[11] This exchange of letters was relayed to Lord Brabazon by M. W. McFarland, head of the Library of Congress's aeronautical section and editor of the Wrights' papers: communication dated 13 May 1958, Brabazon papers, box 26 (1). Brabazon was at that time investigating A. V. Roe's past in an effort to discredit his 'first flight' claim.
[12] P. Jarrett, 'A. V. Roe: the early years', from *Brooklands*, Brooklands Museum Trust 1992.

had he taken up his new duties than the idiosyncratic Scotch inventor, George Louis Outram Davidson, asked A. V. – apparently on the basis of his recent letter to the *Times* – to become his draughtsman instead. Davidson was hoping to build a twin-rotor 'Gyropter', a helicopter-like vertical take-off airliner, with what has been described as 'turbine-like' blades, having somehow persuaded Sir William Armstrong, chairman of the armament manufacturers Armstrong, Whitworth & Co., to back such a project.[13]

This prospect seemed ideal, and Roe had no hesitation in opting for it in preference to the Aero Club appointment, which retained a negative ballooning emphasis. However, he was soon disappointed. Davidson not only refused to revise his utterly impractical ideas in the light of Roe's advice, he also reneged on the agreed remuneration. A. V. had been drawn to his employer's American base at Montclair, near Denver in Colorado, on the promise of a regular salary on top of expenses. He would thus have been able to accumulate sufficient funds to construct his own aeroplane. He was ultimately forced to initiate legal proceedings against Davidson, but even then it was six months before he received his accumulated pay.[14] For practical purposes he therefore found himself both out of work and penniless. In desperation he wrote in November 1906 to B. F. S. Baden-Powell, president of the Aeronautical Society. 'I am very anxious to join someone in making and experimenting with a motor driven aeroplane', he explained, 'and now you are going in for this perhaps we could come to some arrangement, or do you think I could join the staff at Aldershot?'[15] Baden-Powell sympathised, but he had in fact retired from the army and, in any case, unbeknown to Roe he had been out of favour with the Royal Engineers' Balloon School and Factory since his criticisms of the sections in the Boer War. There was thus little he could do, at least in that respect. None the less, he became one of Roe's regular correspondents.

At this point an event was announced that gave A. V. the opportunity both to make his name known to a wider circle and to gain sufficient funds to construct his first full-sized machine. The editors of the *Times* notwithstanding, there was one figure in Fleet Street who did believe in the possibility of heavier-than-air powered flight, and who was prepared to listen to and support would-be aviators, and that was Lord Northcliffe. Realising the publicity value of the new emerging technology, and inspired by Santos-Dumont's autumn 1906 hops with his peculiar '14-bis' carnard-boxkite amalgam, the *Daily Mail*

---

[13] H. V. Roe to G. Dorman, 13 May 1948, AVRO papers, file 56; A. V. Roe, *World of wings*, 28–9; Penrose, *Pioneer years*, 86.

[14] Ibid. 86–7. Davidson's idiosyncratic efforts, already notorious, continued in various guises for some years: P. Jarrett, 'Full marks for trying', *Aircraft Annual '76*, Shepperton 1975, 28–36.

[15] Letter of 18 Nov. 1906, quoted by J. Laurence Pritchard in his obituary of A. V. Roe: *RAeS Journal* lxii (1958), 232. The *Aeronautical Journal* was founded in 1897. The Aeronautical Society received its Royal prefix in 1918 and the publication's title was amended to *Journal of the Royal Aeronautical Society* in 1922. For convenience it is hereinafter cited in the footnotes as *RAeS Journal*, irrespective of date of publication.

– under its proprietor's impetus – announced in late 1906 the modest first of what was to become an increasingly spectacular series of international aviationary awards. Prizes to the value of £250 were offered for model aeroplanes capable of achieving self-sustained flight in public competition. A. V. jumped at the chance and immediately entered a batch of his own devices.[16]

The trials took place on 15 April 1907 at the Alexandra Palace in north London, following a preliminary publicity-seeking exhibition held as part of the International Motor Car Exhibition at the Agricultural Hall in Islington. It was stipulated that flights of at least 50 ft, later widely reported as 100 ft, had to be achieved before the £150 first prize could be awarded. In the event the majority of the 200 entries nose-dived upon release, only Roe's performing with anything approaching a degree of utility. His most successful model was of an essentially Wright-derived pusher-biplane configuration, incorporating a forward-elevator. It had a wing-span of some 9 ft and was powered by a length of twisted elastic (presumably procured from H. V.'s webbing firm) encased in what Roe described as 'a braced frame of triangular section' 8 ft in length. The first prize was plainly his, and with the expected money he could at last set about reconstructing the best features of both this and his other models in a full-size, man-carrying prototype.[17] No sooner had the competition closed, however, than A. V. experienced another of those reverses that were to punctuate his career. The Aero Club judges declared that none of the models entered was of sufficient merit to justify receiving the full financial award. Roe was offered a consolation prize of £75.

A. V. was not a person to harbour resentments but this slight wounded him deeply. He could not, he said, help feeling that it was, to say the least, unfortunate that the judges were more interested in ballooning than in any experimental work with heavier-than-air machines. 'They certainly did not', he insisted, 'realise the difficulties with which one was confronted at that time in making a model aeroplane fly.'[18] Indeed, far from being sympathetic to the needs of genuine pioneers, the trials seem to have been conducted in a very haphazard manner, serious participants being bracketed with the absurd and allotted a minimal amount of time to test and prepare their machines on site.[19]

The Superintendent of the Royal Engineers' Balloon Factory, Colonel J. E. Capper, was given leave to attend the competition, and his dismissive War Office report tends to corroborate both A. V.'s and the judges' misgivings.

---

[16] Jarrett ('A. V. Roe') concludes that Roe entered as many as five models. For the competition's regulations see leaflet preserved in Butler collection of newspaper cuttings, vol. v. 14. Ominously, it concludes that 'The judges reserve the right not to award the prizes if in their opinion the models are not of sufficient merit.'

[17] Roe had written to Baden-Powell as early as February 1907 to say that his model flights had proved so successful that, even with an 8 h.p. engine, which was all he could afford, he was hopeful of achieving full-scale powered flight: A. V. Roe to B. F. S. Baden-Powell, 6 Feb. 1907, RAeS archive.

[18] A. V. Roe, *World of wings*, 33–4.

[19] Jarrett, 'A. V. Roe'.

'Taken all round', Capper commented, 'I do not think we have learned very much from the machines exhibited. . . . The only [model] to my mind that flew with real promise was . . . Mr A. V. Roe's back steering [pusher] machine.' This, he continued, was an adaptation of the Wrights' glider, and it demonstrated that such a device, 'when properly constructed', will float 'under its own power' through the air. The remainder of the event, save for 'a very promising model . . . shown by Mr Weiss' (José Weiss, who was soon to enter into an experimental association with the young Handley Page), was briskly dismissed with a mixture of faint praise and sarcasm.[20] The true nature of the judges' interests was confirmed when the Aero Club's own gold medal – separate from the *Daily Mail* award – went to the Short brothers for a ballooning exhibit. By that time Short Brothers had almost become a subsidiary of the Aero Club anyway.

Despite his disappointment A. V. Roe went ahead with the construction of a full-size, single-propeller, pusher-biplane based on the most successful of his models.[21] For the period immediately before and after the competition he based himself at the home of another brother, Dr Spencer Verdon-Roe, at 47 West Hill, Wandsworth, where there was a sizeable mews, which A. V. was able to convert into a workshop.[22] A neighbouring cycle-shop tradesman, William Daws, undertook the aircraft's more demanding welding–engineering assembly work. The finished machine proved to be a for the most part wooden-framed structure, stiffened by wire-bracing, and with the upper-wing surfaces punctuated by kingposts, i.e. vertical struts, from which the bracing derived. It had an upper wing-span of 36 ft and a lower wing-span of 30 ft, combined with a forward-elevator-cum-steering-plane of about 25 ft span.[23] The pilot's place was in the van of the aircraft's frame, beneath the forward-elevator, and the engine (when obtained) was to be mounted above and behind him, in mid-gap just in front of the mainplanes. The frame itself incorporated what amounted to the first all-purpose control column, or, in effect, joystick, with which the forward-elevator could be warped or have its angle of incidence changed to effect lateral or vertical control. This was placed horizontally in the cockpit area of the machine and culminated in a hand-wheel; if this wheel was rotated the forward-elevator warping was operated, while if it were raised or lowered the angle of incidence was changed. (There was no rudder.)

A. V. had patented this dual-control steering column as early as November 1906. It thus antedated by a year Robert Esnault-Pelterie's primitive two-column system. None the less, the Frenchman was later to claim patent

---

[20] PRO, AIR 1 729/176/4/2/1.

[21] A. V. wrote to B. F. S. Baden-Powell on 16 Apr. 1907, describing his models. The letter is reproduced in *RAeS Journal* lxii (1958), 233.

[22] A. V. Roe, *World of wings*, locates this workshop as Putney (pp. 32, 51). The confusion is understandable. A. V. was now living, at least some of the time, with his parents, who had moved to Oakhill Road, east Putney. This was near the Wandsworth workshop site: the districts merging. For these locations see P. Loobey, *Flights of fancy: early aviation in Battersea and Wandsworth*, Wandsworth 1981, 23–6.

[23] Jarrett, 'A. V. Roe'.

infringement against virtually every aircraft manufacturer, becoming, in the process, a considerable irritant.[24] A. V.'s more immediate concern, however, was to find a suitable environment in which to begin active trials with his aeroplane prototype. On learning that the Brooklands Automobile Racing Club were offering a prize of £2,500 to the first aviator to achieve a circular flight of their recently completed motor track, he decided that that would be his best hope. The 365-acre Brooklands circuit had been constructed in what was effectively a low-lying basin. This created a milieu of still air, which would soon attract many early aviators. Indeed, the *Graphic* and *Daily Graphic* newspapers, controlled by the future aviation publicist and industrialist, George Holt Thomas, were also offering £1,000 to the first aviator who could fly between two given points – a mile apart – on the track, although this prize does not seem to have attracted quite the same amount of publicity.

The BARC flight prize was the brainchild of the well-known automobile advocate, Lord Montagu of Beaulieu, who was at this time the club's vice-president and editor of his own automobiling journal, *The Car Illustrated*. Details were, in fact, announced in *The Car Illustrated* as far back as 27 November 1906, and, through Montagu working in conjunction with his friend and BARC associate, Lord Northcliffe, they appeared simultaneously in the *Daily Mail*. The scheme was to run from the day the circuit opened in July 1907 to the end of the year.

The aim of the prize was obviously to generate publicity for the new track, and in this it was successful. However, that generosity of spirit with which Lord Montagu concluded the organisation's offer, insisting that the club would place at the disposal of 'aerial navigators' 'ample accommodation' for the housing of aeroplanes,[25] Roe soon found to be less than forthcoming. The clerk of the course, an overbearing Austro-Pole named E. de Rodakowski, grudgingly gave him permission to erect a small shed-cum-workshop on the edge of the circuit's internal finishing straight, alongside what was called the judges' box, but without any security of tenure; and his hostility set the tone for the whole of Roe's initial period of residency. It is pertinent to note that Roe was not the first would-be aviator to be granted the use of the club's facilities. An eccentric Frenchman named M. Bellamy had already visited the track with his own bizarre 50 h.p. Panhard automobile-engine 'flying machine'. Unfortunately he was more charlatan than pioneer and quickly departed, leaving many unpaid bills and a considerable accumulation of animosity.[26] Nevertheless A. V. Roe pressed ahead and in September 1907 his first aeroplane – styled 'Avroplane'

---

[24] H. V. Roe memo, 29 Aug. 1931, AVRO papers, file 195; C. G. Grey to H. V. Roe, 8 Sept. 1931, ibid.

[25] 'Wealth for aeroplanists – new £2,500 prize for a three-mile flight', *Daily Mail*, 27 Nov. 1906 (Butler collection, iv. 125).

[26] W. Boddy, *The story of Brooklands: the world's first motor course*, i, London 1948, 54.

– was re-erected on the site; a sprung four-wheeled undercarriage completed the process.[27] By December trials had begun.

It was at Brooklands that A. V. first encountered his contemporary and rival, J. T. C. Moore-Brabazon. Brabazon was already well-known to BARC members as a prize-winning motor-racing driver. Indeed, he had driven in the circuit's opening ceremony in July 1907, some three months before Roe arrived on the site. His early aviational endeavours, undertaken after Roe's arrival, were therefore received with at least a degree of cordiality, although he was later to play this down as his rivalry with Roe developed.

A. V. himself had managed to acquire a 9 h.p. JAP twin-cylinder motor-cycle engine by the time of his transfer to Brooklands, but the power it provided, combined with his own inefficient paddle-blade propeller, proved insufficient for immediate results. He also initially failed to realise that lift derived from pressure on the top-surface of an aeroplane's wing – more so than on the under-surface – and that any exposed struts or ribs would therefore act as drags on the aircraft. So in order to gain experience of how the machine actually handled in flight he sometimes contrived to have it towed into the air by a car, with himself still in the pilot's seat. This was a practice not without danger to all concerned, and the risks were hardly mitigated by the incorporation of a quick release mechanism and the expedient fixture of lightweight wheels to each lower wing-tip, in an effort to protect them against inadvertent ground contact.[28] None the less, by this means Roe successfully discerned his aircraft's limitations. In light of them he planned further modifications and arranged for the loan of a 24 h.p. eight-cylinder Levavasseur-designed Antoinette engine, the most aerodynamically advanced engine-design of its day in terms of power-to-weight ratio. It finally arrived at Brooklands early in May 1908. With its arrival A. V. Roe's career as a pioneer aviator in practice began.

Roe was one of only a mere handful of aviationary enthusiasts on the verge of heavier-than-air powered flight in the years 1907–8, yet, even among his own countrymen, he was still virtually unknown. Generally speaking Great Britain remained uniquely sceptical of the efforts of her own pioneers. Indeed, Brabazon was at this time to reconsider his position and leave for France, where he believed the environment to be more favourable. On the positive side, the War Office was employing J. W. Dunne and S. F. Cody in an effort to create a British army aeroplane, but the work was being conducted in strict secrecy. Britain was not without an entrenched aeronautical community, but it is quite revealing that the renowned American 'prophet' of powered-flight, Octave Chanute, should write to Baden-Powell in May 1907 that while he had heard that there were five or more people building aeroplanes in England he had no idea who they were.[29]

---

27 A. V. Roe, *World of wings*, 39; A. V. Roe to Lord Brabazon, 26 Dec. 1955, Brabazon papers, box 26(1).
28 A. V. Roe, *World of wings*, 40–1; Avro reminiscences; Jarrett, 'A. V. Roe'.
29 O. Chanute to B. F. S. Baden-Powell, 25 May 1907, RAeS archive.

It would, however, be misleading to overstress the situation. Although initially some way behind French and American contemporaries, the somewhat primitive speculations of the first English pioneers were ultimately to develop into the celebrated British-designed aircraft of the First World War. This wider context should be borne in mind. By 1914 the names of Avro, Shorts, Sopwith and de Havilland, together with Fokker in Germany and Sikorsky in Russia, would be dominant, and those of Wright and Voisin little more than history.

J. T. C. Moore-Brabazon

John Theodore Cuthbert Moore-Brabazon was born in London in 1884, of aristocratic Irish descent, and educated at Harrow School. However it is evident that from his earliest days he was devoted to machines and technology to the detriment of any interest in a classical education. This technological bent was not looked upon sympathetically by his father, an officer of the Indian army who had married late and was already sixty years of age when his son was born. Brabazon[30] later remarked sadly that 'although I was devoted to him all my life he could not really share in any of my activities, or indeed in the hopes and probabilities of the changing world. . . . My father, like most of his generation, did not "hold" with motor cars'.[31]

As this suggests, the automobile was the young Brabazon's first great enthusiasm. While still at school he played truant in order to watch the competitors in the 1900 1,000 miles motor trial pass by, and he then had the temerity to lecture the school's scientific society on the subject. Both his housemaster and his father tried to awaken within him the same degree of motivation towards his academic work, but with mixed success.[32] Consideration of what future career the boy might satisfactorily pursue seems to have been the cause of family tensions even from his prep-school days: at that stage he apparently expressed a preference for entering the Royal Navy rather than going on to Harrow.[33] Although that suggestion was firmly quashed, the confrontation between Brabazon's technological inclinations and his father's insistence on some rather more traditional vocation was only postponed. As his school days neared their conclusion the subject blew up again.[34]

---

[30] He was created 1st Baron Brabazon of Tara in 1942, but had long been popularly known simply as Brabazon or 'Brab'. As a boy he was addressed as Ivon, and as such signed his early correspondence.

[31] Lord Brabazon of Tara, *The Brabazon story*, London 1956, 2–3.

[32] Postcard to J. T. C. Moore-Brabazon from his father, Col. J. A. H. Moore-Brabazon, 19 June 1898, Brabazon papers, box 49; H. O. D. Davidson, Brabazon's housemaster, to Brabazon's mother, 2 Aug. 1898, ibid. box 71/3.

[33] J. T. C. Moore-Brabazon to his mother, 19 Oct. 1897, Brabazon papers, box 52(1) (misc. letters); Gilbert Alun, headmaster of The Wick, to Col. J. A. H. Moore-Brabazon, 19 Oct. 1897, ibid.

[34] J. T. C. Moore-Brabazon to his mother, 1 Mar. 1901, ibid.

Reluctantly, Brabazon's parents had to accept that their son, like so many young men with any degree of engineering aptitude who grew to adulthood in the last decade of the nineteenth century – the period of the 'safety bicycle' boom and the birth of the automobile – was captivated by the emergence of the new mechanical locomotive technologies. The question was how to harness their enthusiasms to practical career ends. The effects of this social revolution were still novel by the time Brabazon came to consider his future, and the evident perplexity of his parents at his fervent desire to follow the technological progress of the times was probably not unique. Brabazon's housemaster tried to broach the subject with them as delicately as possible. He recommended that, given the boy's inclinations, the safest thing would be to pack him off to Cambridge to read for the Applied Science Tripos. He should, consequently, 'begin Greek at once'.[35]

The Hon. C. S. Rolls, a future aviationary contemporary, was also shunted off in this direction after leaving Eton, but it is symptomatic of the times that Brabazon should have had to cram Greek in order to win a place at university to read what was in effect mechanical engineering. It was not a prospect that filled him with much enthusiasm, but while at the crammer he at least had the opportunity to ride his first 'motor bicycle', opting for the Singer motor-wheel, designed by Perks & Birch. 'I never loved anything so much', he later recalled.[36]

If those responsible for Brabazon's future thought that by going to Cambridge for three years he would work off in comparative security the excess exuberance of youth, they were to be disillusioned. In his autobiography Brabazon dismisses his university experience in a few throw-away lines, alleging that he never intended to stay for more than a year anyway. He could not recall ever having met his tutor, Professor J. A. Ewing, under whom Rolls had also studied a few years before, and distinguished himself only by helping start-up the university's first motor club.[37] However, it appears from the Brabazon papers that it was C. S. Rolls himself who enticed him away from academe, with heady talk of the motor business he was planning. In an audacious and in some ways rather shocking letter to his father, clearly written while he was still officially at university, Brabazon wrote that he had definitely decided that his future lay with motor cars. Rolls had revealed that he was forming a company, and he (Brabazon) meant to be part of that enterprise. 'What I want you to do', he wrote, 'is, instead of paying £300 a year for me at Cambridge, where the Tripos would be of no use . . . if I took up motor cars . . . is to invest the capital of £5,000 in this business of Rolls – deducting it from whatever I may eventually

35 H. O. D. Davidson to Col. J. A. H. Moore-Brabazon, 7 Oct. 1901, Brabazon papers, box 71/3.
36 Brabazon, *Brabazon story*, 3–4; Montagu of Beaulieu, *Rolls*, 15.
37 Brabazon, *Brabazon story*, 4. For the frivolity of Brabazon's life at Cambridge see J. T. C. Moore-Brabazon to his sister, Crystabel, undated, but written while at Cambridge (1902–3), Brabazon papers, box 52(2).

receive at your death.'[38] One can only imagine the effect that this presumptuous letter must have had upon Brabazon's old father: whatever the response, the Rolls scheme seems never to have materialised. Brabazon's always tenuous tie to academic engineering, however, was now irrevocably cut. Disapproval notwithstanding, his future was with development of the internal combustion engine.

That this situation came gradually to be accepted by his family is evident from the fact that by January 1903 Brabazon was advising his parents and their friends on the purchase of automobiles. Interestingly, the car he recommended above all others was the Lanchester, another motoring name soon to become indelibly associated with the development of early aviation.[39] Having accepted the permanence of their son's engineering aspirations, it was soon agreed that Brabazon should become apprenticed to an established motor firm. A note in the Brabazon papers from September 1903 indicates that the Wolseley Tool and Motor Car Company was approached first,[40] but the enquiry came to nothing. So, after intermittently acting as personal mechanic and sales agent to C. S. Rolls, Brabazon made a similar enquiry of the Darracq Motor Company of Paris. This time the correct strings were pulled.[41] The main attraction of Darracq's for Brabazon, apart, one suspects, from providing the opportunity to put a little distance between himself and his family, was the firm's increasing involvement in the developing world of motor-racing. However, Brabazon was also quick to initiate his own little 'supply' business. This partly involved the purchasing of second-hand Panhard cars, which were then shipped back to England to be resold at a profit.[42] Whether Rolls was involved in this is unclear.

There is evidence that Brabazon was by this time already actively thinking about the possibility of achieving heavier-than-air powered flight. He was writing on the subject for *The Car Illustrated*, and in November 1904 C. S. Rolls provided him with a letter of introduction to the influential French sporting patron, Ernest Archdeacon. Archdeacon was a wealthy Parisian lawyer, whose name had long been synonymous with the promotion of aeronautics and automobiling. His involvement with aviation, on the other hand, was both more recent and precipitant. In early April 1903 Octave Chanute had visited Paris and lectured before the assembled members of the Aero Club de France. In the course of his talk he laid bare the extent of the Wright brothers' recent aviationary progress in America. This galvanised Archdeacon – who had

---

[38] J. T. C. Moore-Brabazon to Col. J. A. H. Moore-Brabazon, undated (c. 1902), ibid. box 71/3.

[39] J. T. C. Moore-Brabazon to his mother, 27 Jan. 1903, ibid. box 52(1). Includes memo on merits of the Lanchester car.

[40] H. Custin, manager, Wolseley Tool and Motor Car Co., to J. T. C. Moore-Brabazon, 5 Sept. 1903, ibid. box 52(5).

[41] C. S. Rolls ('C. S. Rolls & Co., British licensees for Panhard cars') to J. T. C. Moore-Brabazon, 19 Jan. 1903, 11 Mar. 1904; C. H. Selong, A. Darracq & Co. Ltd, to J. T. C. Moore-Brabazon, 4 Oct. 1904, ibid.

[42] Brabazon, *Brabazon story*, 5–6.

recently been elected the club's president – into founding an Aviation Committee of the Aero Club de France with the specific aim of stimulating a co-ordinated European response. Archdeacon was supported in this by such rich and influential figures as the petroleum magnate, Henri Deutsch de la Meurthe, and the French aeronaut, Comte Henri de la Vaulx; but they were building on slender practical foundations as, until then, the Aero Club de France – like its English counterpart – had been almost exclusively concerned with ballooning. Brabazon's own practical aeronautical experience had also been confined to ballooning, an emphasis which was to predominate for some years yet. However, as the letter to Archdeacon reveals, he was already positively seeking to extend his horizons to aeroplane flight by the time he went to Paris as a motor apprentice in 1904: Rolls specifically mentions that Brabazon 'is anxious to make some aeroplane experiments' if at all possible.[43]

A little over three years later, in January 1908, Brabazon's Anglo-French contemporary, Henri Farman, the son of an English journalist domiciled in France, was to win the Grand Prix d'Aviation Deutsch-Archdeacon, worth 50,000 francs, for flying the world's first officially accredited circular kilometre in his own Voisin-built pusher-biplane. Shortly afterwards, Brabazon would begin his own aviation trials at the same Champ de Manoeuvres ground, situated at Issy-les-Moulineaux, on the south-western outskirts of Paris. By then, it is clear, he was to some extent already known in French aviationary circles.

Much of Brabazon's initial period in France, however, seems to have become largely an extension of his frivolous English existence. Unfortunately for him, news of his idle afternoons at the Elysée Palace Hotel and Maxime's found its way back to London and his days as a technological pioneer were nearly brought to an abrupt conclusion. 'You were not sent to Paris to amuse yourself, but to acquire a knowledge of motor engineering', his father wrote angrily, adding that if he did not buck up his allowance would be stopped.[44] But serious though the implications of this admonishment might have been for Brabazon's future, in effect it was little more than a last spasm of parental disapproval. His activities were now to lead on to the world of international motor-racing and a degree of financial independence.

Brabazon first raced as C. S. Rolls's partner during the various speed trials included in the Gordon Bennett-cum-Automobile Club 'Irish fortnight' of July 1903, and the letters he exchanged at this time capture the thrill of being propelled at speed.[45] Later he was to drive in turn for the Mors, Minerva and Austin companies.[46] These activities peaked in 1907 with a celebrated

---

[43] C. S. Rolls to E. Archdeacon, 4 Nov. 1904, Brabazon papers, box 49.
[44] Col. J. A. H. Moore-Brabazon to J. T. C. Moore-Brabazon, 13 Mar. 1905, ibid. box 52(2).
[45] J. T. C. Moore-Brabazon to his mother, 10, 13 July 1903; C. S. Rolls to J. T. C. Moore-Brabazon, 19 Sept. 1903, ibid. box 52(5); Montagu of Beaulieu, *Rolls*, 81–3.
[46] Brabazon, *Brabazon story*, 16–26; S. Larle (Mors Ltd) to J. T. C. Moore-Brabazon, 20 Apr., 5 May 1905, Brabazon papers, box 52(5).

Minerva victory in the Circuit des Ardennes, but by then Brabazon's thoughts had already turned towards the air as the greater challenge. In August 1907, immediately after returning from the continent following his Ardennes victory, he contacted Short Brothers, the balloon manufacturers, and ordered the construction of a full-size man-carrying glider. This was to be built to his own design and delivered to Brooklands. It was to be his and, more controversially, their first aeroplane.

Brabazon's personal progression from automobiling to aviation can in retrospect be seen as symptomatic of a wider technological development, engendered above all by the increasing power-to-weight ratio of the internal combustion engine. Not only Brabazon, but Farman, Rolls and Glenn H. Curtiss had all been motor-racing drivers. For Brabazon, as for the others, the years of automobiling served as an apprenticeship for aviation, in the same way as the years of patient lonely experimentation with gliding flight had been an apprenticeship for A. V. Roe. Unlike Roe, however, whose family, if concerned by his enthusiasm for flight as a means of making a living, were still by no means averse to the idea of his going into the developing mechanical engineering industry, Brabazon had to struggle all the way against the confines of his insulated, anti-industrial upbringing.

Brabazon's comment in his autobiography that the early motorists were 'a community rather disliked by the horsey element'[47] could not have come from someone of Roe's background, where there were no such social tensions. It is ironic that for many years only the wealthy could afford automobiles and that in order to facilitate their acceptance within 'good' society the sociological process of gentrifying their image was begun by such advocates as C. S. Rolls and Lord Montagu of Beaulieu. It meant that the market was initially dominated by the class which disdained those who produced the vehicles. Given the general spirit of hostility to the ethos of technological progress which Brabazon encountered, it is perhaps understandable that he could later claim that both Rolls and himself were exceptional in looking upon the motor car as a stepping-stone to flight;[48] but in actual fact, such a view was by no means unusual among those youths deriving more directly from contemporary industrial society – individuals such as A. V. Roe, Frederick Lanchester and Geoffrey de Havilland.

Brabazon himself lacked any prejudice against working in industry. 'To me', he wrote, 'it has been a privilege to have been one of the first to be "weaned on petrol and fed on nuts".'[49] In his own way he had to overcome almost as many obstacles on the road to pioneer flight as Roe.

[47] Brabazon, *Brabazon story*, 30.
[48] Draft manuscript 'History of early aeronautical experiences', Brabazon papers, box 12/5a.
[49] Brabazon, *Brabazon story*, 32.

The Aero Club and ballooning

In order to gain a fuller understanding of the aeronautical environment at this time, another culturally and administratively important (if technologically insignificant) factor, the development of ballooning, must be examined. Not only did many future aviators gain their first aeronautical experience through an association with the main ballooning bodies of the period, but some major manufacturing concerns, such as Short Brothers, also derived from a ballooning foundation – as did the military's own air service. The principal civilian ballooning bodies, such as the (Royal) Aero Club, the Aero Club de France, the Aero Club of America and the Fédération Aéronautique Internationale to which they all became affiliated, presently became the jurisdictional institutions of aviation. The fact that they originated with a lighter-than-air, ballooning, emphasis was to have a material effect on the development of aviation generally.

Brabazon came to the world of ballooning and its British regulatory body, the Aero Club of the United Kingdom, through Rolls, and it was through his association with the Aero Club that he was introduced to the Short brothers. The foundation and growth of the Aero Club is therefore of crucial importance in analysing the careers of the early British pioneers and the environment in which they worked.

As an executive body the UK Aero Club grew directly out of the (Royal) Automobile Club, in yet another instance of the general evolution from automobiling to aviation. In this case the link was purely administrative, but it would be wrong to assume that for that reason the progression was completely seamless. As so often happens with new ideas the time was ripe for such a development and the inspiration for it occurred simultaneously to two different parties. In this instance those involved were principally Frederick Simms, the motor engineer and founder of the Automobile Club, on the one hand, and the rather more leisured, sport-orientated ballooning clique associated with Frank Hedges Butler, C. S. Rolls and the secretary of the Automobile Club, Claude Johnson, on the other.

Born to an expatriate British family in Hamburg in August 1863, Frederick Simms had served an engineering apprenticeship with the Aktien Gesellschaft für Automatischen Verkauf before undertaking further studies at the Polytechnisches Verein in Berlin. Thereafter he became one of the first men to fully comprehend the commercial implications of the internal combustion engine. In 1889 he made the acquaintance of Gottlieb Daimler, the German engineer who had recently developed arguably the world's first practical petrol engine, and presently negotiated his UK patent rights. This resulted in the establishment of the Daimler Motor Syndicate Ltd of Coventry in 1893. Simms sold the company to the predatory motor financier Henry Lawson two years later, and went on to establish the Motor Carriage Co. Ltd (1898–1900) and the Kilburn-based Simms Manufacturing Co. Ltd, formed in 1900. It was at this time that he designed, in association with

Hiram Maxim (of Maxim gun fame), several armoured motor vehicles, none of which was taken up by the War Office, but for which Simms is probably best remembered today. In collaboration with the German engineer, Robert Bosch, he also helped develop the magneto spark-plug, for engine ignition. From 1899 to 1900 Simms jointly operated the Compagnie des Magnetos Simms-Bosch. Then from 1907 to 1913 he managed the Simms Magneto Co. Ltd, which opened an independent New Jersey sales division in 1910.[50] Of more immediate significance, however, was his founding in 1897 of the Automobile Club of Great Britain and Ireland, within which both UK aero club factions developed. Indeed, Frank Hedges Butler and C. S. Rolls were among the organisation's first recruits.[51]

Simms had the prior claim to found an aero club as such. Not only was he the first to issue a prospectus and canvass support for such a body, he had also been contemplating the construction of his own flying machine since at least 1896.[52] He subsequently co-operated with Hiram Maxim, who had achieved a measure of success with his own enormous steam-driven biplane some two years before, in the design of a petrol-driven aero-engine.[53] As this suggests, Simms from the first associated his aero club scheme with the search for the means of practical heavier-than-air powered flight. This was in marked contrast to Frank Hedges Butler whose primary aim was the cultivation and regulation of 'sport aeronautics', notably ballooning.

Simms distributed his first prospectus in June 1901 and re-emphasised his scheme's utilitarian objective in an open letter to the automobiling press the following October, by which time the issue was in dispute. The Aero Club of Great Britain and Ireland, as he labelled it, was to be predominantly for what was defined as the 'encouragement' and 'development' of 'aerial navigation'. 'I am convinced', Simms affirmed, 'that it only wants a Club or Society on modern lines to bring together the many British enthusiasts . . . interested . . . in Aerial Navigation . . . to ultimately . . . solve [this] great problem.'[54] He was seeking, in other words, to establish an aero club with the aim of stimulating and co-ordinating an indigenous search for aeroplane flight in much the same way that Ernest Archdeacon was to do, some two years later, when he founded the Aviation Committee of the Aero Club de France. As events transpired, he was a little ahead of his time. The Butler clique's separately-registered Aero Club of the United Kingdom won the day, but only after a series of mutually acrimonious exchanges.[55]

Frederick Simms was to be largely shunned by the new Aero Club establishment following this division. This was unfortunate, as the energy he could

[50] DBB, s.v. Simms, F. R.
[51] Montagu of Beaulieu, Rolls, 25.
[52] Simms papers, 11/6–7.
[53] Simms papers, 24/32–8. Simms Aero-Motors Ltd were eventually quite successful, and associated with Voisin: The Aero, 8 June 1909.
[54] Copy of letter to automobiling press, 28 Oct. 1901, Simms papers, 10/13.
[55] Ibid. 10/8–32.

so usefully have contributed to the promotion of powered flight was now effectively lost. By contrast, the ballooning and sporting emphasis that was to be maintained by the victorious Butlerite Automobile Club faction undoubtedly added to the prevalent air of patronising hostility experienced by many modest heavier-than-air pioneers in this country over the next few vital years of aeronautical development. As it was, the ballooning emphasis was to be finally superseded only as the result of events overseas.

Frank Hedges Butler was a wealthy wine-merchant, born in 1855, who found his enjoyment in life through wide-ranging travel and the leisured and well-provisioned pursuit of non-too-hazardous adventure. Ostensibly a partner in the family firm of Hedges & Butler (established in 1667), such responsibilities as his position occasioned were seldom allowed to detract from other more exhilarating interests. Like Rolls, Butler became a pioneer motorist and in 1898 he was appointed honorary treasurer of the Automobile Club of Great Britain and Ireland. His interest in aeronautics, however, was – at least to begin with – purely social.[56] He was to present his particular version of the club's foundation in two publications, neither of which mentioned the Simms controversy or the issues it raised.[57] Thus it would be difficult to argue that he was really a significant figure in the development of heavier-than-air powered flight. Of greater long-term importance was his accomplice, C. S. Rolls, particularly in relation to Brabazon and the Short brothers.

Charles Stewart Rolls's life has been too well chronicled to need much elaboration here. He was born in 1877, the third son of John Allan Rolls (raised to the peerage as Lord Llangattock in 1892) of The Hendre in Monmouthshire. By all accounts the family were cold, distant and cheerless, and these characteristics were passed on to Charles in full. He was sent as a boy to Eton, but appears to have got little out of it, beyond perhaps reinforcement of his natural solitude. Like Brabazon, however, he displayed a marked technological and mechanical bent and was subsequently shunted off to Trinity College, Cambridge, for three years to read for an Ordinary Applied Science degree. It cannot be said that he applied himself with sustained enthusiasm to academic life – he at one stage toyed with and then abandoned the idea of changing to an Honours degree (the Tripos) – but he did at least manage to win a cycling half blue. In October 1896 he also acquired his first motor car, a 3¾ h.p. tiller-steering Peugeot Phaeton, to which a canopy was added. He was tremendously proud of this acquisition which, accompanied by three friends, he immediately drove from London to Cambridge, having obtained the necessary sanction from the Hertfordshire and Cambridgeshire police. Crowds of people awaited the party at Royston and the vehicle was the source of considerable

---

56 For Butler's background see his *Fifty years of travel*, and also pen-portrait 'Celebrities at home: Mr Frank H. Butler at Prince's Chambers', *The World*, 17 July 1906 (Butler collection, iii. 124).
57 *Fifty years of travel*, and *5,000 miles in a balloon*, London 1907.

interest at the university – at least within the student population. This interest was the catalyst that determined the future course of his life.[58]

It was probably through a mutual interest in motor cars that Rolls made the acquaintance of Walter G. Wilson, later co-founder of the Wilson & Pilcher motor firm. Wilson had matriculated the year before Rolls and was at King's. They quickly became active friends, and Wilson was subsequently (c. 1899) to act as Rolls's mechanic during a number of early motoring trials. It just so happened that another of Wilson's close friends and his future business partner was none other than Percy Pilcher, perhaps Britain's most significant practical aviation pioneer of the late nineteenth century. Wilson & Pilcher Ltd was established as an engine construction firm in 1897, but since 1895 Pilcher himself had been indulging a profound interest in monoplane hang-gliding flight, following the example of the German pioneer, Otto Lilienthal. Indeed, Wilson and Pilcher were searching for an engine of suitable power-to-weight ratio to propel such flight when the latter died as a result of a gliding accident in October 1899. There is no evidence that Rolls and Pilcher ever met, but it was almost certainly through Wilson that Rolls was introduced to the embryonic world of aviation.[59]

Rolls's name, however, was first made in the field of automobile dealing. He established the firm of C. S. Rolls & Co. in 1902, and championed the Panhard marque before formally combining his business with that of the Manchester engineer, F. H. Royce, in March 1906. His prospering motoring interests developed simultaneously with his early aeronautical activities.

These, then, were the figures who finally established an aero club in the UK in 1901, and who were to be the dominant establishment of British aeronautics, in all its manifestations, for most of the next decade. For all the grandiose declarations marking its foundation, the club and its activities were initially retarded by the insidious class consciousness of the society which produced it. Stanley Spencer, for example, the aeronaut who initially piloted Butler, his daughter Vera and C. S. Rolls in the ascent that led to their decision to form an aero club, was himself excluded from the body on account of his 'professional' status. The English club was far from unique in maintaining this class emphasis. The spread of gas supplies throughout provincial Britain and Europe in the later Victorian period had made its provision for sporting balloon ascents widely available and consequently the leisurely pastime of aeronautics blossomed throughout Europe, among those wealthy enough to indulge their whims. Unfortunately, the overwhelming desire of most of these sporting amateurs was to keep aeronautics sporting, amateur and exclusive, to the

---

[58] C. S. Rolls to Lord Llangattock, 20 Oct., 23 Nov. 1896; C. S. Rolls to Lady Llangattock, 24 Oct. 1896, Llangattock papers, D361 F/P 867. See also bound family album, 'Extracts, newspaper cuttings, etc., C. S. Rolls May 1895 to Sep. 1899', Rolls papers; and Rolls's scrapbook, including programme of 1896 Cambridge University Bicycling Club meetings, RAeC papers.
[59] For Wilson-Rolls association see G. Bruce, *Charlie Rolls: pioneer aviator*, Derby 1990, 18–19. Wilson & Pilcher Ltd were subsequently acquired by Armstrong, Whitworth & Co.

obvious detriment of those aeronautical enthusiasts whose aim was the true furtherance of powered flight – particularly heavier-than-air powered flight.[60] A similar club had been founded in France in 1898, but before the establishment of Archdeacon's aviation committee in 1903 it took little active interest in heavier-than-air flight. Other associations followed throughout Europe, regulated from October 1905 by the Fédération Aéronautique Internationale.

In earlier decades ballooning had acquired an ignominious reputation as the preserve of showmen and fairground dare-devils, and this probably fuelled the over-compensatory gentrification of the sport when it suddenly became fashionable. In the UK there was also a desire to steer clear of the monopolising commercialism which was the unacknowledged motive behind Henry Lawson's recent Motor Car Club, partly in response to which Simms's Automobile Club had been formed in 1897. One place where this revived society aeronautics was to be less prevalent was the comparatively class-free United States. The rewards of that country's aeronautical endeavours were soon to be reaped in dramatic style.

Butler's Aero Club held its first general meeting at the Automobile Club on 3 December 1901. Among those present were Butler, Rolls, Patrick Y. Alexander and, a little uneasily, Frederick Simms.[61] An Organising Committee was then elected, including Butler, Rolls, Colonel James Templer (Capper's predecessor as Superintendent of the Royal Engineers' Balloon Factory) and John Scott-Montagu, but excluding Simms; and a Balloon Committee was then chosen by, and largely selected from, members of this Organising Committee. The elitism of the club's hierarchy, evident from its disputed foundation, was thus kept intact.

At the opening meeting of the Organising Committee, on 12 December, a resolution was passed pledging club members to contribute £10 each towards a balloon fund. The secretary wrote to Simms requesting his contribution. Not surprisingly, Simms refused on principle to forward any such thing.[62] The Spencer brothers of Highbury were the first aeronautical suppliers patronised by the club. C. G. Spencer & Sons was a long-established aeronautics firm. Indeed its founder, Charles Green Spencer, was the son of Edward Spencer, a Barnsbury solicitor who had become a close friend and associate of the celebrated early nineteenth-century aeronaut, Charles Green. However, through

---

60 For general sociological background note P. Bailey, *Leisure and class in Victorian England*, London 1978.

61 First minute book, minutes of general meeting, 3 Dec. 1901, RAeC papers. For a list of founder members see Aero Club Gazette, No.1, 1901 (Butler collection, i. 132): John Scott-Montagu, Alfred Harmsworth and Sir David Solomons were among the names. The club had been initially registered in October 1901, but legal disputes impeded its immediate practical organisation.

62 Minute book, 12 Dec. 1901, RAeC papers; Simms papers, 10/45. P. Y. Alexander, with his usual generosity, also placed a 35,000 cu. ft Spencer balloon at the club's disposal: Aero Club Gazette, No.1, 1901 (Butler collection, i. 132).

laid for a series of practical experiments.[79] In this way, a full five years after its contentious foundation, the Aero Club began tentatively to address the problems of heavier-than-air powered flight. Little wonder A. V. Roe felt a certain degree of resentment that at the *Daily Mail* model aeroplane competition of April 1907 his well-developed constructions should be patronisingly adjudged inadequate by such figures. By this time only the Wrights, and perhaps Santos-Dumont, had actually achieved flight, and Roe was to have a full-size man-carrying machine constructed by the end of the year. The Wrights were in fact so far ahead of everybody else that once their claims were believed it seems to have blinded many in the British aeronautical establishment to the ungainly efforts of their own pioneers. In March 1906 the Aero Club publicly acknowledged that in their view the Wright brothers had achieved 'a non-stop circular flight of over 24 miles', and they invited them over to England to give a public demonstration of their machine.[80] This invitation was not taken up by the brothers, but assuming some vicarious kudos from their tenuous association with the American pioneers the club's hierarchy proceeded to act as if they had generated the entire achievement.

Brabazon was based in Paris during much of this time, but largely through his friendship with Rolls he kept in touch with the activities of the Aero Club of the United Kingdom, and was sufficiently well-connected to be accepted within the association's inner circle without any difficulties. Indeed, acknowledging that he did not join until 1903, Brabazon was later to remark disingenuously that 'for some unknown reason I was made a member of the [Organising] Committee'.[81] The 'unknown reason' was, in fact, his class. A less socially-advantaged youth would hardly have received such a distinction.

Brabazon was later to claim that he thought of his ballooning experience as 'training for what was soon to come'.[82] Yet the likelihood is that his decision to branch decisively into the incipient world of aviation in 1907 was inspired more by what he had learned in France than by the genteel activities of the Aero Club. Brabazon was for the most part concerned with other matters than faddish ballooning, however cultivated the company.[83] Nevertheless, it was the Aero Club which provided the context within which Rolls and Brabazon

[79] Hurren, *Fellowship*, 32, 45. G. Bruce, C. S. *Rolls: pioneer aviator*, Monmouth 1978, 13. Rolls had joined the Aeronautical Society in December 1901, so he must have been painfully aware of the Aero Club's technological backwardness.

[80] 'Aero Club of the UK: notices', Mar. 1906 (Butler collection, iii. 78).

[81] Brabazon, *Brabazon story*, 43. His Aero Club membership card actually survives among the Brabazon papers (box 49). Dated simply 1903, it is No.73. (It should not be confused with the club's aviator's certificate, the very first of which was to be awarded to Brabazon in March 1910.)

[82] Brabazon, *Brabazon story*, 42.

[83] See letter to J. T. C. Moore-Brabazon from his mother, 5 Apr. 1905, Brabazon papers, box 52(2): 'How disgraceful of you not to have paid your subscription to the Aero Club – with the result that you were posted up . . . as a delinquent'. For a record of Brabazon's balloon ascents see box 71/3/1–5, which contains his balloon log. His first ascent, made as a passenger with Butler and Gen. Sir Henry Colvile, is dated Ranelagh, 23 May 1903.

collaborated in their earliest aeronautical endeavours, and it was as a result of his Aero Club activities that Rolls allegedly 'discovered', in Brabazon's phrase, the Short brothers – or if not discovered, then at least brought them to prominence.[84] In fact the Short brothers later came to rather resent the claim that they had been 'discovered', while the suggestion that it was Rolls and Brabazon who provided them with their first balloon construction order was simply not true.[85] Their meeting with Rolls and Brabazon was, however, undoubtedly their biggest career break, bringing them to the notice of a wider circle than they could ever have hoped to have reached on their own in so short a time.

Short Brothers

The Short family was originally from County Durham, but the father, by profession a colliery engineer, had moved to Nottinghamshire in 1880 to take up an appointment as works manager of the Stanton Ironworks. Here he found himself working under the supervision of the electrical engineering pioneer, R. E. B. Crompton. The three Short brothers who were to make up the aeronautical manufacturing concern were Horace, born 2 July 1872, Eustace, born 26 June 1875, and Oswald, born 16 January 1883. Of these, the most striking was the eldest, Horace, who as a child had suffered from meningitis and hydrocephalus, which was held to be responsible for his cranial disfigurement, manifest in all contemporary photographs. In reality, he suffered from a condition known as *foetus in foetu*: two brains had developed within his head, as was discovered from a post mortem following his death from a brain haemorrhage in April 1917. In spite of this, it was Horace who was the presiding genius of the family in these pioneering years, as Brabazon himself was to testify.[86] His disability, however, made him at first an extraordinarily shy person, and partly it seems to escape ridicule because of his unfortunate appearance he left the country, after periods of training at both Crompton's pioneering electrical works at Chelmsford and the torpedo shop at Chatham dockyard. Eventually he settled in Mexico, where he became the chief engineer of a silver mine.[87] The rest of the family had meanwhile moved on to Derby in about 1890. For some years they maintained only intermittent links with

84 The phrase is taken from Brabazon papers, box 12/5c (Press: aviation articles). Griffith Brewer was another who casually employed the term. Note manuscript, 'The genesis of the flying industry: reminiscences of Short Brothers' (1938), RAeS archive.
85 Oswald Short to J. Laurence Pritchard, 30 Mar. 1948, RAeS archive.
86 Brabazon, *Brabazon story*, 51; 'Lt.-Col. Moore-Brabazon on early aviation', 8, Brabazon papers, box 12/5a. See also G. Bruce, 'Shorts, origins and growth: the sixteenth Short brothers commemorative lecture' (1977), 3–4, RAeS archive, and P. King, *Knights of the air*, London 1989, 64, 485.
87 For a description of his travels and adventures see M. Donne, *Pioneers of the skies: a history of Short Brothers plc*, Belfast 1987, 2–3.

Horace, and this might have remained the situation had not Short senior died in 1891. As it was, a subscription was raised enabling Eustace to visit Mexico in order to implore Horace to return to help the family out of its financial difficulties. This he could not do, but he did provide Eustace with sufficient funds to acquire the coal and coke distribution firm he had joined at New Malden, near Kingston-on-Thames. Eustace and Oswald's early interest in ballooning, dating from about this period, thus originated independently of their elder brother.

Horace's own technological aptitude first found practical expression in the invention of a sound amplifier, subsequently dubbed the auxetophone. This device was patented in 1898, shortly after Horace's return to England where he had managed to find as a sponsor for its production the American entrepreneur, Colonel George Gouraud. Gouraud had served in the Union army during the US Civil War, in the course of which he had won the Congressional Medal of Honor. Subsequently, he moved to England, where he promoted, amongst other things, a variety of Edison's recent technological devices, notably the telephone and phonograph. Gouraud established an acoustics laboratory at Hove, near Brighton, for the production of Horace's amplifier, temporarily restyling it the Gouraudphone, and, following their reunion, the site doubled as the Short brothers' first balloon workshop.

The younger brothers, Eustace and Oswald, had purchased their first aerostat – a second-hand Spencer balloon, christened the 'Queen of the West' – in 1898, and with this they had already established Shorts as an aeronautical concern, acting, for the most part, as showmen-aeronauts, earning money by giving ballooning displays at fairs and fêtes and so on.[88] This was just the sort of 'disreputable' concern that the dilettante aero clubs of Europe were soon to react against – while all the time employing such people for their own instruction.[89] It is clear, however, that the Short brothers never indulged in exhibitionism for its own sake. From the beginning they strove to develop the products of their adopted industry in both a professional and, as far as was possible, a systematic manner, assiduously assimilating the latest technical advances. In 1900 all three brothers with Colonel Gouraud visited the International Exhibition in Paris, Horace ostensibly to demonstrate the Gouraudphone, and Eustace and Oswald to inspect the latest French balloons, specifically the spherical envelopes devised by Edouard Surcouf of the Société

[88] Eustace Short had made a balloon ascent as a fee-paying passenger at the Crystal Palace in September 1897. The 'Queen of the West' – an old, near expended, coal-gas balloon – was purchased for £30 and first used by the brothers the following May. Horace was a passenger on this occasion (a fête in Teddington), but reportedly loathed the experience: Bruce, 'Shorts, origins and growth', 11, 14; Donne, Pioneers, 6–9.

[89] See Rolt, The aeronauts, 234–5. Such was the case, for example, with Butler's own celebrated ascent of 24 Sept. 1901, with Vera Butler and C. S. Rolls, which led to the foundation of his Aero Club. The aeronaut in charge of the balloon that day was Stanley Spencer. In renditions of the tale, however, he (when mentioned at all) is usually accorded the anonymous status of a chauffeur.

Astra. The following year, in the Gouraudphone acoustics laboratory at Hove, they assembled their first indigenous aeronautical artefact, a 33,000 cu. ft Surcouf-derived envelope, which was to act as a replacement for their Spencer-designed 'Queen of the West' balloon. By March 1902 Eustace was writing to the Aeronautical Society from Hove on behalf of what might now legitimately be called the Shorts' aeronautical manufacturing company.[90]

Unfortunately, Gouraud withdrew from the enterprise soon after this, and the Hove laboratory was shut down. Eustace and Oswald were resolved to carry on the aeronautical concern, but Horace again faded from the scene for a time, moving north in October 1904 to work for Charles Parsons, the steam turbine pioneer. Of necessity, Eustace and Oswald fell back once more on the show-men-aeronauts circuit.[91] This was a temporary expedient: their own manufacturing skills were soon sufficiently advanced to win them the first in a series of balloon construction orders.

In his autobiography Brabazon stated that he and Rolls were the first to commission the Short brothers for construction purposes, after directing them to France 'to pick up the latest technique'. However, this version of events gives rather a confused picture since, as just seen, the brothers had already travelled to Paris for this reason in 1900. Nor were Rolls and Brabazon their first customers. The truth, as Oswald Short was to protest even before Brabazon wrote his autobiography, was that by the time Rolls and Brabazon entered the picture the brothers had already produced 'three captive war balloons' for the Indian government – the first being completed as early as January 1904.[92] These had been their first order, and to carry it out Eustace and Oswald had hired a large carriage shed in the London Mews, off the Tottenham Court Road.

Nevertheless, the brothers' association with Rolls and Brabazon did bring the Shorts' enterprise to prominence. The earliest evidence of their meeting seems to date from late 1905 or early 1906, when Rolls was looking for a new balloon with which to challenge for the recently announced Gordon Bennett Cup. The introduction was said to have been effected by Colonel Templer, the Superintendent of the Royal Engineers' Balloon Factory, who had previously been called upon to inspect and approve the balloons ordered by the Indian government.[93] Until then, most official Aero Club-related business had been

---

[90] Eustace Short to E. C. Bruce, 22 Mar. 1902, RAeS archive. The letter was in respect of a 'balloon car', or pressurised cabin, Eustace claimed to have invented, enabling ascents to be made to any height without incurring a 'diminution of atmospheric pressure'.

[91] Bruce, 'Shorts, origins and growth', 19.

[92] Brabazon, *Brabazon story*, 33; Oswald Short to J. Laurence Pritchard, 30 Mar. 1948, RAeS archive. In this letter Oswald also claims to have been the designer of these balloons: 'I designed and Eustace and I built three captive war balloons for the Indian government'. The RE's own Balloon Factory's goldbeaters skin fabric, discussed in ch. 4, had a tendency to deteriorate in tropical climes, and this is presumably why Shorts were commissioned. Their balloons also incorporated a new suspension device. The first was assigned to the Madras Sappers: Bruce, 'Shorts, origins and growth', 20.

[93] Donne, *Pioneers*, 11 (quoting Oswald Short); Penrose, *Pioneer years*, 117.

conducted with the firm of C. G. Spencer & Sons, who had provided virtually all members' equipment in the early years. Rolls, however, rejected this association. Brabazon later recalled that 'for some reason or other [he] disapproved of the firm [of Spencer], who were then about the only [manufacturers] in England dealing with ballooning. . . . I never quite got to the bottom of why he disapproved of this firm', Brabazon continued, 'but he disliked the pear-shaped balloons that they made and was always anxious to get hold of somebody else to make a perfectly spherical model such as had been introduced in France'.[94]

Brabazon then claims that Rolls sent Eustace and Oswald to Paris to study the construction methods of the French aeronaut, Mallet. 'How he managed to get Mallet to consent to this', Brabazon admitted, 'I never have understood . . . but he did, with the result that the two Short brothers returned from Paris confident that they could manufacture balloons as well as anybody.'[95] Rolls thereupon ordered the construction of a 42,000 cu. ft Mallet-model balloon, entitled 'Venus', to be co-owned by Brabazon and Warwick Wright – the brother of the future aeroplane manufacturer, Howard Wright. Other commissions from Aero Club members followed.[96] The 'Venus' was finally completed in May 1906, by which time the Short brothers had also received orders for the 'Zephyr', placed by Professor A. K. Huntington, and the 'Padsop', placed by later recruits Phil Paddon and T. O. M. Sopwith. C. S. Rolls also ordered the 78,500 cu. ft 'Britannia', and this was his entry for the Gordon Bennett competition. With their order books suddenly under this sort of pressure it is scarcely surprising that the brothers were soon looking for new premises. In June 1906 they took the lease of some railway arches adjacent to the Battersea gasworks.[97]

By the following year the association which Rolls had established between Short Brothers and the Aero Club had been officially sealed: at some indeterminate date the brothers were formally appointed the club's aeronautical engineers in succession to Spencer & Sons. It was this affiliation that was ultimately to lead to Shorts' association with the Wright brothers and the establishment of a flying ground on the Isle of Sheppey. Soon Rolls, Brabazon and the Short brothers themselves were to be completely swept along on the current of aviation, and while the relationships forged in the environment of

[94] Draft manuscript 'History of early aeronautical experiences', Brabazon papers, box 12/5a.
[95] Ibid.
[96] Ibid.; Bruce, 'Shorts, origins and growth', 24; 'Aero Club of UK', *Automobile Club Journal*, 26 May 1906.
[97] Brabazon, 'History of early aeronautical experiences'; Bruce, 'Shorts, origins and growth', 24; 'Maps and rate book entries for the Battersea railway arches rented by Short Brothers', RAeS archive. Note also Loobey, *Flights of fancy*, 13, and 'Pastime in Cloudland. Facts gleaned from Messrs Short Brothers of Battersea', *Motoring Illustrated*, 18 Aug. 1906 (Butler collection, iv. 16). For details of Rolls's Shorts-constructed balloon 'Britannia' see *Automobile Club Journal*, 13 Sept. 1906. The result of the Gordon Bennett contest can be found in *RAeS Journal* x (1906) 50–1. It was won by the American, Lt. Frank P. Lahm, with Rolls third in 'Britannia'. His assistant was none other than Col. Capper.

ballooning remained of central importance, the activity itself rapidly became little more than the quaint relic of another age.

One gains an impression of these two spheres overlapping in the last pages of Brabazon's balloon log-book. Here he records the details of a flight in the 'Venus' balloon on 25 May 1907, ascending from Ranelagh and drifting in its course over the 'Weybridge Motor Track'. This was the Brooklands motor-racing circuit where, within a matter of months, both he and A. V. Roe would be endeavouring to fly their first aeroplanes.[98] 'All this time', Brabazon later recalled, 'it must not be forgotten that at the back of our minds the idea of dynamic flight was developing.'[99]

By virtue of his ballooning activities Brabazon had met at Crystal Palace two of the leading figures of British aeronautics even before he went to France as a motor apprentice. These were B. F. S. Baden-Powell, of the Aeronautical Society, and J. W. Dunne. Dunne was soon to be taken up by the War Office in an effort to produce a British army aeroplane, and the particular nature of his innovations and, indeed, his relationship with Baden-Powell, will be discussed elsewhere. But it is interesting to touch on their acquaintance with Brabazon at this time, when the latter was overwhelmingly preoccupied with cars and ballooning, as it provides evidence that the idea of dynamic flight had indeed always been at the back of his mind.

Baden-Powell had constructed a water-chute over one of the lakes in the grounds of the Crystal Palace, and off this contrivance a flat-bottomed boat with interchangeable monoplane and biplane wings was periodically sent hurtling in the hope of achieving some measure of flight before it hit the water. The time Brabazon spent with Baden-Powell in 1904 testing the apparatus seems to have been his first experience of any sort of heavier-than-air flight.[100] He seems to have made a good impression. J. W. Dunne wrote to Brabazon, in October 1904, that Baden-Powell 'has often spoken to me of you'. He then, as always, related details of his own pioneering work in automatic aeroplane stability. Dunne's ideas were far ahead of their time, so what Brabazon made of them one can only imagine, but if nothing else Dunne's correspondence must have further inclined him towards powered flight.[101]

Then, while an apprentice in France, Brabazon would have been caught up in the excitement surrounding figures such as Santos-Dumont – at that time popularly considered the most advanced aviator in Europe. However, in the light of later events, it seems likely that Brabazon's biggest inspiration in deciding, in August 1907, to order the construction of an aeroplane from the Short brothers was the example set by Gabriel Voisin. There is no formal proof

---

98 Balloon log-book, ascent No.27, Brabazon papers, box 71/3/1–5. This is the final log in the book.
99 'History of early aeronautical experiences', 8.
100 Ibid; Maj. B. F. S. Baden-Powell, 'Aeroplane experiments at Crystal Palace', *RAeS Journal* viii (1904), 52–8.
101 J. W. Dunne to J. T. C. Moore-Brabazon, 7, 28 Oct. 1904, Brabazon papers, box 52(5).

of this, but much in the way of circumstantial evidence. For example, when Brabazon went to France as a motor apprentice in 1904 he was, as already noted, given an introduction to the Parisian sporting patron Ernest Archdeacon specifically to discuss the question of aeroplane flight. However, from about this time Archdeacon was himself working in close co-operation with Gabriel Voisin, producing, in 1905, the Voisin-Archdeacon float-glider. This was an important precursive biplane design, incorporating a Hargrave-derived boxkite central wing-section, [102] such as Brabazon would crudely attempt to emulate in 1907.

It is most likely that Brabazon's attention was first drawn to Voisin, by Archdeacon, at this time. Brabazon is not very clear on this in his autobiography, but the fact that after his first aeroplane failed he went straight back to France and to Voisin himself would seem to indicate that he knew him beforehand. Furthermore, it has been stated by some secondary sources that in later years Brabazon gave oral confirmation of a French connection.[103] All this supports the conclusion that it was under the inspiration of Voisin that Brabazon ordered the Aero Club's official aeronautical engineers, the Short brothers, to construct a glider for him to his own Voisin-derived design. To this he then, on his own initiative, added a motor.

Yet the fact that what Eustace and Oswald Short constructed – to Brabazon's design – was simply a glider, and that it was Brabazon who added an engine, was later to lead to some confusion. It was to give the firm of Short Brothers the opportunity to subsequently deny that the aeroplane was ever one of theirs. Once completed, this machine failed to achieve flight. Shorts thus became anxious to refute the claim, highlighted by the publication of A. V. Roe's autobiography in 1939, that the first Shorts-built aeroplane had been essentially inept. 'We were not particularly interested at that time . . . 1906 or early in 1907', declared Oswald Short some years later, denying that he ever knew Brabazon planned to experiment with an engine attached to the craft. This supposed lack of interest is difficult to accept, but the fact remains that this machine of Brabazon's was the first Shorts-built aeroplane, whether by design or not.[104]

Brabazon himself was later to remark that the idea of actually assembling his own glider/aeroplane first came to him in 1906. He entrusted the initial manufacturing work to the Short brothers, he claimed, 'much against their will', but again this is difficult to reconcile with the fact that the brothers themselves were by 1902 already tendering to construct 'flying machines, kites,

[102] That is to say derived from the 'box-kite' design devised by Lawrence Hargrave.
[103] See G. Bruce, 'Shorts aircraft: some new evidence on the early years to 1912' (1977), 3, RAeS archive.
[104] Oswald Short to J. T. C. Moore-Brabazon, 11 Jan. 1940, Brabazon papers, box 26(1). See also O. Short to J. Laurence Pritchard, 30 Mar. 1948, RAeS archive. This controversy led to an acrimonious exchange of letters between A. V. Roe, Oswald Short and Brabazon, in the aeronautical press (for example *Flight* and *The Aeroplane*), in 1940.

etc., from plans'.[105] Be that as it may, the biplane that resulted was extremely crude and, Brabazon remarked, 'of the most singular appearance'. It had a flat mainplane wing-span of 35 ft, a large forward-elevator, rear-mounted rudders, and sat on a dual skid and wheel undercarriage. A 12 h.p. eight-cylinder Buchet engine was fitted and the frame modified by Howard Wright, who – although reportedly undertaking this work at Brooklands in early 1908 – maintained an adjacent workshop to the Short brothers, under one of the Battersea railway arches. The process of adaptation was completed with the attachment of a four-bladed pusher-propeller; 'undoubtedly the most inefficient propeller I have ever seen', as Brabazon described it.[106]

The presence of Brabazon's aircraft at Brooklands did not elicit quite the same reaction as Roe's pusher-biplane. Not only was he an active BARC member, but his machine would not be able to sustain even the most elemental flight-trials, so further considerations hardly arose. Doubtless both these factors helped lay the ground for the future enmity between these two rivals.

What Roe himself did or did not achieve at Brooklands at this time was to cause one of the bitterest disputes in aviation history and will be dealt with later; suffice to say at this point that Brabazon believed that Roe had achieved nothing. He dismissed Roe's biplane as having 'about as fantastic a possibility of flying as my own'. He even went so far as to suggest, in a 1934 memoir, that the authorities at Brooklands were 'quite justified in being rather sceptical as to the desirability of comic inventors being given . . . important pitches on [their] site', presumably forgetting that, in an effort to gain publicity for the new circuit, the Brooklands authorities had actively sought to entice inventors to their pitch in the first place.[107]

Brabazon was less disparaging in his autobiography, published in 1956. In this he maintained that they were both made unwelcome at Brooklands; but following the failure of his aircraft Brabazon was to take leave of the place anyway and head back to France. (In March 1908 three cylinders of his Buchet engine cracked as a result of frost, and this seems to have marked the end of his brief Brooklands sojourn.)[108] No such option seems to have been available to Roe.

---

[105] 'History of early aeronautical experiences'; Short Brothers Ltd, *Seventy-five years of powered dynamic flight*, Belfast 1978 (a brief company history, in which an extract from a 1902 Shorts' catalogue is quoted).

[106] 'History of early aeronautical experiences'. He claims here that the aircraft's frame was 'installed at Brooklands' before the track officially opened, but Brabazon certainly did not start experimenting until after Roe's arrival in September 1907, so this is doubtful. The claim, however, is just one of a number of contradictory reports of the movements of this first Brabazon-Shorts machine. A. V. Roe believed the aircraft to have been delivered to Brooklands 'early in 1908', a statement which Brabazon made no effort to challenge: A. V. Roe to Lord Brabazon, 26 Dec. 1955; Brabazon to A. V. Roe, 28 Dec. 1955, Brabazon papers, box 26 (1). Note also 'Early days of aviation', ibid. box 12/5a, where Brabazon describes the 'paddle blade propellers'.

[107] Ibid. 9.

[108] Bruce, 'Shorts aircraft: new evidence', 3. In 'History of early aeronautical experiences'

Perhaps the most significant thing about this episode, however, is that even Brabazon, who was after all a committee member of the Aero Club, should have claimed in an official Air Ministry memoir, written in 1918, that 'it was very difficult to go on in England'.[109] This would again seem to indicate that the Aero Club's inherent ballooning emphasis did retard the true pioneers of dynamic powered flight. The French Aero Club had been shaken up by Archdeacon. The only comparable figure in England at this time was North-cliffe; but he was scarcely in a position to devote his full attention to the subject. Fortunately for Brabazon and Rolls, they had a practical automobiling back-ground to offset this trend; but the environment in England generally was still disconcertingly unreceptive. 'I cannot stress too much the lack of [an] encour-aging environment in those very early days', Brabazon was to comment, nearly thirty years later.[110]

He arrived back in France some time in mid-1908, to find the country still ablaze with the excitement of Farman's prize-winning circular kilometre flight. Indeed, this may have been one of the inducements to return. Even the London *Times* – which only two years before had disparaged Roe's letter proclaiming the Wrights' achievement – had been obliged to declare this event 'epoch making'.[111] Presumably Brabazon had connections or contacts of some kind with the more prominent French aviators, although to what extent it is difficult to determine. He initially travelled to Châlons, and then on to Issy-les-Moulineaux to join Gabriel Voisin. His first action there was to procure his own Voisin aircraft.

This series of events should again be seen in perspective. The majority of the most promising aircraft in Europe at this time were Voisin-derived. The Voisin brothers themselves were, without question, the primary influence on aviation until Wilbur Wright's first public demonstration flights at Le Mans, in the summer of 1908. Farman's triumph of the first accredited circular kilometre flight simply reinforced this. Thus Brabazon needed no inside knowledge to realise that Issy-les-Moulineaux would probably provide a more encouraging environment for aviation than Brooklands. Robert Blackburn, for example, was another young aviationary-inspired engineer who, studying at workshops on the continent, was drawn to Issy-les-Moulineaux during this period. Gabriel Voisin later recalled – in a passage which could well have been written with Brabazon in mind – that 'Although the Deutsch-Archdeacon victory brought glory to Henry Farman, it brought us, by way of compensation, solid profits. A few days after our [*sic*] success, three or four "sporting types" appeared at the rue de la Ferme and ordered aircraft from us with considerable

Brabazon records that the machine was subsequently taken to Waltham Cross (described simply as 'a place we had in the country'), where he unsuccessfully attempted to propel it into the air by means of an elastic cord. It was never used again.
[109] PRO, AIR 1 724/91/1.
[110] Draft manuscript for 'The Aeropilot', dated 1 Apr. 1935, Brabazon papers, box 12/5a.
[111] *Times*, 14 Jan. 1908.

deposits on the orders.'[112] Brabazon candidly admitted that this direction (i.e. via the Voisins) simply seemed 'the quickest way of getting ahead'. He presently acquired a standard Voisin biplane and began a series of hop-flights over the Issy-les-Moulineaux ground. By November 1908 he was measuring his flight distances in hundreds of yards.[113]

It would seem, in fact, that Brabazon procured from Gabriel Voisin no less than three aircraft during 1908. The first of these was a Voisin tractor-triplane, of which, Brabazon recalled, Gabriel had 'great hopes'. It showed little promise of lifting, however, and seems to have faded from the Voisin catalogue very rapidly. This obscure Brabazon allusion is one of the very few surviving references to this abandoned branch of Voisin development.[114]

Only after this disappointment did Brabazon move on to what he described as a Farman-type Voisin biplane, with a Vivinus engine. It was in this machine that he tentatively became airborne for the first time. It was a machine that often oscillated on the very edge of flight: strenuous efforts had to be made to generate that last spring into the air. Brabazon recalled how he would place only half a gallon of petrol in the tank and simultaneously dispense with his coat and boots in an effort to lighten the load for lift off. This often made the difference between taking wing and remaining humiliatingly earthbound.[115] Then, at the turn of the year, he procured as his third Voisin aircraft a biplane that had originally been earmarked for Farman himself.

Learning to pilot these Voisin machines, when they did manage to become airborne, was a precarious endeavour. All pioneer aviators were by necessity self taught, through the gradual gaining of flight experience. 'Piloting', as Brabazon put it, had to be 'indulged in . . . by "empiric methods".' This meant, in effect, trial and error – and every error meant a smashed undercarriage or worse. Many early accidents were to occur through sheer inexperience in operating the controls.[116] The engines used at this time were also treacherously unreliable. Brabazon considered it a blessing if one of his aero-engines ran for up to four or five minutes without breaking down. And it was a constant struggle to elicit the last few possible 'revs' out of any marque of motor. Again,

[112] G. Voisin, Men, women and 10,000 kites, London 1963, 166–7.
[113] 'History of early aeronautical experiences', 10; 'Early days of aviation'; G. Voisin to J. T. C. Moore-Brabazon, 18 Nov. 1908, Brabazon papers, box 49.
[114] Draft manuscript, 'Early flight', dated Feb. 1930, ibid. box 12/5a. Brabazon himself neglects to mention this aircraft in his 1956 autobiography. Voisin reportedly also constructed triplanes for Ambroise Goupy and Baron de Caters (Sept.–Dec. 1908). Neither was successful: Gibbs-Smith, Aviation, 135.
[115] 'Early flight'; Brabazon, Brabazon story, 53.
[116] Draft manuscript (second version) for 'The Aeropilot', Brabazon papers, box 12/5a. Many early aviators – such as Blériot, A. V. Roe and Grahame-White – were subsequently to set up flying schools in order that their successors should not have to continue learning by such methods.

these last revs often made all the difference between remaining earthbound and becoming airborne.[117]

It would thus be wrong to assume that as soon as Brabazon returned to France he immediately began piloting aircraft with any great facility. Despite the well-publicised success of Farman, the medium was still painfully novel, and even the standard Voisin aeroplane was almost unimaginably primitive. It lacked any form of aileron or wing-warping flight control: thus – pilot experience aside – it was an exceedingly difficult craft to operate, since there was no means of adapting to volatile environmental conditions. It was for this reason that the type was to be superseded by 1910.[118]

Brabazon was to get more out of his third Voisin machine, 'The Bird of Passage'. This aircraft had originally been commissioned by, and built for, Henry Farman. Gabriel Voisin, however, perhaps as the result of some minor disagreement, offered it instead to Brabazon, whom he evidently regarded as his own English protégé. Brabazon, who did not know it had been earmarked for Farman – and who was shocked by his subsequent bitterness – was delighted, as the machine embodied many improvements on the standard Voisin. Farman, on the other hand, was predictably furious: it was this incident that finally induced him to move on to the design and manufacture of his own aircraft. He wrote to Brabazon in February 1909 venting his considerable anger:

> I had ordered an aeroplane from Voisin a few weeks ago giving the distance I wanted between the planes and the position of the front rudder [forward-elevator] and Mr Voisin agreed to build it but at my own risk as far as flying was concerned. This machine was paid for and was ready at Voisin's. . . . Therefore . . . you can judge of my stupefaction to find that Voisin had given you *my* [sic] machine built according to *my* specifications. I am glad the machine flies well, but now you know to whom it is due.[119]

The aeroplane that Farman then went on to construct, the 'Henry Farman III', incorporated ailerons for lateral control, and was again a considerable improvement on its Voisin-built precursor. The first model was to be flown by the English aviator, G. B. Cockburn, of whom more will be heard later. Brabazon, meanwhile, added a 60 h.p. ENV eight-cyclinder engine to his latest Voisin-Farman machine, 'The Bird of Passage', and was immediately rewarded with improved results.[120] By the early months of 1909 he was achieving flights

---

[117] Ibid. See also draft letter to Gabriel Voisin, 24 Sept. 1959, in box marked 'Research, early pioneers 58–68', and Brabazon, *Brabazon story*, 53–4.

[118] Brabazon, *Brabazon story*, 55; *Evening Standard*, 28 June 1930, Brabazon papers, box 12/5a.

[119] H. Farman to J. T. C. Moore-Brabazon, 18 Feb. 1909, ibid. box 49. Note also Brabazon to O. Stewart, 27 Sept. 1963, ibid. box marked 'Research, early pioneers 58–68', where he confesses to having 'inadvertently' been the cause of the Voisin-Farman split.

[120] Among the most prominent early aero-engine manufacturers, ENV were founded as the London & Parisian Motor Co., constructing British-designed engines in Paris. In 1908 the firm patented a 60 h.p. eight-cylinder V-shaped engine, designated 'le moteur en V',

of up to a kilometre and a half.[121] As if symbolising the drift away from Voisin, he had, around the time of this readjustment, arranged to transfer the aircraft to Mourmelon, between Châlons and Rheims, where Farman had chosen to re-establish his operation. Here Brabazon, apparently in conjunction with Farman, began to achieve flights of, reportedly, anything up to five miles, although seldom at heights of above approximately forty feet.[122]

While Brabazon was achieving these successes the Short brothers were securing the honoured position of British contractors to the Wright brothers. As a result, they were to embark upon the manufacture of the world's first 'batch' production of aeroplanes as proprietary articles. Thus, by early 1909, two distinct and contrasting genres of aircraft design were vying for the leading position in aviation. These can be categorised as the 'French' as against 'American' schools of aircraft design,[123] although, ironically, the originators of both – the Voisin and the Wright brothers – were to be almost entirely superseded within the next two or three years: the best features from each genre being amalgamated within new and better aircraft.

Brabazon, in differentiating between the two schools, was to stress the contrasting undercarriages and take-off methods of Voisin as against Wright-derived aircraft; but the differences were more fundamental than that. In the matter of take-off, Wright-derived aircraft relied on a pylon-release mechanism, which propelled the airframe along a mono-rail at great speed, at the end of which the pilot attempted to elevate the machine. Wright aircraft were consequently wheel-less. They landed on skids. The trouble was that this system afforded no opportunity to taxi any such aeroplane, while the prospective pilot familiarised himself with the controls and feel of the craft. The novice was simply propelled along a single-line track. This mono-rail had also to be set facing the wind. If the wind changed direction, as often happened, the device had to be reset. The great advantage of the wheeled-undercarriage was that it left the pilot in control; and Brabazon was fortunate in having learnt

from which the ENV label derived. In 1909 the company was officially renamed the ENV Motor Syndicate and an indigenous factory established at Willesden – distribution rights being acquired by Warwick Wright Ltd. However, ENV motors were soon superseded by the advent of the rotary engine, and the firm subsequently specialised in gear and camshaft production.

[121] Manuscript headed '14/3/40 Aeroplane', Brabazon papers, box 12/5c.

[122] 'Early flight'; 'Lt.-Col. Moore-Brabazon on early aviation', 7. In PRO, AIR 1 724/91/1 Brabazon claims to have actually founded the aviation ground at Camp de Châlons (i.e. Mourmelon) in conjunction with Farman, but there is no corroborative evidence of this and he never claimed it anywhere else.

[123] Again in PRO, AIR 1 724/91/1 Brabazon was to claim that, at this time, he and Farman were the 'only two representatives of the French school of flying, everybody going mad on the Wright type of machine' recently demonstrated at Le Mans. This was an exaggeration, ignoring as it did men like Delagrange. In both these assertions, Brabazon seems to have overstated aspects of his career in anticipation of the official history (Sir W. Raleigh, *The war in the air: being the story of the part played in the Great War by the Royal Air Force*, i, Oxford 1922).

on this second type before moving on, via the Short brothers, to a Wright-derived craft. The Voisin-derived aeroplane, on the other hand, was limited by its narrow margin of lift.[124]

More perceptive critics and experimenters than Brabazon, however, such as J. W. Dunne, José Weiss and the sometime aeronautical journalist, Herbert F. Lloyd, recognised that there was a fundamental difference of concept behind the two types. Dunne and Weiss were both pioneers of inherent stability and their ideas will be examined later. Herbert F. Lloyd, on the other hand, remains a rather obscure figure. Born in 1879, he was a member of the Aeronautical Society who, seemingly on his own initiative, visited France in 1908 in order to investigate the latest aviationary developments. In a series of striking letters written over the summer and autumn months of that year he relayed his observations to the secretary of the Aeronautical Society, Colonel J. D. Fullerton, and in doing so reduced the point at issue to its essence. The real question, he declared, was whether the future of the aeroplane lay with the 'personally-balanced' Wrights' machine, or those precursive 'automatic stability' machines then popularly associated with such French pioneers as Voisin and Farman. The Wrights' machine, he noted, was inherently unstable and its equilibrium derived almost entirely from the skill of its operator. It was thus exceptionally difficult to pilot. By contrast, the French aeroplanes he had seen '[do] really seem to possess a considerable amount of automatic longitudinal stability'. ('I have watched the flights of Farman, Delagrange and Blériot . . . with some attention', he assured Fullerton, as if anticipating some scepticism.) This was predominantly due to the development of the tailplane. The problem was that increased automatic stability left less room for pragmatic lateral control in volatile environmental conditions.[125] It was to be left to the second and third wave of pioneers to synthesise inherent stability and the means of pragmatic lateral control, although the conceptual dichotomy was later to re-emerge – with important consequences for British military aviation.

Brabazon's own sojourn in France was relatively brief – some nine or ten months in all. It was becoming increasingly evident that circumstances were changing in England and he chose to return to his old Aero Club associates, such as Rolls and the Short brothers, at their new establishment on the Isle of Sheppey.[126] Although he would initially use a Shorts' Wright-derived biplane, the Short brothers' own radical change the following year from Wright-derived to Farman-cum-Sommer-derived biplane designs can probably, at least in part, be traced back to Brabazon's French experience. This change was to be the turning point of their careers as aircraft designers; it was to lead directly to an influential series of original Shorts-Admiralty seaplanes.

The ostensible reason for Brabazon's return to England was that the

124 'The Aeropilot'.
125 H. F. Lloyd to Col. J. D. Fullerton, 24 Aug., 15 Sept. 1908, RAeS archive.
126 'Flight in England: Moore-Brabazon's plans', *Evening Standard*, 15 Mar. 1909, Brabazon papers, scrapbook ii.

Aeronautical Society and Aero Club had decided to hold their first joint Aero Show at Olympia in March 1909 and had asked whether 'The Bird of Passage' could be displayed. In the event, it proved to be the only exhibit in the show definitely to have flown. (Neither Roe nor Cody were represented.) Subsequently, it was taken down to the Shorts-Aero Club aerodrome on the Isle of Sheppey, and there Brabazon and the Short brothers began collaboration on a new model of aircraft. During the exhibition the *Daily Mail* had offered a £1,000 prize for the first circular mile flown in an 'all-English' machine. It must have been with a pang of heartache that Brabazon finally cut the tie with France, however. He was cosmopolitan by nature and lacked the provincial roots of someone like Roe. Farman wrote to his former rival on 16 March 1909, wishing him luck and advising caution in future endeavours. 'I hope you will be quite successful . . . flying in England', he remarked, with a hint of sadness; 'and though it is absurd, I will again advise you to be very prudent and try thoroughly the motor with the propeller for hours before flying[,] so to be sure nothing will break.'[127] This was, of course, much easier on the wheeled Voisin-Farman biplane that Brabazon had flown in France: his first 'all-English' machine, by contrast, was to be pylon-released.

On 30 April 1909 Brabazon piloted his Voisin-Farman ('The Bird of Passage') across the Short brothers' Leysdown aerodrome, for what was later to be controversially deemed the first flight in this country by a British pilot.[128] The Isle of Sheppey now became for a time the hub of aeronautical activity, Brabazon either joining or being joined there by such figures as Alec Ogilvie, Maurice Egerton and Francis McClean. Each of these figures was to procure machines from the first historic batch of Short-Wright biplanes. Significantly, they were all men of independent means, indulging their interests, at least initially, in the amateur-sporting spirit of the times.

Such was the stuff of aviation's first paying customers; not the military, or speculating business syndicates, but young, well-to-do adventurers, like Hubert Latham, later Blériot's celebrated cross-channel rival, and Francis McClean, whose philanthropy led directly to the establishment of an Admiralty air arm.

The formation of A. V. Roe & Co.

Having examined Brabazon's disillusionment with the aeronautical situation in England and his return to France, it now remains to pick up the threads of Roe's continued development in the environment from which Brabazon had, temporarily, fled.

---

[127] H. Farman to J. T. C. Moore-Brabazon, 16 Mar. 1909, ibid. box 49. See also 'Early days of aviation', where Brabazon hints at his regret at leaving France.

[128] Brabazon, *Brabazon story*, 58; '14/3/40 Aeroplane'. Here Brabazon confirms that this flight was made in anticipation of the Wrights' planned visit to Leysdown. The flight is usually reported to have occurred over the weekend of 30 Apr.–2 May. In fact, Brabazon flew 150 yards on 30 Apr. and 500 yards on 1 May. The Wright brothers visited Leysdown on 4 May.

By December 1907 Roe was experiencing encouraging 'tow-flights' on his first pusher-biplane. These enabled him to operate the airframe's controls while airborne. The American journal *Aeronautics* carried a report, dated 7 May 1908, indicating the practical value of such artificial flights, and affirming, quite specifically, that they were merely an expedient, pending the long-awaited arrival of a French Antoinette engine. Roe's aeroplane, it stated, was now set for its full trials; but already, towed by an automobile along the finishing straight at Brooklands, it had risen with 'facility' and proved easily controllable.[129]

This means of testing the aircraft had been adopted because Roe's original 9 h.p. JAP engine had proved insufficiently powerful for self-sustained flight. From this *Aeronautics* report, however, the arrival of its 24 h.p. Antoinette replacement can be placed at the first week of May 1908. By coincidence, that was also the beginning of the motor-racing season, and Roe was immediately given notice to quit the track. He was eventually given a temporary stay of execution, provided he made himself sufficiently inconspicuous, but life at Brooklands became increasingly uncomfortable. Roe's small workshop was moved to the rear of the paddock area, on the far side of the clubhouse, and he was told to keep out of sight when club members were *in situ*. So far as the truculent clerk of the course was concerned the period countenanced for the Brooklands flight prize had expired and Roe had no right to be there. This, of course, placed Roe in a quandary, as he had nowhere else to go. But he was not to be defeated. He lay low and countered obstruction with stealth.

Being required to make his flight trials at dawn, when there was little wind and no Brooklands club members to irritate, Roe sought permission to sleep in his shed. This request was brusquely refused; undeterred, Roe did so anyway. At the end of each day he would conspicuously bid 'goodnight' to the gate-keeper, then surreptitiously climb back over the fence and return to his shed for the night, thereby being on the spot at first light. As it was it took him up to two hours to get his machine out of its hangar and on to the track, and another two hours to store it again. This was aggravated by the fact that a five-foot high spiked railing now separated him from the finishing straight. Roe sought permission to make a portion of this barrier detachable; again this was refused, and again Roe went ahead and did so anyway. This was the procedure on normal weekdays: on race-days Roe's shed would be commandeered as a refreshment room and his always delicate aircraft thoughtlessly manhandled over gateposts and across an adjoining dyke-crossed field until it was out of mind, if not sight. The machine seldom escaped unscathed.

It was against this unrelentingly negative background that Roe was obliged to carry out his first full flight trials. To make matters worse, the negativism encountered at Brooklands was echoed in the attitudes of those figures who constituted the aeronautical establishment. The lack of sympathy between A.

---

[129] Report, dated 7 May 1908, from *Aeronautics*, July 1908. See also A. V. Roe, *World of wings*, 40–1.

V. Roe and certain grandees of the Aero Club has already been noted; but Roe was to have perhaps even more reason for resentment at the rebuff he received from the War Office, via the Royal Engineers' Balloon Factory, at about this time.

The Balloon Factory was based at Farnborough, not far from the Brooklands circuit at Weybridge. When he was under notice to quit Brooklands, Roe, who had heard of S. F. Cody's experimental work at the Factory, asked the War Office if he might be permitted to re-establish his operation there. The request was dismissed out of hand. To be so arbitrarily repulsed, or so it seemed, was again a bitter blow, particularly since Cody, who was receiving such favourable treatment, was an American.[130]

The problem of governmental negativism was, moreover, considerably exacerbated the following year when Haldane, with a characteristic regard for what he considered a truly academic and scientific approach to aeronautics, re-established the Balloon Factory along specialist research lines. In pursuit of this scientific approach, mere empiricists (Haldane's phrase) like Roe or the Wrights – or even Cody himself, who was dismissed by the War Office in 1909 – were necessarily disregarded. But Haldane's ideas were pure abstractions, bearing little relation to the reality of the situation. The effect of his initiative was to nearly smother at birth any aeronautical enterprise that might have been developing in this country. Roe's rebuff was the first intimation of what proved to be extremely strained relations between aircraft manufacturers and the War Office over the next eight years.[131] In the meantime, Roe, now with the incalculable advantage of a 24 h.p. Antoinette engine, set out to make the best use of the remainder of his tenure at Brooklands. Aided, on occasion, by two brothers, George and Arthur Halse, he had already refitted his aircraft with a more pronounced triangular-bladed propeller and eliminated the dragging mainplane kingposts. In an effort to gain additional lift, he next extended the top-wing overhang and incorporated two short-span middle winglets within the mainplane structure, thereby creating, in effect, a primitive triplane construction.[132] Full trials began as soon as the new engine was correctly installed. In the event, they were initially hindered by the considerably improved power thereby provided sometimes actually pitching the biplane's propeller off into space. By June, however, this problem seems to have been largely solved. Roe was to claim 8 June 1908 as the date of his first short flight.

The occasion of this first flight was just another solitary dawn trial. No special arrangements had been made regarding witnesses. Much was subsequently made of this fact, but when one recalls that Roe was instructed by

---

[130] Ibid. 45, 51–2. Roe gives no precise dating for this exchange and the original letters do not appear to have survived. However he stated to B. F. S. Baden-Powell, on 16 July 1908 (RAeS archive), that he had recently written to the War Office in an effort to elicit contractual support.

[131] All this is examined in ch. 5. For Haldane's 'empiricists' comment see R. B. Haldane, *An autobiography*, London 1928, 232.

[132] Penrose, *Pioneer years*, 129.

the BARC to undertake his trials at dawn – he had to be off the track and out of sight by between 9.00 and 10.00 a.m. – [133] and that he was, as far as possible, specifically to avoid being seen, it is scarcely to be wondered that there were few witnesses to the event – official or otherwise. The clerk of the course did not want Roe there, let alone hangers-on. Witnesses were later found to some of Roe's subsequent early hops, but detractors like Brabazon refused to accept their retrospective statements as adequate substantiation of Roe's claim. Others, like the Aero Club, refused to accept any hops that Roe may have achieved at this time as in themselves constituting flights.

The problem was compounded by the fact that Roe failed to report any hops or publicly identify any single hop as his first flight until some two or three years later. In his 1939 autobiography he claimed, in particular, a flight of 'about' 150 ft on 8 June.[134] This was rubbished by Brabazon, who had, of course, left Brooklands some months before Roe had both significantly modified his machine and installed the new engine, and who had therefore never actually seen the finalised aeroplane he was so vehemently to denounce over the next half century. He wrote sarcastically to the aviationary press – i.e. *Flight* and *The Aeroplane* – in January 1940, that 'no mention, no claim, not a word was ever said about this "flight" till many years later. Now read Roe's account of his first flight. . . . The claim . . . becomes more detailed as years go by'.[135]

Brabazon was himself exaggerating here. In 1928 the Aero Club had, largely on Brabazon's insistence, initiated an inquiry into the question of who had achieved the first flight in this country, and for a variety of reasons they turned down Roe's claim in favour of Brabazon's (see appendix 1). However, in 1911, some seventeen years before this, the Roe brothers – disturbed by the rumours surrounding A. V.'s 1908 achievements – had themselves tried to instigate a formal investigation into the matter by the Aeronautical Society. To this end those onlookers who had by chance witnessed the June 1908 'flights' (i.e. the head carpenter and the gate-keeper of the Brooklands circuit) were tracked down for signed statements.[136] Unfortunately this investigation does not seem to have materialised in the form envisaged, so the affair dragged on, the old rumours being recycled by those who, for whatever reason, sought to discredit Roe.

The sworn witness testimonies remain, however, as the best independent evidence of what occurred at Brooklands in June 1908. The first, dated 23 May

---

133 Avro reminiscences; L. J. Ludovici, *The challenging sky: the life of A. V. Roe*, London 1956, 52; A. V. Roe to W. Wright, 14 Aug. 1908, Gorell report, Brabazon papers, box 26 (1).

134 A. V. Roe, *World of wings*, 46–7. This was a reiteration of the date he had given in the 'memoir' which he had drafted for the Air Ministry in November 1918, although there he gave no estimated distance: Avro reminiscences.

135 Copy of letter to the press, 10 Jan. 1940, Brabazon papers, box 26(1).

136 H. V. Roe's original letter to the Aeronautical Society on the subject does not seem to have survived, but the society's reply, dated 20 Dec. 1911, indicates what must originally have been suggested: AVRO papers, file 189.

1912, simply states that 'I [W. Boxall] saw Mr A. V. Roe on his flying machine on Whit Monday in the year 1908, rise from the ground of Brooklands Motor Course for about 25 yards [75 ft]. Also on one other occasion I saw him fly for several feet, when his propeller came off and caused him to land.' The other is dated 13 July 1912, and affirms that 'I [E. Harper] saw on a Saturday evening about the middle of June in the year of 1908 . . . Mr A. V. Roe make a flight on his 24 h.p. Antoinette driven aeroplane of about 150 feet at a length of 3 ft high [*sic*] only his propeller blade breaking causing him to come down to the track.'

These statements – and others collected at a later date – clearly do not relate to one particular flight that Roe may or may not have made on the morning of 8 June 1908; they correspond, rather, to a series of hops or flights that Roe made over this period.[137] The presentation of this episode as relating simply to the substantiation of one particular flight claim – which was or was not the first flight in this country – is thus misleading.

Humphrey Verdon-Roe remarked in the April 1910 issue of the *Aeronautical [Society] Journal* that there were a number of such flights that June, most notably on 28 June 1908 when A. V. flew some two feet off the ground for up to 60 yards. This was probably the earliest public reference to such a flight. On 24 May H. V. reiterated the claim in a letter to the editor of the journal *Flight*, but was answered with a request for contemporary press references – of which there were, of course, none.[138] Many years later, after he and Alliott had fallen out, H. V. again championed the June flights. If A. V. Roe's detractors are to be believed, the brothers – who were by then bitterly estranged – conspired to perpetuate the fraud of A. V.'s first flights. This rather stretches credibility.

On 4 July 1908 A. V. Roe was given a fortnight's notice to quit Brooklands and this time there was to be no postponement. 'I was', commented Roe, 'regarded as a . . . liability, rather than an asset.'[139] He left on 17 July and on the 21st was sent a curt note instructing him to sell his shed to the BARC for

---

[137] 'Copy of W. Boxall's certificate', 23 May 1912; 'Copy of E. C. Harper's certificate', 13 July 1912, confirmed April 1928, AVRO papers, file 189. Another chance observer, Mr Herbert Morris, the managing director of Palmers Tyres, also witnessed one of these early 'hops' or 'flights', and emphatically stated so in letters to both Roe and the Gorell committee of 1928–9. As he was unable to accurately identify the date to which his testimony referred, however, other than 'to the best of my recollection . . . the early part of 1908', his evidence was disallowed: Gorell report.

[138] *RAeS Journal* xiv (1910), 90; editor, *Flight*, to H. V. Roe, 24 May 1910, AVRO papers, file 189.

[139] A. V. Roe, *World of wings*, 48; Jarrett, 'A. V. Roe'. Brabazon disputed this interpretation of events and in 1940 contacted the former clerk of the Brooklands course, E. de Rodakowski (by then known as Maj. E. R. Rivers), in an effort to discredit the story. However the latter merely confirmed a personal animosity for Roe: J. T. C. Moore-Brabazon to E. de Rodakowski, 17 Jan. 1940; de Rodakowski to Moore-Brabazon, 16 Feb. 1940, Brabazon papers, box 26(1). De Rodakowski's obituary notices were to corroborate Roe's recollections. *Autocar* subsequently described him as 'somewhat of a martinet' (18 Feb. 1944).

the undervalued price of £15 or remove it immediately.[140] He had little option but to sell it. Thus by August he was without a test-base. By chance the Wright brothers were at this time embarking on their historic demonstration flights, Wilbur travelling to Le Mans, in France. So, in an effort to elicit an invitation to join him, Roe forwarded a cordial letter on 14 August 1908, detailing his latest experiences. Unfortunately the ambiguous wording of this communication only further added to the luckless Roe's troubles. In it he claimed both to have made six trials with his aeroplane after the arrival of the more powerful Antoinette engine ('fast making progress'); and to have made several flights towed by a car ('the power required being very slight'). Roe's critics, like Brabazon and Griffith Brewer, later jumped on this to suggest that the six trials were the tow-flights and hence Roe's subsequent claim to have actually 'flown' in June 1908 was demonstrably a later fabrication, disproved by his own words. This, however, was what they wanted to read into the letter; the text, in fact, could equally well mean the opposite. (The tow-flights had actually been initiated in December 1907: five months before the Antoinette engine arrived.) The Wright brothers – or more particularly Orville Wright, as Wilbur died in 1912 – probably prompted by Griffith Brewer, themselves subsequently read the letter in this negative way. It was to plague Roe for the rest of his life.

Less well known is the fact that A. V. had written to Baden-Powell the month before this, on 16 July 1908, making similar assertions. Obviously in some distress at being ordered out of Brooklands, this letter was written with the aim of interesting Baden-Powell in joining him and unnamed associates in the formation of a proposed aviation company. (He also asked Baden-Powell if he knew of anyone who might like to buy his shed 'cheap' at £38, which gives a better idea of its true value.) In the letter he is keen to stress that his own aeroplane was on the very verge of success: indeed, he had clearly experienced something to excite him already, although unfortunately his prose is characteristically muddled and confusing. He distinguishes between the tow-flights and the six Antoinette trials, and just as to Wilbur Wright he reported that he was 'fast making progress' so to Baden-Powell he wrote that he had already 'nearly' left the ground and that 'next time out' promised 'something interesting', if only de Rodakowski would stop bullying him, but that he would not 'listen to reason'. The implication is clearly that his work had, so to speak, turned a corner. His report of propeller difficulties corroborates the 1912 witness testimonies and, providing corroboration of his later testimony to the Gorell committee of 1928-9, it is evident that he regarded these six trials as no more than preliminaries, carried out in extremely trying circumstances and being both exceptionally difficult to monitor accurately and of no particular uniqueness, given what others had achieved and what he himself was imminently expecting to achieve. The tone of the letter is naturally one of concern

---

140 E. de Rodakowski to A. V. Roe, 21 July 1908, Brooklands Museum, quoted in Ludovici, *Challenging sky*, 53.

for his immediate future, not premature celebration for some undefined and curtailed achievement.[141]

Shortly after his letter to Wilbur Wright, Roe set off on his bicycle to visit him in France. He was not persuaded that Wilbur's aircraft designs were leading in the right direction. Roe's own design development by now owed little to the Wright brothers.[142] Yet he obviously identified himself with their provincial, mechanical, empirical background, out of which they had designed, tested and piloted their own aircraft without the aid of scientific (i.e. academic) or governmental support. Unlike the Wrights, however, Roe's own aeroplane designs were now moving towards the tractor configuration, i.e. with the propeller in front of the engine, pulling rather than pushing the aircraft, which was soon to become universal. The monoplane tractor configuration had already been adopted by both Blériot and Levavasseur – designer of the Antoinette engine – in France.

It was shortly after leaving Brooklands that Roe first dispensed with his aircraft's front-elevator. He then began work on a light-weight tractor-triplane. This incorporated a four-bladed propeller and a 23 ft-long triangular-section fuselage. It has since been designated the Roe I Triplane,[143] although the first intimations of a three-layered mainplane had been evident in Roe's Brooklands pusher-machine. Vertical control of this new aircraft was effected by changing the 20 ft span mainplanes' angle of incidence, while lateral control derived from warping the mainplane-section. The fuselage culminated in a 10 ft span three-layered tailplane, to which a rudder was added for directional control, as the mainplane warping proved insufficient in this respect. (This had, of course, been a Wrights' innovation.) Within the fuselage itself, the machine's petrol tank and pilot's seat were sprung on elastic straps to reduce shock; and finally, to reduce both weight and costs, the largely wooden frame was coated with yellow cotton oiled paper. As a result, the first Roe triplane was sometimes labelled the 'Yellow Peril', although that particular sobriquet is more usually associated with the Handley Page Type E.

By this time Roe's expensive 24 h.p. Antoinette engine had been returned to the manufacturers – he had only had the use of it for about three months – so the new machine was initially forced to carry the cumbersome 9 h.p. JAP first used at Brooklands. The body of the aircraft was constructed in Spencer Verdon-Roe's converted stables at West Hill, Wandsworth, where A. V. had worked before moving to Brooklands. The need to find a new flying ground, on the other hand, remained imperative.[144]

---

[141] A. V. Roe to W. Wright, 14 Aug. 1908, facsimile retained in Brabazon papers, box 26(1); A. V. Roe to B. F. S. Baden-Powell, 16 July 1908, RAeS archive.

[142] A. V. Roe, World of wings, 26–7, 51; Penrose, Pioneer years, 152. In his autobiography A. V. mistakenly dated his visit to Le Mans as 1909, a mistake unfortunately repeated in subsequent biographies. Even Penrose was misled by this.

[143] A. J. Jackson, Avro aircraft since 1908, London 1965, 12. The relevant section of this book has since been revised to incorporate additional information (2nd edn, 1990, 4–8).

[144] A. V. Roe, World of wings, 51.

Having been ejected from Brooklands and refused permission to experiment at Farnborough, Roe next applied to use the military's land at Wormwood Scrubs. His request was turned down. He then applied to use Wimbledon Common, with the same result. In desperation, he finally looked for an area of common ground where no permission would be needed, and none would be sought. In this way Roe alighted upon the Lea Marshes, in Hackney. The site afforded him the opportunity of renting a couple of railway arches under the Chingford branch of the Great Eastern Railway. In these he could accommodate both his machine and an expanding workshop.[145]

While at Lea Marshes A. V. Roe was joined by John Alfred Prestwich, founder of the JAP motor-cycle engine firm. Born in 1874, the son of a successful photographer, Prestwich had already made a number of significant contributions to the development of cinematography. In 1896, for example, he had patented the Greene-Prestwich projector, followed in 1898 by the Prestwich ciné camera. It was Prestwich's devices, indeed, which had recorded for posterity Queen Victoria's Diamond Jubilee in 1897. In 1901 he had then designed the first JAP single-cylinder four-stroke motor-cycle engine. Within two years JAP were working in formal co-operation with the Triumph Cycle Company of Coventry. Thereafter Prestwich's firm became one of Britain's leading motor-cycle engine manufacturers.

At some point Prestwich and Roe had agreed to form a partnership. What its precise nature was is difficult to determine, since for some reason Roe chose not to mention it in his autobiography. H. V. Roe, who clearly had access to additional documentation, dated it from 15 September 1908, with Prestwich putting up capital of £100 and possibly agreeing to supply an improved 35 h.p. engine for an original rectangular-fuselage triplane.[146] Through the offices of Prestwich, Roe then managed to gain his first order for this proposed machine. This came from Charles Friswell, the well-known London agent of the Standard Motor Co. The railway arches were soon a hive of activity. Unfortunately, however, the new engine never materialised and the Friswell deal fell through. The aircraft's frame was sold by public auction in the Friswell sales rooms at Albany Street, realising £5 10s. The Prestwich-Roe collaboration collapsed,

---

145 A. V. Roe, *World of wings*, 52; Avro reminiscences. Wormwood Scrubs was still being used for military training at this time. The following year the Parliamentary Aerial Defence Committee would purchase a Clément-Bayard dirigible for the army and receive permission to house it there, in a hangar provided by the proprietors of the *Daily Mail*.

146 H. V. Roe, 'Pioneers'; Penrose, *Pioneer years*, 148–9. For JAP background see J. Clew, *JAP: the vintage years*, Yeovil 1985. Here it is noted (p. 50) that in October 1908 Prestwich sought permission to erect a shed on Lea Marshes for aviation trials. 'Balance sheet for 1 January 1909', AVRO papers, file 163, also records the building of a shed on the site, but again makes no mention of the JAP partnership. A. V. had evidently been searching in many directions for enterprising partners for various proposed aeronautical concerns. Evidence of these schemes, of which the Prestwich collaboration was just the latest, can be gleaned from his correspondence with B. F. S. Baden-Powell: RAeS archive, 16 July, 16, 18 Sept. 1908.

probably with some recriminations. A. V., however, completed the construction of the Roe I Triplane, and immediately began fresh flight trials.[147]

This time, far from having no witnesses to his activities, Roe's major concern was to keep people sufficiently at bay to enable him to get on with his work. His assistants, like Richard Howard Flanders and E. V. B. Fisher, would cordon off spectators, while Roe attempted to taxi the machine. Flanders later recalled that the Lea Marshes were far from ideal for this purpose, being covered in stumps and divided by fences. However, through continual effort, encouraging results were gained.[148] On 11 May 1909 Roe wrote to B. F. S. Baden-Powell (who had recently established his own *Aeronautics* journal) that he had already succeeded in getting the machine to rise some six inches off the ground.[149] When, at the close of May, a replacement 10 h.p. JAP engine became available, improved hops of up to 50 ft in length immediately resulted.

Work would begin in the early hours, trials being confined to periods of virtually no wind. In the initial runs, helpers would lie on the ground and watch a chosen wheel – shouting when their wheel rose off the ground. If all shouted together, all wheels were off the ground and the machine was airborne. On 13 July 1909 a hop of 100 ft was recorded, and by 23 July a flight of 900 ft was sustained, at an average height of about 10 ft. Once under way, Roe's assistants would chase after the machine with a fire extinguisher (one was evidently carried by bicycle), as fire was an ever present possibility on crash landings,[150] and a crash upon landing was virtually inevitable. If by some chance the machine landed undamaged, it would be taken up again and again until it was damaged. Then it would be hauled back to the arches and repaired. 'The average programme', commented Roe, was 'two weeks work, a 50 yard "hop", a "crash", and then the programme repeated all over again.' But it was soon apparent that Roe now had the makings of a fully practical aeroplane.

This was stressed in a letter of Roe's published in the journal *Flight* on 26 June 1909. By publicly disclosing his triplane's success Roe hoped to elicit a degree of investment, but his words were again subsequently turned against him. Critics later argued that the letter proved that his 'first flight' claim initially related only to June 1909 (see appendix 1). These critics, however, knew nothing of Roe's circumstances at the time and made little effort to find

[147] A. V. Roe, 'Trials, troubles and triplanes', *The Aero*, April 1912, 96; H. V. Roe to C. G. Grey (draft), 11 May 1928, AVRO papers, file 64; H. V. Roe, 'Pioneers'. The latter records that A. V. received £4 19s. from the auction. Penrose records that A. V. and Friswell fell out, Friswell removing the triplane frame from the railway arches, afterwhich it was unsuccessfully converted to a Blériot-derived monoplane by 'a Captain Loveless', and sold off unfinished: *Pioneer years*, 150. This is presumably a reference to Captain J. G. Lovelace, who flew (and crashed) a Blériot at Doncaster in October 1909, and who acted as chief engineer of Humber's aviation department. Neither A. V. nor H. V. Roe made any reference to such a development.

[148] Howard Flanders quoted in *RAeS Journal* lxii (1958), 234–5 (A. V. Roe obituary).

[149] A. V. Roe to B. F. S. Baden-Powell, 11 May 1909, RAeS archive.

[150] A. V. Roe, *World of wings*, 53, 56; Howard Flanders quoted in A. V. Roe obituary. The risk of fire was a constant fear. See also T. O. Smith memoir, Rolls papers.

out. Had they done so they might have realised that the letter was a cry for help, for at this critical stage in his aircraft's development all was threatened by the failure of Roe's partnership with Prestwich and the lack of adequate financial backing. 'I feel confident that [my] machine . . . has reached a stage well worth . . . building in numbers', Roe asserted. No further support was as yet forthcoming, however, and that July the local authorities issued A. V. with a summons and ordered him to quit the marshes.[151]

Now Roe was desperate; here he was on the very edge of success and the ground was again being cut from under his feet. *Flight* reported on 9 October 1909 that he had concluded negotiations with the authorities of the old deer park at Richmond to continue his work there, but nothing came of this. Eventually, through the intervention of the sympathetic motor entrepreneur, Captain Walter Windham, himself an eccentric aircraft-designer and founder of the short-lived Aeroplane Club, Roe re-established his operation at Wembley Park. Windham had reached an agreement with the park's proprietors allowing Aeroplane Club members – of which Roe became one – access to the land together with accommodation rights. A. V. was now employing a second (20 h.p. JAP) Roe I Triplane, originally constructed for the first Blackpool aviation meeting of October 1909. At about the same time as his move to Wembley, in November 1909, the Prestwich-Roe partnership was officially dissolved.[152]

During this period of flux Humphrey Verdon-Roe first seriously entered the fray, in an effort to help A. V. out of his financial predicament. He later commented that he originally joined forces with Alliott as early as April 1909, with the limited aim of helping him find a backer – their partnership amounting to no more than a 'verbal brotherly agreement'.[153] However, H. V.'s account book recorded on 27 April 'cash to clinch the Agreement £1', suggesting something slightly more formal.[154]

By the end of the year, no financier having been found, Roe's father himself refused to lend any more money toward the aeronautical enterprise. Indeed, he was even to prevent another potential backer from investing in it. A young friend of Captain Windham, Lieutenant J. M. Kenworthy RN, had been persuaded to put up £2,000, but as he was only twenty-three years old his lawyer refused to ratify the deal. On learning of it, Roe's father personally refunded the money.[155] It thus fell to H. V. to take full financial responsibility for the Avro scheme or let it die. He decided to take it on and, on 1 January 1910, the firm of A. V. Roe & Co. came into being.

---

[151] 'Mr Roe's triplane', *Flight*, 26 June 1909. For the summons see *The Aero*, 3 Aug. 1909, and A. V. Roe, *World of wings*, 57–8.
[152] H. V. Roe, 'Pioneers'.
[153] Ibid. It may have been under Humphrey's influence that A. V. sent the June letter to *Flight*.
[154] H. V. Roe, 'I financed an aircraft pioneer'.
[155] J. M. Kenworthy (Lord Strabolgi) to H. V. Roe, 12, 19 Oct. 1942, AVRO papers, file 122.

In fact, formal backing of A. V. seems to have been a family decision. The father refused to provide A. V. with any further funds directly, but – prompted by Mrs Roe – would lend money to H. V. for him to finance A. V. This presumably reflected their distrust of A. V.'s business acumen.[156] It was hardly the most auspicious beginning to what was to become, by 1914, perhaps the most influential aircraft firm in England. The agreement left A. V. free to continue inventing, experimenting and constructing, while responsibility for finance, organisation and management was turned over entirely to H. V. If the company were successful, profits would be halved; if it were a failure, then H. V. would underwrite all debts.[157] The fact that H. V. had initially been drawn into his brother's affairs in April 1909 demonstrates how early the Prestwich-Roe partnership must, in reality, have collapsed, even though it was not formally dissolved until November 1909. Presumably H. V. Roe had to end this agreement officially before he was free to establish the firm of A. V. Roe & Co. in January 1910.

By the following March A. V. was back at Brooklands, which had reopened as a dual aerodrome and motor track with a new sympathetic clerk of the course, Major Lindsay Lloyd RE.[158] Here he was presently to begin development of the first Avro tractor-biplane. Roe's belief in this genre of design was soon to be entirely vindicated.

The formation of Short Brothers Ltd

Returning to the Brabazon-Shorts connection, it is now necessary to examine the origins of the Short brothers' association with the Wright brothers and the subsequent founding of Shorts' first aeroplane factory for the production of Wright-derived aircraft.

It was probably the excitement caused within the Aero Club by the Wright brothers' public demonstrations of August and September 1908 that finally induced the Short brothers to unequivocally embrace aeroplane construction. Horace Short was persuaded to return to his brothers that December, his four-year contract with Charles Parsons being about to expire, and by January 1909 their first order from an Aero Club member for a Wright-derived pusher-biplane – the Shorts No.1 – had been received. By then the brothers had already put up £200 each as capital for the new venture: indeed, the reorganised firm had been officially registered as early as November 1908.

The brothers were still based at the old balloon workshop at Battersea at this time. Obviously this was no environment for aeroplane trials, so this branch of activity was immediately moved east, to Leysdown-Shellbeach, on the Isle of Sheppey. This was an isolated marshland area, originally singled out

---

[156] H. V. Roe to C. G. Grey, 18 Jan. 1916, ibid. file 47; A. V. Roe, *World of wings*, 64–5.
[157] H. V. Roe, 'Pioneers'.
[158] C. G. Grey, history of Brooklands, PRO, AIR 1 727/160/2; Boddy, *Brooklands*, i, 54.

by Griffith Brewer, although it is unclear whether he already envisaged it as a suitable location for a Wright aircraft manufactory.[159] The construction of new workshops on the site had begun by February 1909. Here the brothers completed the Shorts No.1 (which never actually flew)[160] for Francis McClean, and the Shorts No.2 (ordered in April 1909) for 'Moore-Brabazon'.[161] Meanwhile the Aero Club acquired the use of a clubhouse on the site. This was Mussel Manor, the property of the local landowner, J. D. F. Andrews, a developer who foresaw commercial benefits from encouraging both sporting and industrial aeronautical facilities in the area.

To establish how Shorts at this point came to be appointed the Wright brothers' British contractors we must go back to the Wrights' public demonstrations in France in the autumn of 1908. There were then several parties chasing the brothers in an effort to secure their patent rights. H. F. Lloyd described the situation to Colonel Fullerton in a letter of 11 August 1908. The Wrights' patents in Europe, he explained, were mainly being negotiated through the offices of the international armaments dealer Hart O. Berg – whom Lloyd describes as the 'American Jew who bought up the Holland submarine boat patents'. Berg, however, had failed to secure control of the UK patent rights, though he was anxious to do so. Wilbur Wright, it seems, felt a personal affinity with England, and preferred to deal with that contract himself.[162] He had already received 'one or two offers' for the British rights, but turned them down because, as Lloyd reports, 'he only desires that [these] should be taken over by a group of people who are interested in aeronautical work for its own sake, and who would make the English company a success'.[163]

In December 1908 another party – this time composed of members of the Aero Club of the UK – travelled down to the Camp d'Auvours, near Le Mans, with the aim of securing the Wrights' British patent rights. Led by Lord Royston, this faction took with them Eustace Short as representative of the club's official aeronautical engineers. As a result of this visit Eustace was first taken up in a Wright Flyer. He was probably already known to Wilbur Wright

---

[159] G. Brewer, *Fifty years of flying*, London 1946, 81–2. Leysdown was the site of the new aerodrome, but in contemporary literature it is also referred to as 'Shellbeach' – and sometimes, inaccurately, as 'Shellness', which was some two miles east.

[160] It was originally fitted with a 40 h.p. Nordenfelt automobile engine and, like the Wright Flyer, rail-launched (Sept. 1909), but to no effect: C. H. Barnes, *Shorts aircraft since 1900*, London 1989, 42.

[161] Delivered 23 Sept. 1909. It initially employed the Vivinus engine from Brabazon's original Voisin biplane: Bruce, 'Shorts aircraft: new evidence', 18.

[162] This must have been a change of heart subsequent to the public unveiling of the Wrights' aircraft, as there was little evidence of such an affinity, in any practical sense, prior to 1908–9. The early abortive Wrights' negotiations are examined in P. B. Walker, *Early aviation at Farnborough: the history of the Royal Aircraft Establishment*, ii, London 1974, and Gollin, *No longer an island*. Both studies cover the same ground and stop short of the Short-Wright collaboration.

[163] H. F. Lloyd to Col. J. D. Fullerton, 11 Aug. 1908, RAeS archive. Lloyd himself tried to raise the necessary support to secure the rights.

through the reports of the Short brothers' patron, C. S. Rolls, who was indeed in France at this time, but the decisive introduction on this occasion appears to have come from another Aero Club member, Griffith Brewer. Brewer was implicitly trusted by the Wrights and was shortly to become their British patent agent.[164] At any rate, although no agreement was reached with the Royston coterie, some rapport seems to have been established between Wilbur and, through Eustace, the Short brothers. Horace rejoined the aeronautical business shortly afterwards.

Wilbur and Orville Wright eventually abandoned the idea of selling their British patent rights to any particular group, and decided instead to sub-contract directly to a suitable indigenous constructor. In February 1909 Short Brothers were selected for this role. By this time they were already based at Leysdown-Shellbeach, where they were working on Francis McClean's Shorts No.1 biplane (Wright-derived, but designed by Horace Short). On receiving the Wrights' proposal, Horace immediately crossed to Pau, in southern France, from where he was able to draft a set of working drawings of the Wright Model A Flyer. By March 1909 a contract for the construction of six Wright Flyers had been agreed.[165]

Despite, or perhaps because of, the novelty of having a batch of half-a-dozen proven aeroplanes on the open market, orders for all these machines were not finally received for some months. The first was earmarked for C. S. Rolls, who had been trying to acquire a Wright Flyer since even before the Short-Wright contract had been agreed. (This was the aircraft that he eventually placed at the disposal of the War Office.) Indeed, in his impatience to learn to fly Rolls had evidently been badgering the Wrights again. On 31 March 1909 Orville wrote to Griffith Brewer that arrangements had been made for Rolls to be taught by Comte Charles de Lambert, a member of the French Wright Company, who had been among Wilbur's first pupils earlier that year, but that this would take some time since he was only sixth in line and de Lambert had only the one machine. Faced with this prospect Rolls seems to have decided to teach himself the rudiments of flying. To this end he ordered a Wright-derived, rail-launched glider from Shorts in May 1909. It was delivered in July.[166]

164 The entire Aero Club party – consisting of Lord Royston, Roger Wallace, Prof. A. K. Huntington, Francis McClean and Eustace Short – was formally introduced to Wilbur Wright by Griffith Brewer: G. Brewer, 'Wilbur Wright', RAeS Journal xvi (1912), 149, but Eustace had already visited the French Wright trials on his own initiative and had known Brewer for some years.
165 Bruce, 'Shorts aircraft: new evidence', 12. The same month as Shorts were selected as Wright sub-contractors C. S. Rolls unsuccessfully proposed to the board of Rolls-Royce that they seek to acquire the Wrights' British patent rights: Bruce, Charlie Rolls, 25.
166 Idem, C. S. Rolls: pioneer aviator, Monmouth 1978, 15–18; O. Wright to G. Brewer, 31 Mar. 1909, RAeS archive. In fact, Orville Wright advised Alec Ogilvie, who presently took the second Short-Wright machine, to do the same; practise first on a glider, then gradually and cautiously take up the powered Flyer: O. Wright to A. Ogilvie, 17 Sept.

The other machines from the batch were not claimed so unequivocally. The second was to be taken by Alec Ogilvie and, as Orville Wright records in his letter of 31 March to Griffith Brewer, by the end of March 1909 an order for the third had been received from Francis McClean. The rest remained for a time unsold. Another order had subsequently come in for No.3, reported Orville Wright, but 'as this [is] already reserved for Mr McClean, we are reserving No.4 for this gentleman [presumably Maurice Egerton] if he wants it. No.5 is therefore the best we can now promise to Lord Montagu'.[167] In the event Lord Montagu was not among the purchasers, but, in any case, delivery of the prescribed Wright-designed Bollée engines was delayed until September 1909, so little could be achieved before that date. Rolls took delivery of the first Short-Wright on 1 October 1909.[168] The last two machines were not finally sold until the turn of the year – one to Percy Grace (brother of Cecil Grace) and a second to C. S. Rolls.[169] They retailed at £1,000 each, with a £200 deposit upon order. From this the Short brothers received just £200 per aircraft – a sum which had to cover construction costs, although the French-built Bollée engines, which would have been the most expensive item, were supplied by the Wrights themselves.[170]

All the aircraft from this particular batch seem to have been in regular use by the early months of 1910, initial difficulties centring on the predictable inefficiency of the Bollée engines rather than on the inherent features of the aeroplanes themselves.[171] For all that, however, the Short-Wright sub-contract was not to be repeated. It was a great boost to the establishment of Shorts as a leading manufacturer of aircraft, and it equally effectively established the Isle of Sheppey as one of the leading centres of aviation, but design technology was moving on and the Short brothers with it. Prompted by Brabazon,[172] and in the wake of the first Rheims meeting, the brothers chose instead to adapt their Wright-derived manufacturing skills to the construction of their own more stable Farman-cum-Sommer-derived aircraft, which were designated S27 Type and which incorporated ailerons and a Sommer mono-tailplane section. It

1909, ibid. (Ogilvie ordered a glider from T. W. K. Clarke, after a dispute with Horace Short: A. Ogilvie to J. Laurence Pritchard, 24 Sept. 1954, ibid.)
167 O. Wright to G. Brewer, 31 Mar. 1909, ibid. Note also Wright to Brewer 18 Mar. 1909.
168 M. J. H. Taylor, Shorts: the planemakers, London 1984, 36. The Short-Wright No.3, purchased by Francis McClean, was ultimately powered by a Bariquand et Marre engine.
169 No.5 was delivered to Percy Grace on 11 Jan. 1910, and first flown – by Cecil Grace – on 14 Feb. No.6 was delivered to C. S. Rolls on 23 Mar. 1910, and flown the following day: Taylor, Shorts, 36. Rolls also acquired, in Apr. 1910 (eighteen months after it was ordered), a French Astra-built Wright Flyer with a Bariquand engine, and it was on this Wright machine – not a Short-Wright – that he was killed in July 1910.
170 O. Wright to A. Ogilvie, 8 Oct. 1909; O. Short to J. Laurence Pritchard, 30 Mar. 1948, RAeS archive. See also Bruce, 'Shorts aircraft: new evidence', 12. The Bollée engines reportedly cost up to £400 each: Donne, Pioneers, 24.
171 Bruce, 'Shorts aircraft: new evidence', 16.
172 Brabazon, Brabazon story, 63.

proved to be a crucial decision.[173] Ogilvie, McClean, Egerton and Percy Grace all joined Griffith Brewer in the creation of a new British Wright Company in 1913, but this was essentially only in an effort to secure compensation for various supposed infringements of the Wright brothers' warping patent.

Meanwhile, utilising a 60 h.p. Green engine,[174] Brabazon prepared his Shorts No.2 (of Horace Short's own Wright-derived design, incorporating interplane 'balancers', i.e. primitive ailerons, and thus not to be confused with any of the six Short-built Wright aircraft) for an attempt on the *Daily Mail* all-English circular mile award. S. F. Cody had actually taken British citizenship during the unofficial Doncaster flying meeting of October 1909 as a preliminary to going for this himself, but he was pre-empted by Brabazon, who took the prize with a flight made at Leysdown on 30 October 1909. However, Brabazon was himself already veering back toward the use of French-derived aircraft at this time. At his request Shorts constructed a S27 variant, designated the S28, for use at the Bournemouth international aviation meeting in July 1910. Following Rolls's tragic accident, however, Brabazon withdrew from the event.[175] The machine was subsequently purchased by Francis McClean, and was to be among the aeroplanes placed at the disposal of the first volunteer naval pilots at Eastchurch.

Development of the Isle of Sheppey and its pioneers

Thus far the inter-related development of the leading figures of the second wave of aviation pioneers has been traced up to the point where they were about to establish what might legitimately be called a productive aeronautical environment. It now remains to outline their future course before turning to that final, third wave of pioneers, who carried the aviation industry into the First World War. Of course, such designations are to some extent arbitrary; much development was simultaneous, with (to change the metaphor) a new crop of pioneers all the time germinating under the surface. However, it gives to the study of so many diverse elements a sense of shape and proportion that would otherwise be lacking.

Of the more influential figures examined in this chapter the most immediately dramatic fate befell C. S. Rolls. Rolls had gained widespread publicity on account of his piloting exploits, notably at the Nice international flying meeting of April 1910, where despite trying circumstances he sustained a total

---

[173] Shorts, in fact, initially designated these new designs 'Short-Sommer' in their order book. Rolls had acquired his own Sommer biplane by March 1910, so the brothers would have been familiar with the design.

[174] Product of the Green Engine Co., Britain's most successful pre-war aero-engine manufacturers. (See appendix 2, *s.v.* Green, Gustavus.)

[175] Brabazon, *Brabazon story*, 63, 65. For the adoption of balancers as the method of lateral control in Brabazon's Shorts No.2 see J. T. C. Moore-Brabazon to E. C. Shepherd, 30 June 1934, Brabazon papers, box 26(2).

of 75 miles over-seas flying (32 miles non-stop), and during his double crossing of the English Channel in a French-Wright Flyer the following June.[176] This was all part of a campaign to publicise aviation in the same way as, a decade earlier, he had sought to publicise automobiling. Following this, he was planning to instruct selected army officers in the use of the Short-Wright biplane he had made available to the War Office. Indeed, a shed had already been erected at Larkhill on Salisbury Plain for this purpose.[177] Surviving evidence also suggests that Rolls was contemplating the formation of his own aircraft company, and plans for what has been termed a Rolls Powered Glider have been discovered.[178] But little had been achieved in this direction by the time of his tragically early death as the result of an accident during an alighting competition at the Bournemouth aviation meeting in July 1910. Approaching the landing target from an excessively steep angle, in the manner made famous by Captain Bertram Dickson, Rolls's French-built Wright Flyer's tailplane broke and he plunged to the ground from a height of some 80 ft.[179] So passed one of the new technology's most consistent and effective advocates. Rolls was a man born to position, but who used it in the cause of mechanical progress rather than its disparagement. The circumstances of his death, however, initially caused a negative reaction, and nowhere more so than in the career of his friend and accomplice Moore-Brabazon.

The Rolls tragedy, Brabazon remarked, 'sickened me . . . of aviation'.[180] He never used the Shorts S28 he had ordered for Bournemouth. Indeed, he temporarily abandoned aeronautics altogether, and was not to fly again until the war. As it turned out, however, he was to remain an important figure in aviation for the rest of his life. After serving in the RFC during the First World War he went into politics, his adoption as a Conservative candidate being sponsored by J. W. Dunne's aeronautical patron, the duke of Atholl.[181] He was elected MP for Chatham in 1918 and became Parliamentary Private Secretary to the Secretary of State for Air and War, Winston Churchill.

Later, as MP for the Wallasey Division of Cheshire, Brabazon was to become, first, Minister of Transport, and then Minister of Aircraft Production in Churchill's wartime Cabinet. He was forced to resign in 1942, after indiscreetly suggesting in public that Nazi Germany and Soviet Russia be left to 'cut each other's throats'.[182] He was subsequently raised to the peerage and in 1946 elected president of the FAI. However it was through the Brabazon Committee

---

176 Bruce, *Charlie Rolls*, 33–4.
177 T. O. Smith memoir, Rolls papers; reminiscences of G. B. Cockburn, PRO, AIR 1 733/192/1.
178 Montagu of Beaulieu, *Rolls*, 206–7; Bruce, *Charlie Rolls*, 35–6. The 'powered glider' seems to have been essentially another Short-cum-Wright-derived design.
179 See Bruce, *Charlie Rolls*, 36ff.
180 Brabazon, *Brabazon story*, 66.
181 Duke of Atholl to J. T. C. Moore-Brabazon, 22 Apr. 1917; J. T. C. Moore-Brabazon to duke of Atholl, 28 Apr. 1917, Brabazon papers, box 49.
182 A. J. P. Taylor, *English history 1914–1945*, Oxford 1965, 641.

on Civil Aviation that his name reverberated loudest, being forever embodied in the huge post-war Bristol Brabazon pressurised airliner prototype.

Two of the leading figures of the Sheppey aerodrome were thus lost in a little more than a year following its establishment. It remained, however, one of the principal centres of aeronautical activity. J. W. Dunne, recently released from his War Office commitments, moved there in late 1909. The following year arrangements were made for Shorts to construct two Dunne aircraft.[183] The site's growing reputation itself fuelled the expansion – and this despite the fact that, over the winter of 1909-10, flooding necessitated transferring the flying area some four or five miles west, to the village of Eastchurch. Here Francis McClean had acquired a more suitable site on the south side of Stamford Hill, which he generously placed at the Aero Club's disposal. He also provided Short Brothers with a more satisfactory three-acre test-site for the nominal cost of £25. This was desperately needed as the company was now constructing wheel-launched rather than Wright rail-launched aircraft. By September 1910 there were eighteen sheds occupied by Aero Club members *in situ*.[184] In December 1910 Thomas Sopwith flew from Eastchurch to win the Baron de Forest prize for the longest flight from England to the continent.

But the most crucial event in the development of the Eastchurch airbase at this time was Francis McClean's offer to the Admiralty of the loan of two Short-Sommer biplanes for the flying instruction of selected naval officers; an offer which included the use of accompanying facilities at the new flying ground. Having no indigenous aeroplane capacity the Admiralty were quick to respond. Receiving formal notification of the proposal in November 1910, a General Fleet Order signalling the service's intention to accept was issued by Admiral Sir Charles Drury, the Commander-in-Chief at Nore, on 6 December.[185] The original agreement was for the loan of two S27 Type pusher-biplanes (S28/S29). In the event, however, through a close association with Short Brothers, trials were undertaken on a variety of models.

Four unmarried volunteers were called for, to go to Eastchurch for flying instruction. More than 200 came forward. The four officers ultimately selected were Lieutenants C. R. Samson RN, A. M. Longmore RN, R. Gregory RN and G. Wildman-Lushington RMA – subsequently replaced by Lieutenant E. L. Gerrard RMLI. In the general spirit of altruism the private aviator, G. B. Cockburn, who had become an unofficial instructor to the early army pilots at Larkhill, then agreed to instruct the volunteers at no cost to the Admiralty. It was, in fact, given to the Aero Club to find an instructor, and they originally

---

[183] PRO, AIR 1 2400/293/1. The first of these was constructed in association with A. K. Huntington, professor of metallurgy at King's College, London. As a result it is sometimes known as the Huntington monoplane – although really of an exaggeratedly staggered biplane configuration. The second was a standard Dunne design (a pusher-biplane with V-shaped swept-back wings) built in conjunction with the Blair Atholl Syndicate.
[184] PRO, AIR 1 2400/293/1; Donne, *Pioneers*, 30.
[185] C. F. Snowden Gamble, *The air weapon: being some account of the growth of British military aeronautics*, London 1931, 149–50.

earmarked for the job Cecil Grace, whom Cockburn himself described as 'probably the best pilot of those days'. Only after Grace's death – he was lost at sea on 22 December 1910, while flying back from an attempt on the Baron de Forest prize – was Cockburn asked to undertake the task.[186] In addition the selected officers were to undergo technical training under the supervision of Horace Short at the Shorts' workshops. For this, however, the Admiralty would pay a fee of £20 per volunteer to Short Brothers.

The four naval volunteers reported to Eastchurch on 1 March 1911, and began work the following day. Their instruction was to last six months, during which time they were responsible to Captain Godfrey Paine, the Commanding Officer of HMS *Actaeon*, at Sheerness, and through him to the Commander-in-Chief at Nore. Under Cockburn's adroit supervision they quickly mastered the rudiments of flight control. On 25 April 1911 Samson and Longmore took RAeC certificates Nos 71 and 72 respectively, and on 2 May Gregory and Gerrard took Nos 75 and 76. The following August, on conclusion of an intensive course of workshop and technical training,[187] their collective report was submitted by Lieutenant Samson, the designated senior officer. 'The rapid progress in the science of aviation is apparent to all who interest themselves in this branch [of activity]', it affirmed. 'Few people now deny that the aeroplane has come to stay.' With regard specifically to naval warfare, it stated that 'the great value of aeroplanes at the present . . . lies in their scouting capabilities'. It recommended that a definite offer be made for McClean's 'Naval biplanes' (i.e. the modified Short-Sommer biplanes on which the greater part of the officers' training had taken place: in the event, models S34 and S38), as these were perceived to be the most suitable for instructional purposes; but it also urged that the Board of Admiralty sanction the acquisition of other 'improved' aeroplane types for what was termed 'experimental purposes'. More important still, it suggested that the board avail itself of the opportunity to maintain a permanent establishment at Eastchurch. McClean had written into the Aero Club's lease that the site should be made available to the Royal Navy should they in future require it. Samson, knowing this, now described its situation as 'most suitable' for naval purposes. It was located 'close to the sea and the sheltered waters of the Swale', and would allow for ground-staff to be trained in the Short Brothers' workshops. Dated 17 August 1911, this report came some two-and-a-half years after Shorts had first established aircraft construction on the Isle of Sheppey. Despite some initial departmental hesitation, it was the first step towards Eastchurch becoming the recognised centre of naval aviation.[188]

---

186 Reminiscences of G. B. Cockburn.
187 See PRO, AIR 1 2469. Note also Lt. Samson's flight log-book, Samson papers, 72/113/1 (28).
188 PRO, AIR 1 2467. M. F. Sueter, Inspecting Capt. of Airships, had already advised Samson of the necessity of establishing a maritime base: Sueter to Lt. C. R. Samson, 26 Apr. 1911, Samson papers, 72/113/2 E (39).

From this point Short Brothers went from strength to strength, but it becomes increasingly impractical to trace in detail the various models and variants of their aircraft, for the company's pioneering work became progressively more interwoven with the development of an Admiralty air arm.[189] One or two developments are of particular note, however. Oswald Short and Lieutenant Longmore, for example, devised various experimental floats for use on Short seaplanes alighting on water, while on 10 January 1912 Lieutenant Samson succeeded in flying Short-Sommer biplane S38 off a specially built wooden trackway erected over the deck of the 15,740 ton battleship HMS *Africa* (a feat which had been accomplished as early as November 1910 when the American pilot, Eugene Ely, flew the Curtiss 'Albany Flyer' pusher-biplane from an improvised platform erected on the deck of the cruiser USS *Birmingham*).[190] This was taken a stage further in May 1912 when the newly-promoted Commander Samson (OC, RFC, Naval Wing) became the first pilot to take off from a moving battleship – HMS *Hibernia* – during the Portland Naval Review.

Support for these innovations derived from the highest quarters. Vice-Admiral Prince Louis of Battenberg remarked in a private letter to Samson in November 1911 that the newly-appointed First Lord, Winston Churchill, had intimated that he meant to 'develop the aeroplane in connection with fleets for all he is worth'. But these specific developments are well-documented, so need not be elaborated.[191]

The previous year Shorts had also experimented with a small series of prototype multi-engine biplanes, the most notable being the S39 'Triple Twin', incorporating a pair of tandem-mounted Gnôme rotary engines driving two adjacent tractor propellers and an additional rear-mounted pusher propeller: but neither this nor its simplified 'Tandem Twin' successor went into production. From Farman-cum-Sommer-derived biplanes Shorts then went on, from late 1911, to devise a series of original tractor-biplanes, culminating in the series of Shorts' Admiralty seaplanes which were to see service during the war. It was on Shorts' tractor-biplanes that much of the Royal Navy's experimental aviationary work was done.[192] To accommodate this trade the company had by 1914 established a new construction site by the River Medway in Rochester.

---

[189] See Barnes, *Shorts aircraft*, passim. Note also photographic record of Shorts aircraft, 1908–1918, PRO, AIR 1 722/47/1.

[190] C. Studer, *Sky storming Yankee: the life of Glenn Curtiss*, New York 1937, repr. 1972, 247–8; C. R. Roseberry, *Glenn Curtiss: pioneer of flight*, Syracuse 1991, 294–6. For the 'Albany Flyer' see L. S. Casey, *Curtiss: The Hammondsport era 1907–1915*, New York 1981, 64–9.

[191] See Gamble, *Air weapon*, 159–60 (where the Samson-HMS *Africa* flight is dated as Dec. 1911); Taylor, *Shorts*, 41–2; and Barnes, *Shorts aircraft*, 59–60. For greater detail note 'Early British naval aviation', Samson papers, 72/113/2 E (39). For the Battenberg letter see Vice-Admiral Prince Louis of Battenberg to Lt. C. R. Samson, 9 Nov. 1911, Samson papers, DS/MISC/100 CRS 1/1.

[192] In 1913 the firm issued a folding wing patent, facilitating the carrying of aircraft at sea,

Development of the Avro venture

For the Roe brothers commercial success came as more of a struggle. With the formal establishment of A. V. Roe & Co. in January 1910 a section of H. V.'s webbing firm's shopfloor, at Brownsfield Mills, Great Ancoats Street, Manchester, was set aside as factory space, following which, in March 1910, A. V. returned to Brooklands to establish the firm's test-base there with an ancillary flying school. Few of the young enthusiasts who came to him could afford full tuition fees, so a system was devised whereby tuition could be paid for by set periods of labour in the Avro workshops. This helped Roe keep down labour costs.[193] Among those who benefited from this scheme was Howard Pixton, a draughtsman with the Simplex Engineering Company of Trafford Park, who was subsequently to win fame as Sopwith's pilot in the 1914 Schneider Trophy contest. Pixton had pleaded with H. V. Roe in May 1910 that he was 'prepared to take absolutely any risk' if A. V. Roe & Co. would take him on.[194] He eventually took his aviator's certificate in January 1911 and became one of Avro's top pilots, before moving on to the British & Colonial Aeroplane Company in June 1911, and from there to Sopwith.[195] By the time he left Manchester for Brooklands the nucleus of A. V. Roe & Co.'s Brownsfield Mills design and construction team was already in being, Reginald Parrott acting as chief draughtsman and, presently, as works manager.[196]

The newly-established firm began work with the construction of a modified 35 h.p. Green engine triplane, the Roe II Triplane, which became known colloquially as the 'Mercury'. Fitted with a two-bladed propeller it was exhibited at the Olympia Aero Show in March 1910, resulting in an immediate order for a second model from Walter Windham. But the process of design modification continued. Following a crash at Brooklands, the original 'Mercury' was reconstructed with ailerons, and the following June a new 35 h.p. (JAP) triplane – the Roe III Triplane – was constructed. In this, the bottom wing of the mainplane-section was shortened, and ailerons were incorporated within the body of the upper wing, being secured to its rear cross-spar, rather than merely hinged to the trailing edge. A larger rectangular rudder was also fixed to the three-layered tailplane. A second 35 h.p. (Green engine) Roe III Triplane was then constructed, with ailerons hinged to the rear spar of the central mainplane, and with the mainplane-section more conspicuously cambered (convexly arched), in order to provide A. V. Roe & Co. with a slower

---

although Oswald Short subsequently claimed that the idea had originated with Horace as early as 1911: O. Short to J. Laurence Pritchard, 30 Mar. 1948, RAeS archive.

[193] H. V. Roe, 'Pioneers'; Brooklands-Avro agreement, 1 Mar. 1910, AVRO papers, file 109. A. V. would also earn money by undertaking 'repair work' for 'amateur constructors'. (*The Aero*, 24 May 1910.) There was no shortage of such work.

[194] H. Pixton to H. V. Roe (A. V. Roe & Co.), 31 May 1910, AVRO papers, file 174.

[195] H. Pixton to H. V. Roe, 5 June 1911, ibid.

[196] H. V. Roe, 'I financed an aircraft pioneer'; 'Pioneers'; A. V. Roe, *World of wings*, 63.

instructional machine. This aircraft was first flown on 9 July 1910, after which – with the 'Mercury' – it was entered for the second Blackpool flying meeting, due to begin the following month. On Wednesday 27 July, however, both these aircraft were lost when the train carrying them to Blackpool caught fire. This was potentially a catastrophe, since A. V. Roe & Co. desperately needed the money offered for competing in this tournament. A. V. returned to his brother in Manchester. They had four days before the start of the meeting: by transferring the necessary components to Blackpool and working round the clock they managed to scrape together a new Green-powered triplane just in time. This was airborne by the afternoon of Monday 1 August 1910. A. V., who had the unenviable task of piloting this untested machine, called it 'without exception, the worst we . . . ever made', but this was essentially because the controls proved exceptionally difficult to manipulate. Such problems were not apparent from the ground.[197]

This episode would suggest, if nothing else, that the firm already had construction down to a well-established routine. Indeed, one of the principal overseas guests at the Blackpool meeting, J. V. Martin of the Harvard University Aeronautical Society, was sufficiently impressed with the extemporised triplane, despite its having to battle against restrictive meteorological conditions, to order another immediately. This was built and despatched within a fortnight. Martin also invited A. V. Roe and fellow Blackpool competitor Claude Grahame-White to participate in the following month's Boston aviation meeting. Both aviators accepted, and Roe despatched for the purpose the same expedient triplane he had flown at Blackpool.[198] Despite this measure of success, however, Roe soon moved on to biplane design. The last of the more primitive triplanes, a 35 h.p. (Green engine) wing-warping model, with a new triangular mono-tailplane section (the Roe IV Triplane), was completed in September 1910. Repaired many times, it was generally used as an Avro instructional machine, and eventually reached heights of up to 750 ft. It was on this aeroplane that Pixton took his aviator's certificate.

At about this time H. V. Roe began to approach a number of larger, more established firms, in the hope that they might help finance expansion of the Avro enterprise. The curt replies he received illustrate only too graphically the low esteem in which aviation was held by the recognised engineering community during this period. Roe initially went through a commercial agency, Wheatley Kirk Price & Co., but this got him nowhere. Having been first contacted in January 1911, the agency was reporting by October that 'the great majority [of financiers] are apt to look upon a business of this description too much in the light of a speculation'.[199] At H. V. Roe's request they continued to search for a suitable investor, but by April 1912 they had to admit defeat.

197 A. V. Roe, 'Trials', 98.
198 Ludovici, *Challenging sky*, 70–1. In America, A. V. succeeded in crashing both the Blackpool and Harvard triplanes – the latter of which he was obliged to replace.
199 Wheatley Kirk Price & Co. to H. V. Roe, 19 Oct. 1911, AVRO papers, file 3.

'In no single instance', they reported, had they 'met with the slightest encouragement.'[200]

Already, however, H. V. Roe had begun to appeal directly to indigenous related manufacturing concerns. He wrote, for instance, to the Sheffield-Simplex Motor Works Ltd (formerly Brotherton-Crockers Ltd, A. V.'s past employer) in March 1912, asking them to consider the benefits of diversifying into aviation. 'We have every facility here for turning out . . . the rate of one aeroplane a week', he insisted, and thus 'fresh capital need not be spent on capital outlay, but solely in the introducing of more business.'[201] However no interest was elicited, so he appealed in the same vein to the Manchester Chamber of Commerce. The need, he said, was for capital reserve. The War Office had provisionally ordered three biplanes (presumably the first Avro 500s), but the firm could not afford the necessary Gnôme engines. They therefore risked losing the order. Surely, Roe argued, an investor could be found 'to help us [keep] this industry in Manchester'. But the response, when it came, was again negative. An undertaking of this kind, wrote the Manchester & County Commercial Agency, 'is outside the usual field of commercial investment'.[202] The problem was that unless someone took the initiative and backed such a venture it would always remain outside the usual field of commercial investment. Forsaking regional pride, H. V. Roe appealed instead to Britain's principal armament manufacturers, Vickers, and Armstrong, Whitworth & Co. Ltd. Neither responded,[203] although Vickers were in the process of establishing, and Armstrong, Whitworth & Co. were soon to establish, their own aviation departments.

Even when he simply attempted to publicise the Avro marque H. V. was rebuffed. In January 1912 he wrote to the Royal Aero Club suggesting that A. V. Roe & Co. be appointed the organisation's official aeronautical engineers. This title, as H. V. was presumably aware, had hitherto belonged to Short Brothers, and the Aero Club was unlikely simply to throw them over now. However, conscious of the increasingly invidious nature of bestowing such a designation on any particular manufacturer, the club judiciously decided that, in future, no firm would be exclusively accredited with that epithet. Since the aeroplane had superseded the balloon the title had become little more than that: an epithet. Minimal direct work resulted. Kudos derived from the fact that since 1910 the club had been issuing official aviators' certificates and that these were recognised by the government as legal pilot licences. The upshot was that Short Brothers were effectively deprived of what they evidently

---

200 Ibid. 22 Apr. 1912. For the agency's advert in Manchester Guardian on Avro's behalf (for 'Capital to cope with expanding trade'), see file 1.

201 H. V. Roe to Sheffield-Simplex Motor Works Ltd, 9 Mar. 1912, ibid. file 7.

202 H. V. Roe to Manchester Chamber of Commerce, 15 Mar. 1912; Manchester & County Commercial Agency to H. V. Roe, 21 Mar. 1912, ibid.

203 Correspondence re. finances of A. V. Roe & Co., ibid.

regarded as their rightful and, in PR terms, beneficial privilege.[204] This will scarcely have improved relations between Shorts and Avro, which had been fraught ever since the Aero Club-*Daily Mail* competition in April 1907. H. V. Roe, however, had simply been seeking another means of attracting investment. So far he had failed to elicit any significant support.

At one point, in November 1911, a Scottish-based engineer named Frank Barnwell, who was later to make his name as the designer of the Bristol Fighter, had proposed investing as much as £2,500 in the Avro firm. But as he was on the point of signing the agreement he apparently received a better offer from elsewhere – presumably from the Bristol-based British & Colonial Aeroplane Company, although this is not specified in his correspondence with H. V. Roe – so nothing came of it.[205] Thus, for the time being, the Roe brothers were forced to continue alone, as best they could.

By now A. V. Roe had abandoned the triplane configuration in favour of the biplane. The first of these, retrospectively designated the Avro Type D, although there had been no A, B, nor C, was constructed in March 1911 with a 35 h.p. Green engine and test flown by Howard Pixton early the following month. It was, again, a two-seater, incorporating wing-warping control. Pixton wrote to H. V. Roe soon after its debut, remarking that compared with previous models this new prototype was 'undoubtedly the better all round machine'.[206] He successfully demonstrated it before the Parliamentary Aerial Defence Committee at Hendon on 12 May 1911, after flying the aircraft in from Brooklands. Later that month Commander Oliver Schwann, of the Royal Naval Airship Tender, HMS *Hermione*, the official quarters of those officers working on the Royal Naval Airship No.1,[207] enquired about its purchase, formally offering £350 for the aircraft on 25 May provided it could be despatched from Brooklands by the end of June. The offer was accepted.[208] 'I hope to be able to get the machine to rise from and alight on water', wrote

[204] RAeC Committee minutes, 6, 13 Feb. 1912 (re. H. V. Roe letter of 24 Jan. 1912), RAeC papers.

[205] F. S. Barnwell–H. V. Roe correspondence, AVRO papers, file 4. Note in particular, Roe to Barnwell, 28 Nov. 1911 and Barnwell to Roe, 20 Dec. 1911.

[206] H. Pixton to H. V. Roe, not dated, but probably May 1911, ibid. file 174. Roe's subsequent detractors – notably Brabazon's friend and ally, the aviation historian Charles Gibbs-Smith – have argued, with little evidence, that in his triplane and biplane designs of this period Roe effectively plagiarised the French Voisin and Blériot-built tractor designs of Ambroise Goupy: Gibbs-Smith, *Aviation*, 135–6, 144. This charge has been authoritatively rebutted by Harald Penrose in his *British aviation: the pioneer years*, 148–9, 291. Evidence of collaboration between Gibbs-Smith and Brabazon can be found in the Brabazon papers, box 26(1).

[207] Subsequently Air Vice Marshal Sir Oliver Swann.

[208] Cdr O. Schwann RN to H. V. Roe, 25 May 1911; H. V. Roe to Cdr O. Schwann, 26 May 1911, AVRO papers, file 168. Sueter subsequently recorded the cost of the aircraft as £700, so £350 may have constituted a deposit, the remainder being settled on delivery: *Airmen or Noahs: fair play for our airmen; the great 'Neon' air myth exposed*, London 1928, 373. However this is not what the contemporary evidence indicates.

Commander Schwann in acknowledgement, 'and to thus lead the way in practical naval aeroplane work.'[209] He had the aircraft delivered to Vickers's Cavendish Dock, at Barrow-in-Furness, where the Royal Naval Airship No.1 was being constructed, and there initiated a series of modifications to turn it into a working seaplane. In particular, a float-undercarriage was added. The Admiralty would provide no financial support for this project. Commander Schwann raised the money for the aircraft himself and a syndicate of naval officers was then formed to finance future experiments. Captain Murray F. Sueter, Commander E. A. D. Masterman and Lieutenant-Commander F. L. M. Boothby were among its members.[210]

The Type D was first tested in its seaplane mode on 2 August 1911, but with little immediate success. H. V. Roe offered to despatch one of the Avro test pilots, R. C. Kemp or, more notably, F. P. Raynham, to Barrow, in order to professionally trial-run the machine, but Schwann, who was not yet a qualified pilot, declined the offer.[211] This was unwise. On 18 November he succeeded in getting the device to lift from the water; however, being unable to control it, the aircraft immediately capsized. Clearly, assistance of some sort would be necessary. The following February Schwann reported that he had secured the services of the Avro-trained test pilot S. V. Sippe, who had recently gained his aviator's certificate on a later Avro Type D model at Brooklands. Schwann himself had meanwhile arranged to take two months on half pay, in order that he might gain his own certificate with the British & Colonial flying school at Larkhill, on Salisbury Plain.[212] Further trials were consequently delayed until April 1912. On 9 April Sippe succeeded in flying the aircraft. Schwann had thus achieved his primary aim of devising a practicable British seaplane. He was not the earliest pioneer in this field – Glenn H. Curtiss is generally credited with having flown the world's first practical seaplane, the Model D Hydro, in America in January 1911 – but the news was a welcome boost for the Roe brothers, who in their advertising proceeded to make much of the Avro D biplane's dual success. Later variants were to follow.[213]

It was in the advancement of biplane design, indeed, that the firm of A. V. Roe & Co. now came into its own. In March 1912 there emerged the Avro 500, the aeroplane which more than any other established A. V. Roe's pre-eminence in aircraft design in the years preceding the war. It was the precursor of arguably the most enduring aeroplane of the era – the Avro 504.

---

209 Cdr O. Schwann to H. V. Roe, 27 May 1911, AVRO papers, file 168.

210 Sueter, *Airmen or Noahs*, 373. A receipt of the syndicate's finances is retained among Sueter related papers in the Fleet Air Arm Museum. Boothby had speculated on forming a partnership with A. V. Roe as early as 1908. Indeed, he was the first person to do so: H. V. Roe, 'A. V. and H. V.'.

211 Cdr O. Schwann to H. V. Roe, 29 Aug. 1911, AVRO papers, file 168.

212 Cdr O. Schwann to H. V. Roe, 26 Feb. 1912, ibid.

213 Capt. Murray F. Sueter, 'Report of experiments carried out with a hydro-aeroplane, with notes and suggestions for further experiments', PRO, AIR 1 2459; Jackson, *Avro aircraft*, 26–32. Newspaper reports of the Barrow trials are collected in Boothby papers, 70/1/5.

Built in compliance with the War Office's military aircraft specification, the Type E prototype, as it was initially known, was first flown on 3 March 1912, powered by a 60 h.p. ENV engine. By the end of the month test pilot Wilfred Parke was able to coax it to an altitude of 1,000 ft in under six minutes and 2,000 ft in thirteen minutes – with a passenger.[214] It was when a 50 h.p. Gnôme rotary engine was added to the new design, however, that the machine revealed its true potential. On 8 May Parke took it up again at Brooklands and climbed 2,000 ft in five minutes.

By early March 1912 the War Office had ordered three of these machines, i.e. the existing one and two more.[215] It was a major breakthrough. A. V. Roe himself recognised this and regarded the Type E as his first fully achieved aeroplane. In acknowledgement of the fact he numbered all future Avro aircraft from this model, beginning with the impressive but arbitrary figure of 500. Further contracts followed. The War Office ordered an additional four machines in November 1912, as well as five single-seater variants, categorised E/S Type or Avro 502s and incorporating a new comma-shaped tail-section, in January 1913. In September 1912 a 500 was also ordered for the Portuguese government. Thus when, in January 1913, A. V. Roe & Co. finally found a backer and was re-established as a limited company, prospective orders for some thirteen of the new Avro 500 class biplanes were on the books – twelve of them from the War Office. Later that year six more were ordered by the Admiralty, to be stationed at the Aero Club's old aerodrome at Eastchurch. Lateral control of these service machines was again by means of wing-warping, although ailerons were subsequently incorporated into replacement outer wing-panels.[216]

However, all this was in the future. Development of the Avro 500 in March 1912 did not immediately solve the firm's cash flow problems. 'We carried on with very slender finances', admitted H. V. Roe.[217] Indeed, the brothers relied to a large extent on such sums as could be gleaned from the various competitions and exhibitions characteristic of this period. Since 1910 A. V. Roe & Co. had also run an aircraft components service, engagingly styled 'The Aviator's Storehouse'. This commercial penury must be borne in mind as the constant backdrop against which Avro's new design successes were developed.[218]

In April 1912 A. V. Roe designed the world's first enclosed cabin aeroplane, the Avro Type F monoplane. This was followed, more successfully, by the

---

[214] Jackson, *Avro aircraft*, 38.

[215] H. V. Roe, 'Pioneers'. H. V. Roe dates this order as 7 March 1912, the first new Type E being dispatched from Manchester to Brooklands, for trials, on 27 April.

[216] Jackson, *Avro aircraft*, 40–2; H. V. Roe, 'Pioneers'. (For the background to the Portuguese order see AVRO papers, file 181.)

[217] H. V. Roe, 'Pioneers'.

[218] Ibid. For the Aviator's Storehouse catalogue see RAeS archive; for details of A. V. Roe & Co.'s early finances see AVRO papers, file 163. Prize money from, for instance, Brooklands events, could be both lucrative and regular during the season. Note F. Lindsay Lloyd (BARC) to H. V. Roe, 24 Apr., 27 May, 15 June 1911, ibid. file 109.

construction of a 60 h.p. Green engine cabin biplane, the Avro Type G, intended for that August's military aeroplane competition on Salisbury Plain. It was here, in this machine, that Avro pilot Wilfred Parke became the first man to make an authoritatively documented pull out of a spin. It was from this event, known as 'Parke's Dive', that a technique was derived for dealing with this potentially fatal problem.[219] On 24 October 1912 another of A. V.'s former pupils, F. P. Raynham, also briefly established a new British duration record of 7 hrs 31 mins in the machine; but the military trials had revealed that it had a poor rate of climb and it never threatened the superiority of the Avro 500.[220]

Despite this, Alliott could still write to Humphrey Verdon-Roe from the Larkhill competition ground that he had been told, off the record, that 'both [the] Army and Navy now consider [that] we are at [the] top of [the] tree . . . and have nothing to fear . . . from Shorts'.[221] Such was the impression made by the firm's new biplanes – particularly the Avro 500: an impression soon reflected in material orders. By this time Alliott had the invaluable practical support of a young personal assistant named Roy Chadwick. Chadwick had joined A. V. Roe & Co. in September 1911 as a locally recruited junior draughtsman previously employed with the British Westinghouse Electrical & Manufacturing Company of Trafford Park. At Westinghouse he had been a near contemporary of Arthur Whitten Brown, subsequently one half of the Alcock and Brown Atlantic-crossing team. Chadwick was to be made Avro's chief draughtsman in May 1914 and presently became the firm's chief designer also. In this role he would later design the Lancaster bomber of World War II.[222]

At the close of 1912 Humphrey Verdon-Roe's commercial lobbying finally bore fruit: some three years after the brothers had formally established the firm, a financial backer was found for A. V. Roe & Co. This was James Grimble Groves, philanthropist, chairman of Groves & Whitnall, brewers, and former Conservative MP for South Salford. It is perhaps significant that even now it was a local figure who took up the Avro venture, despite its recent successes.[223] Established manufacturing firms still shied off such an investment. But the brothers had at least elicited sufficient capital to expand. On 1 January 1913

---

[219] See *Flight*, 31 Aug. 1912. H. V. Roe was quick to utilise this incident in his impromptu sales pitches. Note, for example, H. V. Roe to sec., War Office, 20 Sept. 1912, AVRO papers, file 121. In the event of Parke not being available for the military trials, C. G. Grey had suggested that his place be taken by Robert Smith Barry, subsequently the originator of the wartime Gosport-Avro system of flight training. Grey described Smith Barry to H. V. Roe as a 'magnificent' flyer: C. G. Grey to H. V. Roe, 22 July 1912, AVRO papers, file 7.
[220] Jackson, *Avro aircraft*, 50.
[221] A. V. Roe to H. V. Roe, 18 Aug. 1912, AVRO papers, file 7.
[222] H. V. Roe, 'Pioneers'; A. V. Roe, *World of wings*, 141. See also H. V. Roe to J. G. Groves, 2 May 1914, AVRO papers, file 10, reporting that Chadwick had been made chief draughtsman. For Chadwick's background see H. Penrose, *Architect of wings: a biography of Roy Chadwick – designer of the Lancaster bomber*, Shrewsbury 1985.
[223] For Groves's background see obituary article, *Manchester Evening News*, 24 June 1914, AVRO papers, file 11.

A. V. Roe & Co. was officially reconstituted as a limited company. To show their good faith the Roe brothers agreed to Groves receiving preference shares, while they themselves accepted less remunerative ordinary shares.[224]

The first practical step following these events was to transfer the construction side of the business from H. V. Roe's webbing firm's somewhat outdated workshop, still situated in the former cotton factory, Brownsfield Mills, to a more commodious site. This entailed H. V. finally giving up direct control of the webbing firm, Everard & Co., in order to concentrate on the growing Avro enterprise. In April new premises were opened at Clifton Street, in the Miles Platting district of Manchester.[225] War Office orders for the Avro 500 biplane meant that the reconstituted company began with a not inconsiderable amount of work in hand. Groves became the organisation's chairman.

The precise extent of A. V. Roe & Co.'s capital at this time is difficult to determine. H. Montgomery Hyde, compiling the A. V. Roe entry for the *Dictionary of business biography*, states that the limited company had an authorised capital of £50,000; but how much its initial subscribed capital amounted to, or exactly how much Groves invested in the company, is not clear. Regarding the latter, H. V. Roe acknowledged merely a 'few hundred pounds'.[226] Other evidence suggests that Groves guaranteed extra capital whenever required.[227] A major cash flow problem was caused by the extended time it took the War Office to pay for their machines. H. V. Roe's letters to Groves over the spring and summer months of 1913 continually harp on this and the problems that resulted.[228] H. V., in line with Groves, saw it as symptomatic of their reluctance to deal with small self-made firms: they would have preferred, he contended, to have dealt with established armament manufacturers, like Vickers.[229] But it seems more likely that it reflected War Office inexperience of their needs, since a system of advances was eventually introduced, although not until well into the war.

Work had begun in November 1912 on a successor to the successful Avro 500 biplane and development continued throughout the following year. This successor proved to be the first of the Avro 504 series. The aileron-controlled prototype was constructed amid great secrecy, to be publicly unveiled at the

---

[224] H. V. Roe, 'Pioneers'. The Groves family took 10% cumulative participating preference shares, while the Roe brothers accepted ordinary shares, 'which were to receive two-thirds of the remaining profit, if any, after the interest on the preference shares had been met': A. V. Roe, *World of wings*, 140.

[225] H. V. Roe, 'Pioneers'; A. V. Roe, *World of wings*, 139. See also H. V. Roe to C. G. Grey, 11 May 1928, ibid. file 64. According to one former employee, Harry Goodyear, the construction staff at this time amounted to between 30 and 40 men: Penrose, *Architect of wings*, 22.

[226] *DBB*, *s.v.* Verdon-Roe; H. V. Roe, 'I financed an aircraft pioneer'.

[227] J. G. Groves to H. V. Roe, 13 July 1913; H. V. Roe to J. G. Groves, 14 July 1913, AVRO papers, file 9. See also H. V. Roe, draft list of correspondence with James Groves, 1913–14.

[228] 8 Mar., 12 June, 11 Aug. 1913, ibid.

[229] H. V. Roe to J. G. Groves, 13 Aug. 1913, ibid.

Hendon Aerial Derby of 20 September 1913.[230] In competition with faster monoplanes it only came fourth on that occasion[231] and further modifications proved necessary, but this preliminary model – piloted by F. P. Raynham – was the forerunner of one of the most accomplished aeroplanes in aviation history.

On 24 November Raynham flew the prototype to Farnborough to undergo its official tests. In these, employing an 80 h.p. Gnôme engine, and carrying a pilot, passenger and fuel for three hours, it climbed to 1,000 ft in 1 min. 45 secs. It also registered a top speed of 80.9 m.p.h. – outpacing Farnborough's own BE2. The following February Raynham climbed to 15,000 ft in the aircraft, surpassing the existing British solo altitude record by nearly 2,000 ft. As this flight had not been officially observed, however, he followed it within the week by carrying a passenger to 14,420 ft – another record. The machine's responsive handling qualities were shortly to become well-known. Roe himself subsequently suggested that the Avro 504 was to the aircraft industry what Henry Ford's Model T was to the automobile industry, and this is not unreasonable. A generation of airmen were to be trained on it. The first War Office order for a dozen of the type was issued on 1 April 1914.[232]

Such was the company's growing reputation that unsolicited offers to help launch Avro on the continent soon arrived on H. V. Roe's desk at Clifton Street. An Avro 503 seaplane had been purchased by the German government in June 1913. It had been test flown in England by a German naval officer named Captain Schultz and despatched to Germany by the end of the month. Following these negotiations Schultz again contacted A. V. Roe & Co., this time proposing that he become their agent in Europe, representing the firm in Germany, Austria and Russia. 'As we have nobody in these parts', wrote H. V. Roe to James Groves, 'we think it as well to close with him.'[233] However, nothing seems to have come of this proposal. Schultz became Fokker's agent in England instead.[234]

---

230 H. V. Roe to J. G. Groves, 16 Sept. 1913, ibid. file 10. H. V. Roe had become increasingly security conscious over this last year. See, for example, H. V. Roe to unspecified recipient, 3 July 1913, ibid. file 9, where he warns, in particular, of losing design secrets to Sopwiths. He seems to have seen Sopwiths as Avro's main rival. See also letter to H. Lutwyche, 19 Jan. 1914, ibid. file 12.

231 *Flight*, 27 Sept. 1913.

232 A. V. Roe, *World of wings*, 122. The Type had a limited combat career as (procurement restrictions apart) it was – like the BE – flown from the rear seat, leaving the observer/gunner hemmed in by the upper wing and its central section struts. It came into its own as a dual-control training aircraft, particularly when incorporating a 100 h.p. Gnôme Monosoupape engine. Maj. Robert Smith Barry's celebrated Gosport-Avro School of Special Flying was initiated in August 1917: J. M. Bruce, *The aeroplanes of the Royal Flying Corps (Military Wing)*, London 1982, 112–17. See also Jackson, *Avro aircraft*, 56–8; H. V. Roe to C. G. Grey, 15 July 1922, AVRO papers, file 195. A single-seat fast scout biplane, with swept-back wings – the Avro 511 'Arrowscout' – also emerged in early 1914, but had little opportunity to establish itself before the outbreak of war.

233 H. V. Roe to J. G. Groves, 21 June 1913, ibid. file 9.

234 A. Fokker and B. Gould, *The flying Dutchman: the life of Anthony Fokker*, London 1931, 111.

Some three months later, in September 1913, a Frenchman named Donnet proposed establishing a French Avro company. This, in essence, would involve letting A. V. Roe & Co.'s French patent rights to an independent organisation. In return, the parent company would refrain from competing in the same market. Humphrey Verdon-Roe was again initially keen on this idea, but it seems that Groves was against it, being reluctant to segregate the firm's responsibilities.[235] It too, therefore, came to nothing.

In June 1914 Avro's chairman, James Groves, died. In a sense, this removed the barrier that had separated the Roe brothers and kept the tensions inherent in their relationship in check. A. V. had already begun agitating to get the company's works moved south, to Shoreham or Hamble, where seaplanes could be tested and his dream of a new garden city realised; but H. V. Roe, with Groves's backing, had firmly vetoed this.[236] (The main Avro flying school had already, in October 1912, been transferred from Brooklands to Shoreham.)[237] Now there was no buffer, and an increasingly acrimonious dispute developed within the company. Its immediate effects were temporarily mitigated by the advent of war and the enormous expansion of business which resulted; but by the following year tensions had become intolerable.

In fact, even in the later months of 1914 the conflict between A. V. and H. V. was only too evident behind the façade of unity engendered by the serious-ness of the hour. With the onset of war an immediate increase in the firm's manufacturing capability became imperative. Avro's neighbours, Mather & Platt, had recently completed a large expansion to their Miles Platting works, so H. V. Roe arranged to rent this area for the duration of the emergency – thereby securing an expeditious shopfloor expansion without dangerously extending A. V. Roe & Co.'s holdings in the event of a post-war slump. A. V., however, fiercely opposed temporarily renting additional premises in Man-chester. He chose instead to resurrect his semi-utopian Hamble scheme, again advocating the construction of a unified works and employee garden city complex on the south coast. This, he insisted, would in itself provide long-term benefits as after the war there would be a developing market for flying-boats with which to link the empire.[238]

The intensity of this division was exacerbated by the War Office's manipu-lation of the market at this time. As will be shown, it had developed a procurement bias in favour of its own Royal Aircraft Factory designs, leaving private manufacturing firms, with their own design teams, heavily dependent on Admiralty orders. Thus, on purely pragmatic grounds, A. V. urged the necessity of establishing a coastal research and manufacturing base. It was in

[235] H. V. Roe to J. G. Groves, 22, 26 Sept. 1913, AVRO papers, file 10.
[236] J. G. Groves, A. V. Roe, H. V. Roe correspondence, July 1913, ibid. file 9.
[237] Jackson, *Avro aircraft*, 31–2. One of Avro's instructors there was John Alcock, a Manchester mechanic, who was subsequently to achieve fame as one half of the 1919 Alcock & Brown Atlantic-crossing team.
[238] A. V. Roe, *World of wings*, 144–5; H. V. Roe, 'Pioneers'. See also H. V. Roe to C. G. Grey (draft), 11 May 1928, AVRO papers, file 64.

his view an essential prerequisite to the development of vital Admiralty war contracts. The brothers were, by now, joint managing directors of the company, and initially neither would give way.[239] The Hamble scheme, however, eventually won the day and in August 1917 Humphrey Verdon-Roe washed his hands of both his family and the business and left to rejoin the army. (He joined the Royal Flying Corps and the following year became an observer in Trenchard's newly-inaugurated Independent Bombing Force.)[240] Ironically, the greater part of the contentious Hamble plan was then thwarted by a war-time restriction on available building materials.

Alliott was to resign from the company which carried his name in 1928, by which time he had himself lost control of the organisation to the unsympathetic motor industrialist, John Siddeley. The Manchester-based firm of Crossley Motors, which had formed the basis of one of the government's three national aircraft factories in 1918, had already gained a majority shareholding by May 1920.[241] A. V. subsequently acquired a controlling interest in S. E. Saunders Ltd, the boat-builders to whom many Avro 504s had been subcontracted in the war, and this firm became Saunders-Roe, the manufacturers of flying-boats.

Sadly, the family estrangement was never healed; rather, the old wounds festered with time. Humphrey Verdon-Roe was, for example, exchanging the most painful and recriminating letters with his family on the eve of his death.[242] It was a distressing end to so courageous an enterprise.

Conclusion

The decade following the emergence of aviation saw the progressive development of a productive aeronautical environment. During this period the foundations of an industry were laid. The main market to be targeted, however, was initially somewhat vague. After the first band of well-to-do adventurers had purchased their machines, these embryonic firms and their advocates were obliged to turn increasingly – indeed solely – to the military for support. But if a sustained market were to be created here the aeroplane's utility in war had to be demonstrated. Successive new aircraft designs were consequently drawn up with this specifically in mind.

The military were not always responsive, preferring to develop any new aeronautical devices through their own research and development establishment, the Balloon Factory – or Royal Aircraft Factory as it became. This was later to

[239] A. V. Roe to Col. J. G. Groves (James Groves's son), 30 Apr. 1915; H. V. Roe to Col. J. G. Groves, 2 June 1915; A. V. Roe to Col. J. G. Groves, 26 Apr. 1916, ibid. file 12.
[240] H. V. Roe to his family, 21 July 1917, and Alliott's reply, 23 July 1917, ibid. file 111; H. V. Roe to C. G. Grey, 1 July 1917, file 47; H. V. Roe to W. H. Bell, 11 Aug. 1917, file 159.
[241] The nature of these national aircraft factories will be examined in ch. 5.
[242] E. V. Roe to H. V. Roe, 12 Jan. 1945, and H. V. Roe to E. V. Roe, 15, 22 Jan. 1945, AVRO papers, file 111.

cause an outcry, but to begin with – particularly while the Factory concentrated essentially on airship development – there was no perceived conflict of interests. At this early stage it was more a question of persuading the authorities to accept the feasibility of aeroplane flight at all. To this end private enterprise was largely dependent on forming coalitions with interested parties who were in a position to promote their cause; links with figures like Rolls, for example, who was able to exploit his social position in order to foster interest; or that of Shorts and Francis McClean with the Admiralty, which was less restrictive in its procurement programme. The example of Lord Northcliffe might also be cited. He was able to bring pressure to bear through his newspapers – in turn boosting their sales through an enormously successful series of prizes, awards, competitions and so on. It was only after this acceptance of the aeroplane had been achieved that the public/private sector dichotomy arose.

Humphrey Verdon-Roe was probably correct in surmising that the War Office was by then hesitant about procuring aeroplanes from private enterprise because technology was advancing so rapidly that expensive models could become outmoded virtually overnight.[243] On the other hand, it was in the industry's interest to persuade those in authority that, for military purposes, continual upgrading of aircraft was essential. These firms and their projected market, principally the British army and the Royal Navy, had therefore to strike a balance between availability and accelerating efficiency.

It should always be borne in mind, however, that the scale of operations in these years was tiny as compared with events after 1914. Grudging acceptance, on the one hand, of the practicability of the product from established industry, and, on the other, of the utility of the product from its potential market, the military, meant that the permanence and commercial potential of this new industry was only gradually recognised. It also, of course, accentuated the value to be placed on each individual aircraft. None the less, the individuals examined thus far had together managed to gain a firm foothold in the door of commerce and industry. They achieved this largely through their success in persuading the naval and military authorities of the worth of their product. This was despite the sporting or ballooning ethos of the contemporary aeronautical establishment, which initially hindered rather than aided the new pioneers. This inherent tension derived largely from the ambivalent social position of a growing and highly-motivated mechanical class.

Through the indefatigable efforts of individuals like A. V. Roe, Brabazon and the Short brothers, any such ambivalence had been swept aside. It now remained for the next, the third wave of pioneers, to further develop the increasingly accepted practicability of aviation and to exploit the opportunities created by their precursors.

---

[243] H. V. Roe to J. G. Groves, 22 Sept. 1913, ibid. file 10. This fear was ultimately Treasury led. The Permanent Secretary to the Treasury, Sir R. Chalmers, made precisely this point during the CID Standing Sub-Committee discussions of 18 Dec. 1911, preceding the drafting of plans for the RFC: PRO, CAB 16/16, 38.

3

The emergence of a military–industrial complex

Industrial foundations

It was the pioneers, entrepreneurs and advocates of the third phase of aviation's development who ultimately carried the industry into the crucible of full-scale war. It was they who transformed it into what was essentially an extension of the armaments industry. Indeed, in terms of the increasingly sophisticated nature of its products – involving high risk capital expenditure, with the possibility of failure or rejection after months or even years of research investment – it is possible to discern the beginnings of that military–industrial complex so familiar in the defence market of the later twentieth century. The stick and string days of aviation were now largely over. Instead, the lone pioneer quickly found himself embroiled in a restrictive net of production and commerce. In place of the empiricist evolved trained design staffs; in place of the inexperienced entrepreneur, company directorates. As will become clear through an examination of the aeroplane manufacturers' deputation to the then Under-Secretary of State for War, J. E. B. Seely, in December 1911, the industry now demanded government subsidy – in, for example, the form of conditional contracts – if firms in the private sector were to survive in what was becoming an increasingly expensive and internationally competitive market. A powerful lobbying force was formed in parliament and the press to articulate and disseminate such views to those in authority. This again presages the future trend of defence procurement.

The military, conscious of the risks inherent in their growing dependence on the private sector in this area – they seemed to imagine that total dependence would mean that the industry could virtually dictate terms, despite the balancing effect of competition – increasingly sought to divert the technical development of military aviation to their own research and development establishment at Farnborough. This was to have potentially disastrous consequences, apparent when war broke out. What began as an attempt to standardise service production and procurement, with orders for Farnborough designs being progressively sub-contracted to the private sector, ended by virtually throttling design initiative when it was most needed.

Finally, the question of public perceptions of the new technology also became of increasing importance. Moulded and manipulated by the newspapers and journals of the period, and by publicists such as George Holt Thomas and Claude Grahame-White, it is clear that attempts were being made by those with a vested interest in the developing industry to by this means influence the manner and extent of governmental procurement.

Handley Page

The earliest of the third wave pioneers was probably Frederick Handley Page, whose company was officially incorporated in June 1909. Handley Page was born in Cheltenham in November 1885. His father ran a small upholstery business in the town and doubled as a lay evangelical preacher with the nonconformist Plymouth Brethren sect. It was a staid, lower middle class background, somewhat at odds with the general run of life in the fashionable Victorian spa: yet within this conventional environment the young Handley Page became captivated by advances in locomotive technology. Unlike his wealthier contemporaries, however, he was unable to give expression to his technological bent in the purchase or development of any of the standard automobile types then beginning to appear on the highway. Instead, he developed an active interest in electric rail and tram traction, and at the age of seventeen somehow persuaded his reluctant parents to let him leave Cheltenham Grammar School and enrol on a three-year electrical engineering course at Finsbury Technical College in north London.

Institutions such as this, unpretentious and largely disregarded by fashionable society, significantly advanced the locomotive revolution of the period through the practical education they offered to many mechanically-minded young men. A contemporary of Handley Page's at Finsbury was Richard Fairey. Indeed, they shared the same tutor, Sylvanus Thompson, although as an employee of the Jandus Electric Company of Holloway Fairey attended night school. Geoffrey de Havilland likewise studied at the south London Crystal Palace Engineering School.[1] It was at Finsbury that Handley Page first became interested in the possibility of powered flight. The need to earn a living intervened before anything practical could be achieved at this stage, however, and in 1906, in keeping with his original interest in electric traction, he joined an established firm of electrical machinery manufacturers, Johnson & Phillips Ltd, of Charlton. Even so, his thoughts constantly veered back to aviation, and the following year he was elected a member of the Aeronautical Society.[2] In his earliest correspondence with the society Handley Page emphasised his belief in the possibility of achieving inherent aerial stability.[3] By chance, an older member, José Weiss, was already experimenting in this area. The encouragement he provided was to induce Handley Page to begin his own practical, rather than merely abstract, research.

---

[1] de Havilland, *Sky Fever*, 36 passim.
[2] The Aeronautical Society retained more of a scientific emphasis than the Aero Club. Handley Page became one of the organisation's protagonists against the latter body, and was subsequently instrumental in reforming its council: C. H. Barnes, *Handley Page aircraft since 1907*, London 1976, 8–9; J. Laurence Pritchard, ' "H.P." Sir Frederick Handley Page, CBE', *RAeS Journal* lxvi (1962), 738.
[3] Handley Page to sec. (Col. J. D. Fullerton), 5 Nov. 1907, RAeS archive.

Weiss was a Parisian, born in 1859, but long domiciled in England. He was a talented aerodynamicist, but, in common with most of the older generation, he seems to have had little practical experience of the internal combustion engine. His early aeronautical activities were consequently centred exclusively on gliding flight. From his home at Houghton, near Amberley in Sussex, where he apparently earned a living as a landscape painter, he organised the testing, from about September 1906, of a series of full-sized, man-carrying, monoplane gliders, built on the perceived aerodynamic principles of soaring (i.e. gliding) birds.[4] As pilot he engaged the young Gordon England, who had been born in Argentina in 1891 and who was in due course to be recruited by the British & Colonial Aeroplane Company of Bristol. Contemporary reports record considerable success with these early aircraft, glides of anything up to 400 ft altitude and a mile in length being documented.[5] But their chief feature was their inherent stability, and it was this quality – resulting from Weiss's distinctive wing design – that primarily interested Handley Page. These wings were crescent shaped, swept back from a curved leading edge culminating, on either side, in upturned tips. Unbeknown to the English experimenters the Austrian pioneer, Igo Etrich, was simultaneously developing wings of much the same type, later to be incorporated within the design of the German Rumpler Taube reconnaissance monoplane of World War I.[6]

It was on Handley Page's initiative that the pair met. He requested an introduction, through the Aeronautical Society, as early as January 1908.[7] At this stage he evidently lacked confidence, having come to doubt the originality of his own ideas.[8] His desire to meet Weiss may therefore have reflected the need to find a contemporary against whom he could gauge the significance of his own research. They first met on 8 February, at Weiss's home in Sussex. Two days later Weiss was reporting to Colonel J. D. Fullerton RE, secretary of the Aeronautical Society, that 'Mr Page . . . is an exceedingly nice and intelligent man'.[9] Soon the pair were experimenting together on the aerodynamics of wing design.[10] In June Handley Page became a shareholder in Weiss's recently established Aeroplane and Launcher Syndicate. By July Weiss was cautiously predicting that this syndicate would produce a channel-crossing aircraft by the

4   For the dating of these first full-scale glider flight trials see J. Weiss to E. S. Bruce, 3 Sept. 1906, ibid. Here he states that he hopes to begin full-scale experiments in a fortnight. He also discusses some of his early models in a letter to Baden-Powell of 21 Mar. 1907. For Weiss's background see B. Talbot-Weiss, 'The centenary of José Weiss', West Sussex Gazette, 22 Jan. 1959, reprint in RAeS archive.
5   C. G. Grey, notes on early aviation, PRO, AIR 1 727/160/1; reminiscences of J. Weiss, AIR 1 728/161/1.
6   PRO, AIR 1 727/160/1. See also J. H. Morrow, Building German airpower, 1909–1914, Knoxville, Tenn. 1976, 75–6.
7   Handley Page to sec. (Col. J. D. Fullerton), 28 Jan. 1908, RAeS archive.
8   Ibid. 5 Nov. 1907, 28 Jan. 1908.
9   Weiss to Fullerton, 10 Feb. 1908, ibid.
10 Weiss to Fullerton, 28 Feb., 17 Mar. 1908, ibid. As a member of the RAeS's Wings Committee Handley Page also collaborated for a time with Maj. (retd) R. F. Moore RE, who

end of the year.[11] It is difficult to say whether he really believed this: there seems to have been little substance behind the claim. In fact Weiss was never to incorporate any engine successfully into his airframes.[12]

In the meantime Handley Page had lost his job as chief designer with Johnson & Phillips Ltd through misappropriating the firm's materials for aeronautical experiments. He was replaced there by Archibald Low, subsequently to become Vickers's chief aircraft designer. Unabashed, Handley Page rented a shed and office at 36 William Street, Woolwich, from where he proceeded to operate as an aircraft constructor. As a result of his stability experiments he had already received a commission from the idiosyncratic amateur aviation enthusiast, G. P. Deverall Saul, to build one of the latter's self-styled inherently-stable quadruplanes. This was essentially an opposed staggered, tandem-winged (i.e. with two sets of wings), single-seat biplane, with the upper front wing forward of the lower front wing and the upper rear wing further back than the lower rear wing. In the event the finished machine, which Handley Page had somehow had to fashion out of Saul's rough plans, proved practically useless, but this was not altogether unexpected. Immediate practical success was not the salient point of the commission, or it would never have been accepted. Nor, presumably, would the finished machine have been powered, as it was, by an ineffectual 8 h.p. motor. More importantly, the order – by merely constituting work – effectively launched the Handley Page aircraft firm. Saul, whose occupation was variously described as 'Gentleman' and 'Engineer', was to become a major shareholder in Handley Page when the organisation became a limited company the following year.[13]

In December 1908 Handley Page and Weiss decided to combine their efforts for the following March's Olympia Aero Show. In return for making Weiss's (unsuccessful) 12 h.p. Anzani twin-propeller pusher-monoplane the centrepiece of his display, Handley Page, who had in any case modified its wing configuration, was to be granted the use of the Weiss wing patent in his own designs. Weiss declared himself 'absolutely convinced of the immense superiority of the monoplane',[14] and, at this stage in his career, Handley Page was content to agree. It would not always be so. Before any machine could be properly tested, however, a suitable flying ground had to be found.

---

was something of an ornithopter pioneer (i.e. one concerned with flapping rather than fixed-wing aircraft): however, it is doubtful that he derived much of any practical value from this association.

11 Weiss to Fullerton, 21 July 1908, ibid.

12 He had acquired a 12 h.p. Anzani engine the previous February (Weiss to Fullerton, 10 Feb. 1908, ibid.), and this was incorporated into the Weiss monoplane exhibited at Olympia in March 1909, but it was ineffectual: *RAeS Journal* xiii (1909), 61–2. The following year Gordon England attempted to fit both a tailplane and a 35 h.p. ENV engine to the design, with mixed success.

13 G. P. Deverall Saul to T. O'B. Hubbard, undated, but *c.* 1908, RAeS archive; Register of members, directors and shareholders 1909–18, Handley Page papers, box 73/7–9. For details of the Deverall Saul 'quadruplane' see Penrose, *Pioneer years*, 170.

14 Weiss to Fullerton, 12 Nov. 1908, RAeS archive.

Handley Page was initially drawn to the so-called Colony of Aerocraft or Essex Flying Ground, recently established at Fambridge by the as yet little known entrepreneurial figure Noel Pemberton Billing – the future founder of Supermarine, whose rabid wartime demagogy would presently place him on a par with Horatio Bottomley. Here a tract of marshland had been placed at the disposal of pioneer aviators: indeed José Weiss was already *in situ*. But after studying the terrain in more detail Handley Page decided that it did not suit his particular needs. (Weiss was himself eventually forced to employ a Wright-derived pylon-release mechanism, so rough was the ground.) Instead he learnt that the Aeronautical Society had, after prolonged investigation, recently acquired its own test-site at Dagenham, nearer London. Again this was far from ideal;[15] however, anxious to get started on his own designs Handley Page took the lease of some adjacent marshland, with flying rights extending along the north bank of the Thames from Barking Creekmouth to Dagenham Dock – a passage flanked by the London, Tilbury and Southend Railway. A workshop was swiftly erected on the site, and this became Handley Page's first genuine factory.[16] The commercial importance of the location was underlined the following January, when Handley Page offered the Aeronautical Society £40 for their three sheds at Dagenham.[17] These he had re-erected next to his workshop, and from them he supplemented his meagre income by retailing aeronautical components.

The marsh near Barking Creekmouth had fortuitously been built up by the dumping of clay excavated during the construction of the London Underground. This enabled Handley Page to test jump a primitive canard (i.e. tail-first) monoplane-glider off gentle man-made slopes. His craft consisted essentially of a three-dimensional triangulated frame mounted on a tricycle-wheeled undercarriage. This carried mainplane and foreplane-cum-elevator wings of the Weiss crescent pattern. It was not a success.

Meanwhile the Deverall Saul quadruplane, completed in May 1909, had also, as anticipated, proved a failure. Undeterred, Handley Page re-established his operation as a limited company on 17 June. The firm was subsequently to contend that this made it the first company 'to be constituted exclusively for the design and manufacture of aeroplanes'.[18] This is perhaps debatable. Registered under the title Handley Page Ltd it had an authorised capital of £10,000, through the creation of 500 £20 shares. Initial subscribed capital, however, amounted to no more than £500. Handley Page himself became the firm's managing director, and he was joined on the board by his brother, 'Arthur Page', who, as an Indian civil servant, must have been a predominantly

15 Handley Page to Fullerton, 17 Jan. 1909, ibid. For the background to the acquisition of this site see *RAeS Journal* xiii (1909), 55–6. Both Handley Page and José Weiss were, in fact, among the list of subscribers to the Aeronautical Society's test-site fund.

16 *Flight*, 21 Aug. 1909; *The Aero*, 31 Aug. 1909.

17 Handley Page to B. F. S. Baden-Powell, 13 Jan. 1910, RAeS archive.

18 Handley Page Ltd, *Handley Page: forty years on*, London 1949. See also F. Handley Page to Lord Brabazon, 29 May 1959, Brabazon papers, correspondence (Handley Page).

absentee director, and two London engineers, Francis Neale Dalton and Walter Magdalen.[19] A former Johnson & Phillips employee named Herbert Tucker, who had doubtless aided Handley Page in his original illicit aeronautical experiments, and who Handley Page had then engaged as his embryonic firm's foreman and patternmaker, also subsequently became a shareholder – presumably in part payment for his labour.

Three further manufacturing commissions rapidly followed: for a two-seater Deverall Saul quadruplane, a W. P. Thompson (Planes Ltd) 'pendulum stability' Voisin-derived biplane (in which the pilot was situated below the lower wings), and for what has been described as a direct-lift monoplane, designed by a Alexander Thiersch – although in this last case Handley Page seem to have been primarily responsible for the manufacture of component parts.[20] Thus established, the firm then began work on its own first full tractor-monoplane, the Handley Page Type A, otherwise known, after its coloured cotton coating, as the 'Bluebird'. Powered by a 20 h.p. Advance air-cooled engine – really only a motor-cycle engine – the 'Bluebird' re-employed Weiss-style crescent wings. No means of lateral control was incorporated into the machine, however, owing to its presumed inherent stability.

It was exhibited, still uncompleted, at the Olympia Aero Show of March 1910, but extensive alterations proved necessary before the aeroplane finally became airborne the following May. Soon it had been so comprehensively modified that it was redesignated the Handley Page Type C.[21] (One of the earlier commissions, the pendulum stability biplane, designed by W. P. Thompson and amended by Handley Page, had been labelled the Type B.) In this revised form it incorporated a warping facility for lateral control and a 25 h.p. Alvaston engine, but the aircraft remained capricious in performance. Indeed, another structural overhaul, centred upon the installation of a 50 h.p. Isaacson radial engine, was soon required. By the time it was completed Handley Page was already concentrating on a new monoplane, the Type D. This incorporated an adapted mahogany monocoque, i.e. single-shell, fuselage, but otherwise resembled its predecessor. A handwheel mounted on the elevator lever controlled the warping mechanism and a foot tiller-bar controlled the rudder. After being displayed at the April 1911 Olympia Show, during which it was fitted with a borrowed 35 h.p. Green engine, it was reconstructed with a new fabric-covered fuselage in order to accommodate the Type C's 50 h.p. Isaacson engine. It was then transferred to a somewhat smoother flying ground which Handley Page had established approximately six miles north of Barking, at Fairlop.

As a means of publicising this improved model Handley Page had registered it for the Daily Mail's Circuit of Britain race, scheduled for 22 July 1911. It

---

[19] Register of members, directors and shareholders 1909–18; Barnes, Handley Page aircraft, 6.

[20] Ibid. 6–7; Penrose, Pioneer years, 176.

[21] See Barnes, Handley Page aircraft, 50–4.

never made that event, however, for on the machine's maiden flight at Fairlop, on 15 July, the company's adoptive test pilot, Robert Fenwick (W. P. Thompson's former assistant, who had test flown the Type B pendulum stability biplane), crashed the machine. There was no hope of repairing it in time for the race. Handley Page was furious and promptly dismissed Fenwick, who was soon replaced by another young enthusiast, Edward Petre, who seems to have been employed on a more formal basis.[22] Edward Petre generally worked in association with his brother, Henry. Indeed, following the recent Sidney Street siege, involving the cornering of a notorious gang of anarchists in London's East End, the brothers had acquired the joint nicknames of 'Peter the Painter' and 'Peter the Monk', to the evident enjoyment of contemporary journalists. With more enthusiasm than experience the pair had recently constructed their own pusher-monoplane, which Henry Petre had subsequently crashed at Brooklands. This seems to have convinced them that they could not go it alone. Fortunately Handley Page found a place for both of them at Barking, and work was begun on a tandem two-seater alternative to the Type D. This became the Handley Page Type E, otherwise known colloquially, and as a derivative of the reconstructed lanoline-coated Type D, as the 'Yellow Peril'. The Type E retained the distinctive swept-back, tip-twisted, Weiss-derived wings, although with multiple wire-bracing and extended flexible trailing edges, and was powered by a 50 h.p. Gnôme engine – one of the stock acquired from Horatio Barber's defunct Aeronautical Syndicate. It was the Handley Page company's first genuinely practical product.

There can be little doubt that Handley Page's work at this time benefited greatly from his free use of the facilities of the Northampton Institute, in Clerkenwell, south London. Since late 1910 he had been head of the aeronautical division of the mechanical engineering department, having previously earned a little extra money teaching evening classes at Finsbury Technical College. In this capacity he not only had access to a fully-equipped aeronautical laboratory, complete with wind tunnel, which he had been instrumental in acquiring, but he was also provided with a regular salary.[23] Handley Page always viewed his academic posts as complementary to his manufacturing work, rather than as an end in themselves. Indeed, it might be said that he rather exploited his position, as he had previously exploited his employment with Johnson & Phillips. His commitment was first and foremost to the development and maintenance of his own aeroplane company. His mind was continually focused on that end. The same might be said of his forays into the clubs and offices of London's West End. Everywhere, whether in the Northampton Institute's laboratory or in the Aeronautical Society's lobby, he was either modifying his own aircraft designs or seeking some degree of financial support for his firm.

22 Penrose, *Pioneer years*, 304.
23 Barnes, *Handley Page aircraft*, 9. Among those who later attended classes in aeronautical engineering at the Northampton Institute was the future Sir George Dowty, founder of the Dowty aircraft components firm: L. T. C. Rolt, *The Dowty story*, London 1962, 10.

With a view to encouraging potential investors he relinquished his original Woolwich premises during this period and, with the aid of a £600 loan from his uncle, took an office at 72 Victoria Street, London SW1 – near St James's Park.

The Type E 'Yellow Peril' was completed and first flown in April 1912. In July it was transferred to Brooklands, where Handley Page had taken over the Howard Flanders flying school. (Brooklands was by this time the hub of commercial–industrial test flying.) Back at the soon-to-be-superseded Barking workshops, meanwhile, construction began on a new monoplane for that August's military aeroplane competition, details of which had been announced the previous December. This design incorporated capacious side-by-side seating for pilot and observer thereby, it was hoped, improving its reconnaissance potential. But for all the expedient innovations the Type, labelled F and powered by a 70 h.p. Gnôme engine, was dogged by misfortune, both before and during the official trials. Significant delays in its assembly were caused by preparations for that September's removal of the company's works from Barking to a more commodious site at Cricklewood Lane, off the Edgware Road near Hendon. Then there was confusion over the machine's final construction, certain members of the workforce, specifically Edward Petre's brother Henry, at one point seeking, unbeknown to Handley Page, to replace the Weiss-derived wings with those of a more orthodox configuration. Finally, in the trials themselves, the company's test pilot, Edward Petre, was obliged to make a forced landing, resulting in considerable damage to the aircraft. This effectively knocked it out of the competition.

Handley Page Ltd were subsequently awarded War Office orders, but not for their own designs. As part of their muddled policy of both strangling and aiding private enterprise at one and the same time the War Office were beginning to sub-contract orders for the construction of their own Farnborough-designed BE2 aircraft to the private sector and Handley Page eventually won a contract for five of them. Given the dearth of genuine orders at this juncture one would have thought that the company would have taken great pains to complete the contract satisfactorily; in the circumstances, they had been fortunate to receive it. Indeed, it was partly on the strength of it that Handley Page Ltd finally moved their works to Cricklewood Lane. The contract, however, was not a success. The War Office's dossier on the episode amounts to a damning indictment of the firm's manufacturing capacity at this time, and cannot have made those in authority any more inclined to pursue procurement through the private sector.

Criticisms derived from the report of a Royal Aircraft Factory inspector's visit to the Cricklewood and Barking works on 25 and 26 September respectively. Efforts to arrange an official visit to the Handley Page workshops before then, commented the Superintendent of the Factory, Mervyn O'Gorman, in introduction, had met with no response. Consequently they were obliged to send an inspector to view the sites without prior notice. It was evident from his submission that no serious works yet existed. 'The old stable and riding school that [the inspector] found [at Cricklewood] are not [in the least] equipped', remarked O'Gorman disappointedly.[24] Indeed, 'he could not find 1% of the parts necessary to complete the 5 machines which [Handley Page] has on order, the first of which was due for delivery on the 21st [September], this being the extended date'. It was, he concluded, with some justice, 'a very bad state of affairs'.[25] This was, of course, a period of flux for Handley Page, with the removal of the works to a new site, but the inspector visited both locations and found each equally inadequate. To the War Office it appeared that the firm had tendered for the contract under false pretences. After further negotiations it was recommended that the order be cancelled.[26]

It seems likely that Handley Page was at this stage more concerned with the continued development of his own designs than the fulfilment of however promising a sub-contract. Even here, however, there were serious reverses. On 15 December 1912 Lieutenant Wilfred Parke, the naval pilot previously associated with A. V. Roe & Co., was killed with his passenger, Alfred Hardwick, Handley Page's assistant manager, when 'guest flying' the renovated Handley Page Type F. Soon after taking off on a flight from Hendon to Oxford the failing engine stalled while the aircraft was subject to the lee of a wooded area in Wembley Park. In the resulting downwash the machine spun and crashed to the earth. It was one of a number of fatal monoplane accidents during this period.[27] The result was already evident in a War Office edict banning the flying of all such machines by the Military Wing of the RFC. The Naval Wing, under the political control of Churchill, refrained from so drastic a measure; but the implications for a manufacturer like Handley Page – who was totally absorbed in monoplane design – were serious enough.

Despite these setbacks, however, he continued to develop the reasonably established Type E 'Yellow Peril' monoplane. With the aid of his new chief designer, George Volkert, who had recently graduated from the Northampton Institute, ailerons were now incorporated into the design for lateral control. The aircraft was first flown in this revised form in May 1913, piloted by the company's new test pilot, Ronald Whitehouse. But in reaffirming his belief in the monoplane Handley Page was swimming against the strong tide of contrary opinion – a tide being inexorably drawn forward by the biplane successes of competitors like Avro and Sopwiths. So gradually, reluctantly, he too came round to designing biplanes.

During the summer of 1912 Handley Page had, in fact, already constructed

---

[24] The site consisted of a 20,000 sq. ft redundant riding-stable complex.

[25] PRO, AIR 1 729/176/5/70.

[26] Barnes records that Handley Page was unhappy with the paucity of the order, from which he could scarcely recover his capital costs, such was the expense of the necessary components: *Handley Page aircraft*, 65. He also records that three of the projected BE2s were eventually delivered – the last in 1914!

[27] RAeC reports on aeroplane accidents, No.8, PRO, AIR 1 733/199/7. Edward Petre was killed on a Martin-Handasyde monoplane within days of Wilfred Parke's death; Robert Fenwick had already lost his life as a result of a monoplane accident at the military trials.

a heavily staggered two-seater tractor-biplane, to the design of a Japanese engineer named Sonoda. The following year, in the wake of the monoplane controversy, came the first Handley Page-Volkert tractor-biplane, the 100 h.p. (Anzani radial) Type G. This still employed Weiss-derived wings and ailerons, but it also incorporated a tandem two-seater cockpit, probably derived from the BE2, the design details of which both Handley Page and Volkert would have known from the firm's recent government contract. In addition the fuselage was suspended on struts, midway between the two mainplane wings. The expense of such radical revisions to the company's customary designs was offset by the recruitment of several new investors to the firm at this time. They included a metal merchant named Douglas Russell and a diamond merchant named Walter Dunkels.[28] Even so, there was some difficulty in attracting a buyer for the new aircraft. The Type G prototype had originally been planned as a seaplane, with the Admiralty obviously in view; but when they expressed no interest in the design it was amended and eventually constructed as an exhibition landplane on account of its ability to take off and alight in restricted areas. In this form it was sold to Rowland Ding, acting on behalf of the Northern Aircraft Company (previously the Lakes Flying Co.). Ding had learned of the design when a pupil at one of the Hendon flying schools, during which time Handley Page persuaded him to diversify beyond the employment of seaplanes. The type first flew in November 1913 and was finally released the following April.[29]

This largely involuntary change of emphasis from monoplane to biplane design, although no-one knew it at the time, was crucial to Handley Page's future. However it would be wrong to make too much of the constrained aspect of this decision. Given the increase in size of Handley Page Ltd's aircraft over the next year the change would probably have occurred anyway, regardless of external pressures. Handley Page had already come to the conclusion that, where specifically larger and heavier aeroplanes were concerned, the biplane genre was inherently more stable. This was the gist of a paper he delivered, comparing the two types of aircraft, before both the Aeronautical Society and the Royal United Services Institution in January 1913.[30] The trend was confirmed in the summer of 1914 when, as an enlarged derivative of the Type G, a 60 ft span, 200 h.p. Canton-Unné Salmson engine two-seater tractor-biplane – designated the Handley Page Type L – was designed as a commission for one of the Northern Aircraft Company's most enthusiastic patrons, Princess Ludwig von Löwenstein-Wertheim (née Lady Anne Saville, daughter of the earl of Mexborough). Built with the *Daily Mail*'s £10,000 transatlantic flight prize in view, and incorporating side-by-side seating, dual controls and fuel

[28] Register of members, directors and shareholders 1909–18. (Summary of share capital and shares, Dec. 1912.)

[29] Barnes, *Handley Page aircraft*, 16, 66–7.

[30] 'The comparison of monoplanes and biplanes, with special reference to the stresses in each type', repr. *RAeS Journal* xvii (1913), 49–58.

tanks capable of carrying up to 350 gallons of petrol, this aircraft was the indirect precursor of the first makeshift Handley Page wartime bomber design. Indeed, although originally intended to carry the princess and Ding as co-pilots, war intervened before it was completed and military variants were immediately proposed.[31]

When war came it was the Admiralty which saw the potential in the latest Handley Page designs. The War Office had effectively cut all dealings with the firm following the disastrous BE2 sub-contract. Commodore Murray F. Sueter, Director of the Admiralty's Air Department, was looking for a long range heavy patrol bomber capable of attacking naval bases and Zeppelin hangars inside enemy territory: capable, in other words, of pre-emptive strikes. To this end he had already sanctioned the construction of a J. Samuel White 115 ft span, twin-fuselage, twin-engine, float-mounted seaplane.[32] This was strikingly la-belled the AD 1000, apparently by way of a sardonic reference to Igor Sikorsky's huge 'Ilia Mourometz' four-engine biplane, which took its name from a legen-dary tenth-century Russian hero. Designed by one of the NPL's young Cam-bridge recruits, Harris Booth, this experimental prototype was essentially a revision of Howard Wright's recent White-built twin-fuselage biplane. It ultimately failed to satisfy Sueter's requirements, however, and he remained open to alternative designs. Handley Page was quickly forward with plans for a suitably amended twin-engine biplane in both land and seaplane modes. In this Weiss-pattern wings were finally eschewed in favour of straight (although staggered) mainplane wings of 70 ft span. The aircraft was to have been powered by two tandem-mounted Salmson radial engines driving parallel outboard propellers, and was predictably designated the Type M; however Sueter rejected it as well: 'what I want is a bloody paralyser not a toy', he is reputed to have told Handley Page,[33] and in December 1914 work began on an even larger twin-engine successor.

A draft design for this, labelled the Handley Page Type 0/100, was completed by February 1915, but it was the following December before the prototype, incorporating a 100 ft upper wing-span and a 62 ft long rectangular-section

---

[31] The princess was never to lose sight of the transatlantic goal, however, and, indeed, lost her life in 1927 after sponsoring and acting as passenger aboard a Jupiter-powered Fokker F.VII being piloted across the Atlantic in a westward direction by Lt.-Col. F. F. Minchin and Leslie Hamilton. (The aircraft vanished without trace: H. Penrose, *British aviation: the adventuring years 1920–1929*, London 1973, 516–17.) Canton-Unné engines were the product of the Société Anonyme des Moteurs Salmson, of Billancourt, France, but they were manufactured under licence by the Dudbridge Ironworks Ltd, of Stroud, Glos.

[32] J. Samuel White & Co., of East Cowes, Isle of Wight, were established Admiralty constructors – notably of diesel engines and torpedo-boat destroyers. The firm had instituted an aviation department, under the management of Howard Wright, the previous year (1913). See M. H. Goodall, *The Wight aircraft: the history of the aviation department of J. Samuel White & Co. Ltd 1913–1919*, London 1973.

[33] Penrose, *Pioneer years*, 545. Cdr C. R. Samson (OC, Naval Air Unit, Ostend, 1914) had himself signalled to Sueter that what they required was 'a bloody paralyser to stop the Hun in his tracks': Barnes, *Handley Page aircraft*, 18.

fuselage with armour underplating, made its first test flight, and it was October 1916 before the RNAS received any deliveries. The aircraft's first operational flight then occurred in March 1917. Its employment thereafter depended upon the production of revamped Rolls-Royce Eagle aero-engines. Starting at 320 h.p., these were to be considerably refined and improved over the coming months, but they were manufactured in frustratingly small numbers, owing to Rolls-Royce's reluctance to sub-contract. As an alternative, 320 h.p. Sunbeam Cossack engines were used, but they were of an inferior power-to-weight ratio. As its *raison d'etre* the Type 0/100 fuselage was equipped to carry a 1,792 lb bombload in sixteen 112 lb bombs, and with this emphasis an ominous precedent was seemingly set. Yet the emergence of the multi-engine bomber can in retrospect be seen as the culmination of the whole trend of aeronautical development – almost from its very inception. Given that military utility provided the essential dynamic for design development then aviation effectively reached its apogee with the evolution of the long-range bomber. In time, Handley Page bombers – particularly of the improved Type 0/400 class, employing 360 h.p. Rolls-Royce Eagle VIII engines and carrying a potential 2,000 lb bombload – were to become the mainstay of the independent strategic bombing force and its precursors which were to play so important a part in the foundation of the Royal Air Force. Furthermore, and ironically given the fact that with the onset of aerial combat the concept had been widely discredited, such enormous machines had largely vindicated Handley Page's original emphasis on stability, for the employment of twin-engines meant that inherent stability became a requisite feature – lest one of the outboard motors fail. The Germans followed much the same design in the development of their own Mercedes-powered Gotha bomber, used in raids on London in 1917 and 1918.

With the war expected to continue into 1919 huge prototype 13½ ton four-engine 'Super-Handleys' – or V/1500s, as they were officially designated – were painstakingly constructed specifically for heavy bombing raids on Berlin. Two machines and their crews from 166 Squadron were in readiness for an initial sortie from 27 Group's Bircham Newton airbase in Norfolk when the armistice intervened. By that time 160 such aircraft were on order. Although only about a quarter were completed they presaged future air policy.[34]

[34] For the development of all these Handley Page bombers see Barnes, ibid. 74ff; Bruce, *Aeroplanes of the Royal Flying Corps*, 261–8; and C. Bowyer, *Handley Page bombers of the First World War*, Bourne End 1992, passim. The Gotha company (Gothaer Waggon-fabrik) began constructing aeroplanes as early as 1913, but made little impact before the war: Morrow, *Building German air power*, 52. In Russia, however, Igor Sikorsky had been developing unprecedented extended-fuselage four-engine tractor-biplanes, incorporating four-rudder tail-sections for inherent stability, since 1912: I. Sikorsky, *The winged S: an autobiography*, New York 1938, 73ff.

Sopwith

Unlike Handley Page, Thomas Sopwith emerged not only from a wealthy background, but from one already rooted in the engineering world: his father was a successful civil engineer, tragically killed in a sporting accident in 1898. Sopwith, like Brabazon, had all the advantages of a private income, but, unlike him, did not have any of the attendant anti-industrial family prejudices. He was born in Kensington in January 1888 and educated at a small private school, Cottesmore, at Hove in Sussex, and at a private engineering college, Seafield Park, at Lee, near Portsmouth. The latter provided practical engineering instruction to the less academic sons of the upper middle classes. Of all the young pioneers who grappled with the new medium of flight in these years, Sopwith came nearest to being born with a silver spoon. He had both finance and a natural mechanical aptitude.[35]

Certainly his family's wealth was sufficient for him to be accepted socially by the grandees of the Aero Club. The young Sopwith gained his earliest aeronautical experience with ballooning associates like Frank Hedges Butler and C. S. Rolls – he made his first ascent on 24 June 1906, as a passenger in C. S. Rolls's 'Venus' balloon – [36] and like Rolls and Brabazon he bought his own gas balloon from Short Brothers. This was designated 'The Padsop', as it was jointly owned with Phil Paddon, another member of the Butler-Rolls set, with whom Sopwith had already established a London-based automobile retail business under the title of Paddon & Sopwith of Albemarle Street, Piccadilly. He thus had many early connections with those pioneers who were subsequently to establish an aerodrome on the Isle of Sheppey.

Yet Sopwith was a comparatively late convert to aviation itself – perhaps as a result of being some years younger than most of the other pioneers. It was the autumn of 1910 before he purchased his first aeroplane, a Howard Wright 40 h.p. (ENV) 'Avis' monoplane, designed by W. O. Manning. It cost him £630. Howard Wright was the brother of one of Sopwith's Aero Club ballooning associates, Warwick Wright, and was well-known through his maintenance of a workshop next to Short Brothers in Battersea. Having already secured some slender experience as a passenger on a Farman biplane hired out at Brooklands by Hewlett & Blondeau, Sopwith made his first tentative solo aeroplane flight in the 'Avis', at Brooklands, on 22 October 1910. A beginner's smash delayed his progress for a week or two but unlike Handley Page, who soon realised his limitations as a practical aviator, Sopwith quickly proved himself a most adept pilot. In November he acquired one of the new Howard Wright biplanes (a Farman-derived pusher-biplane, powered by a 60 h.p. ENV

[35] B. Robertson, *Sopwith: the man and his aircraft*, Letchworth 1969, 11; A. Bramson, *Pure luck: the authorised biography of Sir Thomas Sopwith, 1888–1989*, Wellingborough 1990, 14–17.
[36] Ibid. 18.

water-cooled engine) and on the 21st, just a month after his first flight, he gained his aviator's certificate (RAeC No.31).

Almost immediately Sopwith entered for the Michelin Cup, awarded each year for the longest closed-circuit flight. By the end of the month he had flown a circuit-accumulated 107 miles in 3 hrs 12 mins, taking himself to the front of the competition. S. F. Cody subsequently flew a greater total distance, but by that time Sopwith's name was made, for he had secured – against keen competition – the £4,000 Baron de Forest Prize for the longest flight from England to the continent in an all-British aeroplane. His modified Howard Wright biplane, incorporating upper wing extensions and an enlarged petrol tank, left Eastchurch on 18 December 1910 and travelled 169 miles over the channel, before landing some 3 hrs 40 mins later at Thirimont, near Beaumont in Belgium.[37]

At this stage Sopwith seems to have seen aviation primarily as a sport, similar to his equally expensive passions for both yachting and motor-racing. When exactly he began to think in terms of developing his own designs is difficult to determine. Certainly he and his mechanic, Fred Sigrist, would have tinkered with, and probably modified, their various machines (Sopwith had continued to purchase aircraft throughout 1911); and in the course of their discussions they doubtless schemed improved designs of their own. If so, then this interim period can perhaps be seen retrospectively as a time of hidden apprenticeship, prior to what was later to seem an extraordinarily rapid rise to eminence within the industry.

In February 1912, after a successful American tour in which he flew Gnôme-powered Blériot monoplanes, Sopwith founded a flying school at Brooklands. In terms of the gathering together of staff and holdings this was the first step towards the establishment of a Sopwith aviation company. Former Avro pilot F. P. Raynham was appointed chief instructor, leaving Sopwith himself free to pursue other interests. In June 1912 Sopwith won the first *Daily Mail* Aerial Derby at Hendon, flying a 70 h.p. Gnôme Blériot. Raynham left soon after, to be replaced by yet another former Avro pilot, Copland Perry. Under him the Sopwith school continued to flourish. (This trail of former Avro pilots – Sopwith was soon joined by Howard Pixton as well, albeit from British & Colonial – presumably reflected the older firm's inability to pay competitive wages, symptomatic of their wider financial difficulties.) Among the first pupils successfully taught by Perry was the then Major Hugh Trenchard. Soon after, a man who was subsequently to play a decisive role in the development of Sopwith aviation, Harry Hawker, also passed through the school.[38]

It was in association with his successful flying school that Thomas Sopwith began to construct his own aircraft. During his 1911 American tour he had acquired a Burgess-Wright (i.e. Wright brothers) pusher-biplane. This he amended by installing side-by-side seating for instructional purposes, and by

[37] RAeC Committee minutes, 3 Jan. 1911, RAeC papers.
[38] M. Hawker, *H. G. Hawker, airman: his life and work*, London 1922, 31.

affixing a small forward nacelle. In contemporary literature this modified design is sometimes dubiously designated a Sopwith-Wright: it was on this machine, powered by a 40 h.p. ABC engine, that Harry Hawker secured the 1912 Michelin Cup following his setting of a new British duration record of 8 hrs 23 mins on 24 October 1912. Taking this process a stage further, Sopwith then incorporated a set of wings derived from his American Burgess-Wright into an otherwise conventional Sopwith-designed two-seater tractor-biplane. This became known as the Sopwith Hybrid and is usually considered the first authentic Sopwith aircraft.[39] Powered by a 70 h.p. Gnôme engine, it was first flown at Brooklands in July 1912. By October, when it was tested at Farnborough, it could climb to 1,000 ft in under three minutes – with a passenger. Major F. H. Sykes, by then OC, RFC, Military Wing, praised its qualities in a report to the Director of Fortifications and Works,[40] but, in the event, the Naval Wing were to acquire it first.

Meanwhile, in association with the boat-builders, S. E. Saunders Ltd – subsequently to become Saunders-Roe after the firm's takeover by A. V. Roe – Sopwith and his team designed and constructed a biplane flying-boat, powered by a 90 h.p. Austro-Daimler engine. This aircraft incorporated a reconstituted boat-hull, rather than being merely an adapted landplane. It became known as the Sopwith Bat Boat, and bore many similarities to prototypes simultaneously being tested by Curtiss in America: both used the pusher-biplane format. It was later recognised as an early, if somewhat primitive, example of the type of aircraft ultimately to culminate in the multi-engined flying-boats pioneered most notably by Short Brothers in the inter-war years.[41]

Once Sopwith started on design and construction the flying school quickly faded into the background and out of it emerged the Sopwith aviation company. In June 1912 Hawker was taken on – ostensibly as test pilot, but in reality as general adviser; and in October R. J. Ashfield was appointed the firm's chief draughtsman. (The future sports-car manufacturer, Donald Healey, was also presently taken on as an articled pupil.) In November 1912, through the intervention of Commander Oliver Schwann, who had been instrumental in development of the Avro Type D seaplane, the Sopwith Hybrid was purchased by the Admiralty for £900.[42] It was delivered to the old Aero Club ground at Eastchurch, by now the centre of naval aviation. By contrast to the War Office, whose procrastination in honouring their contracts caused such serious

---

[39] H. F. King, *Sopwith aircraft 1912–1920*, London 1980, 16–18. Sopwith's first Burgess-Wright had, in fact, ditched into the sea near Coney Island, so the 'Sopwith–Wright' amendments related to a second acquisition: Bramson, *Pure Luck*, 37.

[40] F. H. Sykes to DFW, 17 Oct. 1912, PRO, AIR 1 801/204/4/1099.

[41] Ibid. AIR 1 733/203/1; AIR 1 733/203/2. The aircraft was subsequently temporarily fitted with a 100 h.p. Green engine, in which 'all-British' mode it won the £500 Mortimer Singer prize for amphibious aircraft on 9 July 1913. Refitted with the Austro-Daimler engine it was then sold to the Admiralty, only to be wrecked the following month when left moored at sea. A replacement, known as the Bat Boat II, was ordered: Penrose, *Pioneer years*, 451–2.

[42] Ibid. AIR 1 727/152/7.

cash-flow problems for Avro, the Admiralty paid for their machine promptly. This enabled Thomas Sopwith to take possession of a redundant skating rink at Kingston-on-Thames, which he then converted into the firm's workshops.[43] The company was officially registered at the end of the year.

By now Sopwiths had begun work on a new tractor-biplane, incorporating staggered mainplanes and a modified fuselage, culminating in a comma-shaped tail-section similar to that of the Avro 502. This became known colloquially as the Sopwith Three-Seater. With the military market in mind, it was designed to carry two observers, side-by-side, well forward in the fuselage, with the pilot situated directly behind. He then controlled the aircraft by means of a Roe-derived joystick, consisting of a steering wheel mounted on a vertical elevator. Lateral control was initially by wing-warping, but ailerons were incorporated into later models. To aid the observers, windows – large, non-inflammable, celluloid panels – were incorporated into the sides of the forward fuselage.[44] Powered by an 80 h.p. Gnôme engine, the type was first flown at Brooklands in February 1913.

This prototype was again taken by the Royal Navy. A second model, which the Admiralty had also ordered, was completed by May. During official tests at Farnborough, on 8th and 9th of that month, it climbed to 1,000 ft in 2 mins 22 secs and recorded a speed range of between 40.6 and 73.6 m.p.h. On 31 May Hawker then established a new British altitude record of 11,450 ft in it. On 16 June he climbed to 12,900 ft solo, and 10,600 ft with two passengers.[45] These were such consistently good performances that even the War Office, which was notoriously indifferent to the advances of private manufacturers, was soon obliged to take notice. On 20 June 1913 Major Sykes sent a memo to the War Office requesting permission for one of his officers to test fly the new aircraft.[46] Captain A. G. Fox of No.3 Squadron was then despatched to Brooklands on 29 June to undertake both solo and passenger trials. His subsequent report, dated 1 July 1913, was exceptionally favourable. The machine, he wrote, got quickly off the ground and climbed well, 'without forcing'. The aircraft's controls were 'sensitive', yet the aeroplane corrected itself 'laterally' in heavy gusts. It thus combined manoeuvrability with a degree of inherent stability. In flight it turned with ease, taking up 'its natural bank of its own accord'; and finally, it landed with equal deftness. 'I consider it a better machine than the BE', Fox concluded; adding that, in fact, he liked it 'better than any machine' he had yet flown.[47]

Fox forwarded this report to Sykes via his Squadron Commander, Major Brooke-Popham. It was clearly what he expected to hear. Sykes forwarded it

---

[43] Ibid. AIR 1 727/152/7; AIR 1 733/203/2. See also Sir Thomas Sopwith, 'My first ten years in aviation', *RAeS Journal* lxv (1961), 242.

[44] PRO, AIR 1 727/152/7.

[45] Robertson, *Sopwith*, 34; King, *Sopwith aircraft*, 24; Bruce, *Aeroplanes of the Royal Flying Corps*, 491–2.

[46] PRO, AIR 1 822/204/5/24.

[47] Ibid. AIR 1 822/204/524.

on to David Henderson, Director General of Military Aeronautics at the War Office, with a recommendation that no less than nine such machines be ordered for immediate use with the Military Wing of the RFC. 'I may say', he added, 'that I have myself been taken [up] as a passenger in one of these machines, and from a passenger's point of view can corroborate Captain Fox's favourable report.'[48]

The order was duly placed, and Sykes was reporting delivery of the first aircraft by November 1913.[49] All this occurred no more than a year after the emergence of the Sopwith Hybrid. Already, however, in this RFC report, can be discerned those features which were to make the company's aircraft the outstanding fighting machines of World War I: stability for reconnaissance combined with manoeuvrability for combat. It was this discernible military potential that found Sopwith his market.

By the time these machines were delivered to the RFC Sopwiths were already working on a new compact dual cockpit-seating biplane known as the 'Tabloid'. The prototype underwent trials at Farnborough on 29 November 1913 and revealed an exceptional versatility. Powered by an 80 h.p. Gnôme engine, and designated for the purpose the Sopwith Scout, it gave a speed range of between 36.9 and 92 m.p.h., and climbed at a rate of 1,200 ft per minute – with a passenger and sufficient petrol for 2½ hours flying. A later wartime report described its performance on this occasion as 'a sensation'.[50]

Hawker, who according to Sopwith had been instrumental in the Tabloid's development,[51] was given permission to demonstrate the prototype in Australia. In his absence, Sopwith engaged former Avro and Bristol pilot, Howard Pixton. Production of the Tabloid-Scout began in earnest in January 1914, following despatch of the last three-seater to the RFC. By then a batch of nine single-seater Tabloids had been ordered by the War Office, and three more were to be ordered in March. Deliveries to Farnborough began in April and May of 1914; but the later models incorporated modified and strengthened undercarriages, leading Major Sykes to recommend that all Sopwiths be so amended and that none be flown until then. This inevitably delayed their entry into service.[52] The type none the less proved to be a major influence on the development of aviation prior to the war, setting the standard by which nearly all manufacturers measured their achievements. It also marked the emergence of Harry Hawker as one of the great aviators of the day, both as pilot and co-designer. Born the son of a blacksmith in Australia in 1889, Hawker had

---

48 F. H. Sykes to Director of Aeronautics (*sic*), War Office, 5 July 1913, ibid.
49 Sykes to DGMA, 8 Nov. 1913, ibid.
50 'The Sopwith Aviation Company', report dated Jan. 1916, PRO, AIR 1 733/203/2. These statistics are also given in AIR 1 727/152/7. The prototype Tabloid employed the wing-warping method of flight control, but this was soon amended in production to ailerons: Sopwith, 'My first ten years in aviation', 244.
51 Quoted in Robertson, *Sopwith*, 37. See also Hawker, *H. G. Hawker*, 119ff, and Sopwith, 'My first ten years in aviation', 251.
52 Bruce, *Aeroplanes of the Royal Flying Corps*, 495.

left school at an early age and – in a now familiar pattern – served his technological apprenticeship in the motor industry. Inspired by the idea of flight, he then joined forces with fellow motor mechanic, Harry Busteed, and came to England in May 1911. Busteed quickly joined the British & Colonial Aeroplane Co., but it was another year before Hawker found a place with Sopwiths.[53] He started out with no advantages in life, save his mechanical aptitude; in a sense, the Tabloid was his vindication.

Soon after the emergence of the Tabloid Sopwiths were commissioned by the Admiralty to build a special scaled-up side-by-side twin-seater biplane specifically to carry the First Lord – at that time Winston Churchill – over the Fleet. In contrast to the Tabloid, this incorporated larger two-bay wings and a 100 h.p. Gnôme engine to facilitate take-off – apparently on account of Churchill's already fulsome weight – and was to be flown by Lieutenant Spenser Grey RN. In February 1914 Churchill visited the Sopwith works in person to witness the final stages of its construction. Known alternatively as the 'Sociable' or 'Tweenie',[54] the finished machine was test flown by Howard Pixton, and delivered to the Royal Navy on 19 February. Some of Sopwiths' competitors were greatly riled by this sign of official patronage. H. V. Roe, in particular, instructed his own staff to wake up to such opportunities: 'We ought not to let Sopwith pick out these little plums', he complained.[55] An official commission on behalf of the First Lord was about as good an advertisement as any firm could hope for in those precarious days.

Sopwiths having clearly secured a profitable corner of the market with its latest successes, the firm was reconstituted as a limited company in March 1914. Soon foreign interest was also evident. The Greek navy, in particular, commissioned the company to construct a new design of pusher-Gunbus, with a machine-gun mounted in the nacelle. Six of these were ordered and trials begun in March 1914.[56] Then the following month Sopwith Aviation spectacularly confirmed its international reputation. Entering a single-seater Tabloid biplane on floats, with a 100 h.p. Gnôme Monosoupape engine, the company won the second of the recently inaugurated Schneider Trophy seaplane contests. Howard Pixton piloted the aircraft on this occasion, on account of Hawker's continued absence in Australia. Winning this event had enormous prestige value as, due to the wide field of entry, it was considered a benchmark of design efficiency. No British competitor had entered the first contest, which had been won the year before by Maurice Prevost on a 160 h.p. Gnôme Deperdussin float-monoplane, and Sopwiths' victory in the second, held again at Monaco, caused something of a sensation. Yet it really only

[53] Hawker, H. G. Hawker, 27–8; 'Mr H. R. Busteed: pilot-instructor' ('Men of moment' series), Flight, 20 Dec. 1913.

[54] Penrose, Pioneer years, 507–8; King, Sopwith aircraft, 64–5.

[55] H. V. Roe to H. Lutwyche, 19 Jan. 1914, AVRO papers, file 12.

[56] They were commandeered by the Admiralty upon the outbreak of war, before completion: PRO, AIR 1 727/152/7.

confirmed what was already generally known in informed aeronautical circles – that English-designed aircraft were rapidly eclipsing the French and all comers.[57] More immediately, Sopwiths' Schneider victory was seen as further evidence of the extraordinary capabilities of the Tabloid-Scout biplane. Production of the single-seater model continued at Kingston-on-Thames.[58]

With the development of shipborne aircraft the need for folding wings became apparent. This idea had been devised by Short Brothers as early as 1911 and patented by them in 1913. Sopwiths were therefore obliged to come to terms with Shorts before they could incorporate the device into their own designs. This done, a new two-seater 100 h.p. Gnôme Monosoupape seaplane – the Type 807 – was marketed, with the Admiralty obviously in view. The first was completed in July 1914, during which time a large, four-bay, experimental 'torpedo-dropping' float-biplane, designated the Sopwith Type C, was also under construction.[59] The firm was certainly wise to cultivate the Admiralty connection; despite recent orders War Office procurement was growing increasingly restrictive and with the onset of war it was primarily as naval contractors that Sopwiths were to be employed, Sopwith aircraft serving predominantly with the Royal Naval Air Service.[60] As aerial operations developed progressively into aerial combats, however, the qualities offered by Sopwith aeroplanes became of rapidly escalating importance. Sopwiths were quick to realise the implications of what was happening and turned their attention to the design of aircraft specifically as combat machines.[61]

It is no part of this study to trace the evolution of the aeroplanes which followed, aircraft which have since become part of the popular legend of World War I; suffice to say that from the Tabloid developed the Sopwith Strutter, Pup and Triplane in 1916, and Camel in 1917. The Strutter was a two-seater reconnaissance machine, designed to defend itself. It had both a fixed propeller-synchronised forward-firing gun for the pilot, and a movable rear gun for the observer. (The synchronising device was devised by Hawker's compatriot, H. A. Kauper; formerly foreman of Sopwith Aviation's fitters.)[62] As with Avro, however, the unprecedented demand resulting from the continuation of the war led to much of the firm's later construction work being sub-contracted to other manufacturing concerns.[63]

The cessation of military orders following the armistice, coupled with punitive taxation, specifically the excess profits duty, forced the Sopwith Aviation Company into voluntary liquidation in September 1920. It re-

---

[57] For evidence that even French constructors recognised Sopwiths' pre-eminence see PRO, AIR 1 783/204/4/515; and AIR 1 727/152/7, 8.

[58] Ibid. 11.

[59] Ibid.; King, Sopwith aircraft, 67–74.

[60] PRO, AIR 1 727/152/7. See also AIR 1 733/203/2, for details of early Sopwith–RNAS operations.

[61] PRO, AIR 1 727/152/7, pt II.

[62] King, Sopwith aircraft, 94–5, 155.

[63] Robertson, Sopwith, 150.

emerged almost immediately as the Hawker Engineering Company, subsequently to form the basis of the Hawker-Siddeley conglomeration. Harry Hawker himself, however, died shortly after the launching of the new company in his name. After winning enormous publicity for his dramatic and near-fatal May 1919 transatlantic flight attempt aboard the Sopwith 'Atlantic' biplane, accompanied by his navigator, Commander Mackenzie Grieve RN, he was killed while test flying Henry Folland's Nieuport Goshawk at Hendon on 12 July 1921. A post mortem revealed tuberculous of the spine, which had resulted in extensive pre-death hemorrhaging. He had agreed to pilot the Goshawk in that year's Aerial Derby.

British & Colonial (Bristol)

Most early aeronautical enterprises were established with some precariousness by young engineering enthusiasts, with little capital and less business experience. It has been seen through, for example, Avro's development, the immense difficulty such technological entrepreneurs had in attracting significant investment from established industry. Likewise the desperate plight such fledgling firms could fall into without rapid remuneration for capital outlay has been demonstrated. In about 1910–11, however, once a potential market had been identified, this situation began to change. Established businesses, with large capital reserves, began to take an active interest in the new industry. Foremost among these were the Bristol aeroplane company (i.e. British & Colonial, originally a scion of Bristol Tramways) and the Vickers corporation.[64]

The Bristol organisation owed its existence to one man, Sir George White. White's career was in the mould of the classic entrepreneurs of Victorian tradition. Obituary notices of his death in 1916 rang with phrases emphasising the self-made accomplishments of his life: 'From Office Boy to Millionaire' ran the *Morning Post's* headline; 'Architect of his own fortunes' stressed the *Financial News*; and there were many more.[65]

Born in Kingsdown, Bristol, in 1854, the second son of a painter and decorator, George White was educated at St Michael's School in the city, but left early to take up a position as junior clerk in the offices of Stanley & Wasbrough, a local firm of solicitors.[66] The Bankruptcy Act of 1869 resulted in a flood of insolvency cases for solicitors at this time and White was entrusted with the task of superintending dozens of liquidations. Through this work he was able to gain, in an inverted way, valuable experience of business methods, which he would later employ to great effect.

[64] Amongst others, there were also the Coventry Ordnance Works, and later (from 1913) Armstrong, Whitworth & Co. Ltd.

[65] *Morning Post*, 24 Nov. 1916; *Financial News*, 24 Nov. 1916 (Bristol papers, Cuttings Album AC 76/23).

[66] Particulars from obituary article in *Western Daily Press*, 23 Nov. 1916.

From bankruptcy work White moved on to the preparation of parliamentary private bills. In this capacity, between 1872 and 1874, he helped conduct the Bristol Tramways Bill through parliament. Subsequently, in 1874, when still only twenty-years-old, he was appointed part-time secretary of the Bristol Tramways Company. The following year he set himself up as a stockbroker and was elected to the Bristol Stock Exchange in 1876.[67] From this time White's astuteness in various locomotive and transport industries became a byword. Perhaps as a result of his earlier experience with bankruptcy cases he succeeded in revolutionising[68] the railway map of western England by obtaining control of, and restructuring, several struggling lines. These he would then generally sell off to their more powerful competitors, moving to divest himself of his accumulated shareholding at a considerable profit. The Bristol Port Railway & Pier Company, the Taff Vale Railway Company, the Bristol and North Somerset, the Severn and Wye, the Severn Bridge Railway Company, were all grist to the mill in this regard.

It was through the introduction of electric street traction to many cities, however, that George White became a national figure. He pioneered electric traction in the north of England, in Dublin and in London, as well as in Bristol. He gained, for example, control of Imperial Tramways and electrified their systems.[69] He also transformed the derelict West Metropolitan Tramways into the successful London United Tramways – the capital's first electric line. Even after the establishment of the British & Colonial Aeroplane Company it was as the 'tramway millionaire' or 'tramway king' that Sir George White was popularly remembered.[70] In Bristol itself, whilst maintaining a fierce opposition to trade unionism, he was also known for his public works and many acts of philanthropy, for which he received a baronetcy in 1904. He was, C. G. Grey later commented, the first of the big businessmen to see the potential of aviation. 'I have always regarded Sir George', he added, 'as one of the three or four men I have met who could justly be called great.'[71]

It was, then, a figure of some stature, supported by considerable capital, who now turned his attention to the commercial exploitation of aviation. Sir George White had visited the aerodrome associated with the Wright brothers at Pau, in the south of France, sometime in 1909 and thereafter began to investigate French developments in this new field of technology. He first publicly announced his intention of diversifying into aeroplane manufacture at a shareholders' meeting of the Bristol Tramways & Carriage Company – of which he was by then chairman – on 16 February 1910. Within the week a

---

67 Ibid. For a detailed examination of Sir George White's early business activities see C. Harvey and J. Press, 'Sir George White: a career in transport, 1874–1916', *Journal of Transport History* ix (1988), 170–89.
68 The term used by the *Financial News* in its obituary article, 24 Nov. 1916.
69 Ibid.
70 *Chronicle*, 24 Nov. 1916; *Daily Sketch*, 28 Nov. 1916. Even the journal *Flight* had to admit this fact: 30 Nov. 1916.
71 C. G. Grey, 'Something about Bristol's', *The Aeroplane* lxxviii (1950), 220, 222.

new firm was registered under the trading title of 'The British & Colonial Aeroplane Company Limited'. The aim was that the company should become the principal manufacturer of aircraft for Britain and her empire – hence its name. The label 'Bristol' was always adopted, however, to denote the company's products.[72] Yet capital for the new enterprise came not from Bristol Tramways as such, but directly from Sir George White and his immediate family, for Sir George was shrewd enough to realise how speculative such a venture would initially appear to most investors. Trading began with capital raised through the sale of 25,000 £1 shares: of these, 10,000 were taken by Sir George White himself, while 10,000 were taken by his brother, Samuel, and 2,500 by Sir George's son, G. Stanley White. These three figures became the company's directors, with Sir George as chairman.[73] Sir George then made his nephews, Henry White Smith and Sydney Ernest Smith, who were the remaining shareholders, secretary and manager respectively.

In other ways than finance, however, Bristol Tramway connections did facilitate British & Colonial's entry into the aviation market. As well as electric trams the Bristol Tramways & Carriage Company also provided an omnibus and taxi service. The taxis they employed were built by Charron in France, and in securing them Sir George worked through a French motor agent named Émile Stern. Having been impressed by the Voisin-derived pusher-biplanes displayed at Rheims in 1909, Sir George indicated that his immediate interests lay with that genre of aircraft. Consequently Stern was able to place him in contact with the Zodiac aircraft company of Paris, which manufactured such machines. A British licence for the construction of Zodiac aircraft was then duly negotiated, the agreement being ratified at British & Colonial's first directors' meeting, on 28 February 1910. Stern was subsequently appointed the firm's agent in France for the procurement of aero-engines.[74]

At the same meeting it was reported that the Bristol Tramways omnibus depot at Filton, some four miles north of Bristol, was soon to fall vacant, so it was decided to appropriate the site as the aircraft company's works.[75] The Bristol Tramways Company also provided Sir George with a pool of skilled labour, which he drew on for the basis of his new workforce. George Challenger,

---

[72] Statement on company history (1915), B&C papers. In fact, on 19 February 1910 Sir George White simultaneously established three other companies as well: the Bristol Aeroplane Co., the Bristol Aviation Co., and the British & Colonial 'Aviation' Co., with a nominal capital of £100 each. In 1920 the trading company (the British & Colonial Aeroplane Co.) went into voluntary liquidation as a means of circumventing excess profits duty, and its assets were transferred to the Bristol Aeroplane Co.

[73] 'Memorandum and articles of association, British & Colonial Aeroplane Co. Ltd', George White papers, 35810/GW/T/30; 'List of first subscribers of capital', B&C papers. In the event the company made comparatively quick returns: from 0.87% profit on turnover at the end of 1911, to 15% in 1912, 18% in 1913, and 20% in 1914: statement on company history (1915), B&C papers.

[74] Company minutes (cited hereinafter as minutes), 28 Feb. 1910, B&C papers.

[75] The agreement for this was only officially ratified in January 1911, however: rider on minutes, 18 Jan. 1911, ibid.

son of the tramway company's general manager, was appointed the aviation firm's chief engineer and works manager. He had previously been in charge of the tramways company's Brislington carriage works and from there he brought with him a selected team of carpenters, fitters and coachbuilders. Among these was a former tramway's apprentice named Collyns Pizey, who became Challenger's assistant engineer.

In the event the French contract proved a disaster. Powered by a 50 h.p. four-cylinder Darracq engine the first Bristol-Zodiac biplane was tested at Brooklands in May 1910, piloted by the French aviator, Maurice Edmond, and it could barely be induced to hop. Five further models were under construction at Filton before the total inadequacy of their design was fully comprehended. They were all written off. The Zodiac agreement was terminated and in August 1911 the company commenced legal proceedings for the recovery of costs. Zodiac settled out of court.[76] Marking this episode down to experience British & Colonial turned instead to the development of their own designs. George Challenger, having gained some practical experience through construction of the abortive Zodiacs, was given responsibility for the company's first indigenous aircraft. He produced the Bristol Challenger Boxkite, a Farman-derived pusher-biplane. Utilising one of the first export 50 h.p. Gnôme rotary engines to come on the market, the main prototype, aircraft sequence No.7, was successfully test flown in July 1910.

Prospects for the success of the Bristol venture were further improved when the company, through the agency of Émile Stern again, went on to secure (temporarily as it turned out) British rights to the Gnôme engine. The priority now was for the firm to somehow ingratiate itself with the military authorities. As early as May 1910 Sir George White had visited Haldane in Whitehall and offered to place the 'entire resources . . . of the company' solely at the War Office's disposal, 'abstaining from all business with Foreign Powers'. Clearly he was angling for some sort of semi-official contractual position with the War Office. Haldane, however, refused the bait. 'The Minister', it was stated, would prefer that the company develop 'business relations in the fullest way with all foreign countries without restriction.'[77]

Exchanges like this were to become familiar over the next four years. Undeterred, British & Colonial sought other ways of manoeuvring themselves into the military's sights. Also in the spring of 1910 successful negotiations were entered into with the War Office for flying-rights over some 2,000 acres of army land at Larkhill, on Salisbury Plain. Here the first army aviators, led by Captain J. D. B. Fulton, were already carrying out trials. In June 1910 the company erected three iron hangars on the site.

Both at Larkhill and at Brooklands Bristol flying schools were established:[78] Larkhill being situated so conveniently near Bulford and Tidworth, however,

---

[76] They paid 15,000 francs in compensation: minutes, 21 Aug. 1911, 16 May 1912, ibid.

[77] Rider on minutes, 24 May 1910, ibid.

[78] Minutes, 24 May, 2 Sept. 1910, ibid. B&C had already ratified an agreement with the

drew, as planned, curious soldiers to the assembled office and workshops like so many moths to a flame. There had been considerable excitement at the launching of the first Bristol Boxkite on the site on 30 July 1910 for instead of preliminary hops, in anticipation of which witnesses had situated themselves in prone positions on the ground, the aircraft immediately climbed to an altitude of 150 ft. Topographically the area was also ideal. Salisbury Plain, commented C. G. Grey in a later report to the Air Ministry, was 'by far the best stretch of country in England for the training of aviators'.[79] In time the Royal Flying Corps would be stationed in sheds adjoining the Bristol hangars, while nearby would be situated the Central Flying School. Here also, in August 1912, the military aeroplane competition would be held. The directors of British & Colonial had been particularly astute. By just being on site they were bound to be noticed by local army units. Their policy of assiduously pursuing favour with the military was unquestionably the correct one from a commercial point of view, and to facilitate the process Sir George White would entertain officials at Larkhill in a way redolent of today's arms market.[80]

Yet what the Bristol organisation really wanted was some practical link with the army, like that Shorts had with the Admiralty; or alternatively, some useful intermediary. In the event they found the latter, in the person of Captain Bertram Dickson of the Royal Horse Artillery. Dickson was arguably the most successful pilot to have emerged from the British army in these pre-Flying Corps years. By the time of British & Colonial's move to Larkhill he had already acquired a considerable reputation as an exhibition aviator. He had won first prize at the Tours meeting in May, flying a 50 h.p. (Gnôme) Farman. This was the first important international prize won by a British pilot. At the same meeting he was also awarded the original Schneider Cup – a forerunner of the seaplane trophy – which was subsequently to be presented to RAF College, Cranwell, to be awarded annually to the most proficient pilot. Dickson had then gone on to both win prizes and establish records at Anjou and Rouen. His setting of a new world's duration record for a passenger flight, when he flew for two hours and covered 98.75 km (61 miles) at Anjou on 6 June, was the first world record established by a British pilot. Such was his success that he was said to have accumulated as much as £10,000 from international flying exhibitions over the 1910 season.[81] As this might suggest, while he was still,

---

BARC for the lease of 'sheds' at Brooklands (30 Apr. 1910: reported minutes, 24 May 1910). By the close of May hangar No.17 was occupied, the site being described as the firm's 'London HQ': *The Aero*, 24 May 1910. An aerodrome was also established at Durdham Downs, on the outskirts of Bristol.

[79] PRO, AIR 1 727/160/5.

[80] See C. G. Grey, 'Bristol's', 222; and minutes, 18 Jan. 1911, B&C papers, where arrangements for a press visit are detailed.

[81] Penrose, *Pioneer years*, 239; 'A successful British airman – Captain B. Dickson', *Daily Mail*, 17 May 1910 (Butler collection, xi. 87); 'A British winner abroad', *The Aero*, 17 May 1910.

officially, a serving officer at the time he commenced his flying activities, he was in no way confined to regular duties.

Born in Edinburgh on 21 December 1873, Bertram Dickson had been commissioned as a 2nd Lieutenant in the artillery in November 1894, and promoted Lieutenant three years later. In December 1899 he was a member of General William Gatacre's column during its unsuccessful attempt to capture Stormberg railway junction in Cape Colony – the first reverse of what became known as the 'Black week' of the Boer War. He was subsequently a victim of the typhoid fever which broke out among the poorly sanitised army camps at Bloemfontein the following March and April and was invalided home, but was sufficiently recovered to command the Royal Horse Artillery detachment incorporated within the duke of York's escort for the state opening of the Australian federal parliament later that year. Service in the Boer War and his other duties brought him the Queen's Medal (with two clasps) and, in November 1900, promotion to the rank of captain.[82] Thereafter he enjoyed a career of varied activity such as was perhaps uniquely available to officers of the Victorian–Edwardian period. What was categorised in the artillery officers' list as 'special service' followed in Mombassa (1901) and with the Boundary Commission in South America (1902–3), after which Dickson explored and surveyed Patagonia and then indulged his hazardous passion for Big Game hunting in such diverse locations as East Africa and Kurdistan. His last major overseas assignment came in July 1906 when he was appointed a vice-consul in Asia Minor. This saw Dickson stationed predominantly on the Persian frontier. The appointment was officially to run until April 1910, but it ended sometime before then for by the summer of 1909 Dickson was in Europe and had begun to take an active interest in aviation. Stopping in Paris on his extended return journey to visit his sister, he had taken the opportunity to attend that August's much publicised flying meeting at Rheims. Soon after, he procured his own Farman biplane which he learned to fly at Mourmelon. He qualified on 19 April 1910, being awarded Aero Club de France certificate No.71 on 12 May. Although no longer on the active list he was the first British serviceman to gain an aviator's certificate. By March 1910, aided by a French mechanic, he was successfully participating in continental flying exhibitions.[83] Having for some time been on sick-leave as a result of his consular service, he officially retired from the army, on half-pay, on 3 August 1910. He was listed on the reserve from the following month.[84]

[82] *Kane's list of officers of the Royal Regiment of Artillery*, ii, Sheffield 1914. For Dickson's background see R. L. Munro, 'Flying shadow: Captain Bertram Dickson', unpubl. manuscript *c*. 1991, Museum of Army Flying Library.

[83] 'A new British flier', *The Aero*, 1 Mar. 1910; PRO, AIR 1 727/160/5; Munro, 'Flying shadow', 8.

[84] *Army list*, Sept. 1910. C. G. Grey subsequently recorded that his friend Dickson had in fact taken up aviation in the hope that clear air and exercise might improve his precarious health: PRO, AIR 1 727/160/8/2, extract from *The Aeroplane*, 2 Oct. 1913. Munro records that he had been suffering from a fever, 'probably malaria': 'Flying shadow', 4.

Dickson first met Sir George White at the Bournemouth aviation meeting of July 1910 and seems almost immediately to have become a trusted advisor, although it was January 1911 before he was officially designated British & Colonial's London and Continental representative.[85] Consequently, when the War Office permitted the inclusion of aeroplanes in the 1910 autumn manoeuvres, Dickson, on behalf of the Bristol organisation, was able to place his services at the army's disposal. As a result, British & Colonial were brought to the attention of an even wider and more influential circle than that normally frequenting Larkhill. Two Bristol Boxkites were employed during the manoeuvres.[86]

The year after, in October 1911, the company offered to undertake, and complete within six months, the tuition of 500 army and naval officers at specially advantageous rates. This was another attempt to coax the military into the aviation market. Both services declined the offer.[87] Prospective pupils, they rejoined, were free to enrol at the various flying schools on an individual basis. Many were already doing so. Among the first was a Captain H. F. Wood, of the 9th Lancers, who was successfully taught at British & Colonial's Brooklands school under the supervision of Archibald Low. Both these figures were subsequently to play major roles in the development of Vickers's aviation department. Unfortunately, however, tragedy overtook Dickson himself. In October 1910, shortly after the British army manoeuvres, he was involved in the first recorded mid-air collision while participating in an international air meeting at Milan. He was demonstrating a silent vol-plané gliding descent, with the engine of his Farman biplane switched off, when his aircraft became entangled with a fast descending Antoinette monoplane piloted by René Thomas.[88] Although severely injured and permanently marked he was eventually to fly again, but he was never the same man as he had been in the summer of 1910. He eventually succumbed to his ailments in September 1913.[89] In the intervening period he continued in the service of Sir George White and British & Colonial.

Having thus belatedly established themselves in the aviationary field British & Colonial's capital was increased in November 1910 to £50,000, through the

---

[85] Minutes, 15 Mar. 1911, B&C papers. Penrose, *Pioneer years*, 234–5, suggests that Sir George White sought Dickson's advice following the Zodiac fiasco, the latter recommending the acquisition of Farman's British patent rights (which, in fact, Holt Thomas was already negotiating). But as Challenger's Bristol Boxkite design was drafted in June 1910, and the prototype completed by the close of July, it is doubtful Dickson had much influence on immediate post-Zodiac developments. Penrose is also mistaken in supposing that Edmond was engaged as test pilot only after the Zodiac affair.

[86] The manoeuvres themselves are examined in ch. 6.

[87] Rider on minutes, 20 Dec. 1911, B&C papers.

[88] *The Aero*, 19 Oct. 1910; Munro, 'Flying shadow', 22.

[89] Died Lochrosque Castle, Ross-shire, 28 Sept. 1913: *Kane's list*, ii, 1914. It is generally reported that Dickson died as a result of his aviation injuries, but while these may have exacerbated his condition they were not the primary cause of his early death. He died of a stroke and was buried at Achanalt, Ross-shire.

creation of 25,000 new £1 shares.[90] The following month, in keeping with the firm's avowed aim of becoming the principal manufacturer of aircraft for Britain and her colonies, aviation commissions were despatched to Australia and India, and later an agency was also established in South Africa.[91] In India, three company machines were placed at the disposal of the military authorities for reconnaissance demonstrations. These demonstrations, or more accurately trials, were led by Farnall Thurstan and company pilot-instructor Henri Jullerot, and supervised by Captain Sefton Brancker, a spirited artillery officer with a limited degree of ballooning experience who was at that time Deputy Assistant Quartermaster-General to General Sir John Cowans, Calcutta Brigade, but who subsequently became the GSO responsible for the early expansion of the RFC. It was this speculative Bristol mission that first awoke Brancker's active interest in aviation. Indeed, when an aeroplane was incorporated within the Deccan cavalry manoeuvres, attended by, among others, Sir Douglas Haig, he took some leave and participated as an aerial observer. Jullerot acted as pilot, and on 16 January 1911 a successful twenty-seven-mile dawn reconnaissance was made. Next morning, however, the Boxkite was unable to attain take-off and crashed. The machine was written off, but no great harm came to the crew. The following month a second machine flew during the Presidency Brigade's training at Midnapore, near Calcutta, but the flights were not so successfully co-ordinated with ground operations. From this point, nevertheless, aviation remained the essential dynamic of Brancker's career.[92]

Soon afterwards the company adopted the roguish figure of Graham Gilmour as test pilot and, as a result, found themselves in receipt of much unwelcome publicity. Flying a Bristol Boxkite Gilmour became probably the most daring aviator of his day. Unfortunately he was too ostentatious and, if brought to book, pugnacious, for his own good. His disregard for petty flying restrictions repeatedly brought him into conflict with the authorities. Indeed, his career as a Bristol pilot is best illustrated by a succession of official reprimands in the Royal Aero Club's minutes for 1911.[93]

Serious trouble ensued in July when, flying his firm's new Prier racing monoplane, Gilmour caused pandemonium by deliberately diving low over the Henley Regatta. He had already disrupted the University Boat Race in this way, in a Boxkite. His flying certificate was immediately suspended, and he was banned from competing in that month's *Daily Mail* £10,000 Circuit of Britain

90 Rider on minutes, 25 Nov. 1910, confirmed at shareholders' meeting, 18 Jan. 1911, B&C papers; George White papers, 35810/GW/T/30. Harvey and Press suggest that this and subsequent capital increases owed much to the fact that Sir George White was able to borrow substantial amounts from the Western Wagon & Property Co., of which he was also chairman: 'Sir George White: a career in transport', 185.

91 Minutes, 25 Nov. 1910, 18 Jan. 1911, B&C papers.

92 Statement on company history (1915), ibid.; N. Macmillan (ed.), *Sir Sefton Brancker*, London 1935, 14–20.

93 RAeC Committee minutes, 4 Apr., 9, 30 May 1911, RAeC papers.

race.[94] Gilmour contemptuously dismissed the decision as 'a great stew' about nothing; Sir George White, however, began legal proceedings against the club and eventually terminated Gilmour's contract.[95] Gilmour moved on to Martin-Handasyde, but was killed shortly after, on 17 February 1912, when his monoplane reportedly broke up in the air above Richmond Park.[96] It was one of a number of monoplane accidents over this period.

Meanwhile the company had been encouraged by the securing of its first significant governmental orders. The breakthrough came in November 1910 when Émile Stern successfully negotiated the sale of eight Bristol military biplanes (i.e. modified Boxkites) to the Russian government, via the Russian attaché in Paris.[97] There was a certain frisson within the aviationary community when knowledge of this contract became public. Haldane, who had little regard for private British manufacturers, had had his remark about the company 'developing business relations in the fullest way with all foreign countries' thrown back at him. To what extent he and his advisors were influenced by this Russian contract, and to what extent by reports of the Boxkites' use in recent manoeuvres, is difficult to determine, but by March 1911 the War Office had ordered four of them.[98] (At this point the Royal Engineers' Air Battalion was coming into being.) The first two deliveries, of aircraft Nos 37 and 38 in May 1911, were to be powered by 50 h.p. Gnôme engines; the second two, of modified nacelle-fronted Nos 39 and 42 in July–August 1911, by 60 h.p. Renault engines.[99] An order for an additional four standard Boxkites, two with 50 h.p. Gnôme engines and two as spare airframes, was received as the last two were delivered.

Yet already the Boxkite was beginning to appear out-dated. Its heavy, slow, handling made it suitable for 'school' work, but ultimately little else.[100] So Challenger, who had designed the Bristol model, and Archibald Low formed a new experimental research department at Filton, with the aim of devising a

---

94   Ibid. 18 July 1911.

95   Minutes, 20 Dec. 1911, B&C papers; RAeC Committee minutes, 16 Jan., 20 Feb. 1912, RAeC papers. The Aero Club lost the action and paid £300 in compensation.

96   All the warp and wing wires were found to be intact when the debris was examined: *RAeS Journal* xvi (1912), 61.

97   Minutes, 25 Nov. 1910; correspondence between Russian Purchasing Commission and British & Colonial, B&C papers. The frames of these aircraft incorporated enlarged fuel tanks and triple rudders: there was little else to distinguish them from 'civil' models. An additional biplane (No.32) was purchased in April 1911, after being displayed in St Petersburg: Barnes, *Bristol aircraft*, 50.

98   Minutes, 15 Mar. 1911, B&C papers. Farman chose this moment to claim patent infringement against British & Colonial for their Challenger Boxkite design. The company successfully resisted the charge: minutes, 24 May 1911.

99   Bruce, *Aeroplanes of the Royal Flying Corps*, 148; Barnes, *Bristol aircraft*, 50–1.

100   For technical criticism of the design see Penrose, *Pioneer years*, 286–8. A total of 76 boxkite biplanes were ultimately constructed by the company, although some were reconstructions of existing models which subsequently received new sequence numbers. A demand for replacement parts ensured that component production continued until 1914.

more adaptable successor. With the active support of other members of the company's staff they speedily drafted blueprints for both a small tractor-biplane and a slim triangular-section tractor-monoplane. (The Short brothers, at this time, found themselves in a similar position, resulting in their successful adoption of the tractor-biplane genre.) Both types were displayed at that year's Olympia Aero Show. Unfortunately, however, neither performed adequately in subsequent trials. Indeed, little more than hops were achieved, and both models were quickly damaged. Yet, as with the abortive Zodiac biplane, valuable practical experience was gained from what were ostensibly failures.

Anxious to get a new product on to the market quickly British & Colonial next turned to the French aviator, Pierre Prier, whom they seem to have head-hunted directly from the Blériot school at Hendon, where he was chief instructor. He officially joined the company in June 1911. Prier was best known for the first non-stop flight from London (Hendon) to Paris (Issy-les-Moulineaux) on 19 April 1911, but he was also a skilled engineer and draughtsman. In view of this he was invited by Sir George White to design a light and fast monoplane for Gilmour to pilot in that year's Gordon Bennett competition, to be held at Eastchurch. In the event the resulting wing-warp-operated machine was not ready in time, but otherwise was to prove reasonably successful. Powered by a 50 h.p. Gnôme engine it had a top speed of 70 m.p.h., and from the racing prototype there evolved a military two-seater, later modified in conjunction with Bertram Dickson. This secured for British & Colonial a number of new orders, initial interest and accompanying sales again deriving from the continent. Evidently strong representations were made to the War Office on behalf of this new Bristol monoplane, but in vain. In particular Major Bannerman, Commandant of the Air Battalion, complained of the 'tendency to "boom" this machine'.[101] It was eventually given a trial, but Prier, while test flying it with Captain J. D. B. Fulton (OC, Aeroplane Company, Air Battalion) in September 1911, was compelled to make a forced landing, only narrowly averting disaster. Fulton consequently reported that the aircraft was 'completely unsuitable' in its present condition.[102] Such active lobbying, however, did exemplify an increasingly prevalent trend in the aviation market: although established armament manufacturers, like Vickers, were already well-versed in these procedures.

It was during this period, indeed, that Vickers began its own diversification into aviation. Captain H. F. Wood, who had been one of the first successes of the Bristol schools, was appointed manager of Vickers's aviation department in March 1911. Given the task of recruiting a technical staff from scratch, he returned to his Bristol associates and enlisted both George Challenger and Archibald Low. With Prier also expected to leave the firm British & Colonial

---

[101] 'Report on Bristol Monoplane – September 1911', PRO, AIR 1 1609/204/85/45.
[102] Ibid. A Prier Bristol monoplane was subsequently procured by the Air Battalion in February 1912, at a cost of £850, but it crashed twice and was struck off the army's lists in little over a year: Bruce, *Aeroplanes of the Royal Flying Corps*, 151–3.

were very soon in need of new blood.[103] Fortunately a number of equally able replacements were rapidly found, notably Harry Busteed, who had come over from Australia with Harry Hawker; Gordon England, who had previously acted as Weiss's test pilot and more recently flown a Hanriot monoplane at Brooklands; Howard Pixton, recruited – at a fee of £250 per annum – from British & Colonial's poorer, if more talented, competitor, Avro; and Henri Coanda, the quarrelsome[104] son of the Rumanian War Minister whom Sir George White had been introduced to by Émile Stern at the 1911 Paris Aero Salon. As at the time Coanda was specialising in monoplane design (he was, in fact, a wide-ranging and innovative aeronautical engineer) he was engaged specifically to replace Prier. In January 1912 he was formally appointed the firm's chief technician.[105]

In addition, a further – secret – design office was established for the development of hydro-seaplanes. This was done in conjunction with, and largely under the supervision of, a Royal Naval officer named Lieutenant Charles Burney, the son of Admiral Sir Cecil Burney. British & Colonial thus came out of this transitional period concentrating on three new projects: Gordon England's adaptive biplane designs, Coanda's monoplane designs and Burney's hydro-plane designs.

Gordon England first came to British & Colonial as a pilot, in which capacity he flew Boxkites. He gained RAeC certificate No.68 at the firm's Brooklands school on 25 April 1911. When, however, a later variant of pusher-biplane, the single-seater nacelle-incorporating Type T, the first of which had been constructed for the firm's chief competition pilot – Maurice Tabuteau – to use in the June 1911 Circuit of Europe tournament, did not prove wholly satisfactory, one of the frames was given to him to revise. England, who was to have piloted a Type T in the July 1911 Circuit of Britain, then installed a tractor-propeller, as he had done with the Weiss monoplane the year before. Driven by a 60 h.p. ENV engine, it was first flown in this amended form in November 1911, but again, with minimal success. England, however, was deemed to have demonstrated sufficient aptitude to be given the opportunity to design his own aircraft. He accordingly developed a small series of conventional tractor-biplanes in anticipation of the forthcoming military aeroplane competition. Two models of the second design of this series – the GE2 –

103 None the less, the company's capital was still increased at the turn of the year 1911/12 to £100,000, through the creation of 50,000 new £1 shares: rider on minutes, 20 Dec. 1911, B&C papers; George White papers, 35810/GW/T/30.
104 'Most amusing', was how C. G. Grey described him: 'Bristol's', 224. Others were not so sanguine in their assessment.
105 Minutes, 30 Jan. 1912, B&C papers. The exact phrasing is 'chef technique'. Elsewhere he is described as British & Colonial's 'technical director' (for example in *Flight*, 24 May 1913, where he featured in the 'Men of moment' series, his photograph revealing a young, dark, handsome man, with long hair and a dilettante air). For Coanda's origins see Penrose, *Pioneer years*, 330–1. Munro speculates that he may have been acquainted with the Dickson family, who had Rumanian social connections: 'Flying shadow', 31.

employing, respectively, a 100 h.p. Gnôme and a 70 h.p. Daimler-Mercedes engine, were presently entered for the trials in August 1912. In each the fuselage was raised above the lower wing to facilitate air-flow, and pilot-operated ignition, side-by-side seating, dual control and detachable wings were incorporated for military expediency. Due to engine trouble, however, only the Gnôme-powered model, piloted by England himself, participated in the prescribed tests, and even this failed to complete the course.[106] As a result of interest from the Turkish authorities there followed two modified 80 h.p. GE3 biplanes, each with a circular-section fuselage, tandem seating and a long-range three hour duration capability, but England did not remain with the company for long and, following some unsuccessful seaplane modifications, the type was abandoned. The Italian blockade of Turkish ports prevented any deliveries to the Turkish government.

Lieutenant Burney, on the other hand, was inspired from the first by the idea of using hydrofoils as a means of facilitating seaplane launches directly from the water's surface. This was essentially an extension of the idea of using hydrofoils to lift motor boats above the water's surface, in an effort to reduce drag at high speeds.[107] He interested Sir George White in his theories and as a result British & Colonial agreed to develop a prototype. The Admiralty was also prepared to help subsidise the project, but in return imposed conditions of strict secrecy. Work in the designated section of the design office at Filton (X department), which began at the close of 1911, was thus maintained in as much isolation as was expedient.[108] British & Colonial themselves delegated thirty-one-year-old Frank Barnwell to the department. Barnwell had wavered for a time between backing the penurious Avro venture and joining Bristol. Born in Lewisham in 1880 and educated at Fettes School in Edinburgh, and subsequently at evening classes at the Technical College and University of Glasgow, where in 1905 he obtained a BSc in naval architecture, Barnwell had begun his working life as an apprentice engineer with the Fairfield Shipbuilding Company of Clydeside, of which his father was a partner. With his brother, Harold, he had then formed the Grampian Engineering & Motor Company of Stirling, under whose auspices the brothers began the design and construction

---

[106] Bruce, *Aeroplanes of the Royal Flying Corps*, 14; Barnes, *Bristol aircraft*, 67–8.

[107] In contemporary jargon the term 'hydro-plane' was commonly used for 'seaplane'. This makes for confusion when describing specifically hydrofoil seaplanes. In fact, the common usage was a misnomer: see Gibbs-Smith, *Aviation*, 263. In July 1913 the First Lord, Winston Churchill, stated in parliament that the Admiralty had adopted the term seaplane in preference to hydroplane to describe its aeroplanes; and in September 1913 the War Office decided to end further confusion by following suit in respect of its own water-borne aircraft: PRO, AIR 1 791/204/4/689; Gamble, *Air weapon*, 154.

[108] Minutes, 20 Dec. 1911, B&C papers. A Burney-Bristol patent was taken out as early as January 1912: ibid. 30 Jan. 1912. For details of the work of 'X department' see British & Colonial Aeroplane Co.: correspondence and reports concerning experimental work on hydro aeroplane, RAeS archive.

of their own biplanes and monoplanes.[109] It was Barnwell who advocated a boat-hull monoplane design for the Burney hydroplane, rather than merely a GE modification, as originally intended. Employing an 80 h.p. Canton-Unné engine, to drive both air and water propellers, tests were begun on the resulting prototype, the X-2, at Milford Haven in May 1912.[110] It was soon evident from them, however, that substantially more research than had been anticipated would be necessary if anything approaching practicability was to be achieved. Testing was consequently resumed in September, after further analysis. No breakthrough ensued. Burney returned to naval service in December 1912. It was decided to revive experiments on a revised prototype, the X-3, using an Admiralty-owned 200 h.p. Canton-Unné engine, in March 1913, but after protracted disappointments the whole project was finally – reluctantly – abandoned in July 1914.[111] That it was allowed to continue so long was due, one suspects, to the Admiralty subsidy.

Coanda, meanwhile, had begun his career at Bristol by designing a two-seater school monoplane, first tested at Larkhill in March 1912. The design derived in some measure from the amended Prier-Dickson machines and was manufactured in both tandem and twin-seater modes. Two modified versions, intended to increase the crew's field of vision, were subsequently entered for that August's military aeroplane competition on Salisbury Plain. They were powered by 80 h.p. Gnôme engines and incorporated wing-warping for lateral control, in addition to which each undercarriage comprised a four-wheeled frame, on which the aircraft sat 'tail high' when at rest.[112] Piloted by Busteed and Pixton, these machines succeeded in gaining third prize for the company.[113] Many orders for the Bristol military monoplane, as the modified design had been expediently redesignated – including further overseas orders – followed. The Italian government ordered as many as fourteen, while Coanda's own country, Rumania, facing the possibility of war with Bulgaria, ordered ten: an impressive demonstration of how military tensions could boost sales.[114]

Then tragedy struck. Shortly after the reopening of the Larkhill school in September 1912, following the military trials, Bristol's exceptionally successful

[109] Penrose, *Pioneer years*, 318, and *British aviation: the ominous skies 1935–1939*, London 1980, 220. See also 'The Barnwell aeroplane', *The Aero*, 10 Aug. 1909.

[110] Minutes, 16 May 1912, B&C papers.

[111] Ibid. 13 Dec. 1912, 28 Mar. 1913, 7 Aug. 1914.

[112] Bruce, *Aeroplanes of the Royal Flying Corps*, 153; Barnes, *Bristol aircraft*, 71–2.

[113] Minutes, 30 Aug. 1912, B&C papers. In fact, they won two of the three third prizes (worth £500 each) in the category open to British aeroplanes: Bruce, *Aeroplanes of the Royal Flying Corps*, 155.

[114] Minutes, 19 Feb. 1913, B&C papers; Barnes, *Bristol aircraft*, 73. Of course as the military market evolved many machines were designated 'military type' irrespective of their origins – as Geoffrey de Havilland complained, when sent to report on the 1912 French Aero Salon. 'Although a large number of machines are described as "military type" ', he commented, 'it is difficult to determine what constitutes this type [as] any machine capable of taking a passenger may bear this title.' He included Bristol aircraft in this survey: PRO, AIR 1 729/176/5/69.

instructor, Edward Hotchkiss, was attached to the Royal Flying Corps reserve as a volunteer pilot. As part of the autumn manoeuvres, on 10 September 1912, he was taking one of the new military monoplanes cross-country when on reaching the first staging-post of the journey he began to lose control. The aircraft plunged to the ground near Port Meadow, outside Oxford, and both Hotchkiss – who had very little experience of the new machine – and his passenger, Lieutenant C. A. Bettington, were killed instantly. This example of yet another monoplane fatality considerably alarmed the War Office. Only days before Captain Patrick Hamilton and his observer, Lieutenant Athole Wyness-Stuart, had suffered a similar fate on an RFC-owned Deperdussin monoplane when on reconnaissance duty with the cavalry in Hertfordshire, and Captain E. B. Loraine and his passenger had died on a Nieuport on 5 July.[115] Soon afterwards Lieutenant Wilfred Parke was to be killed on a Handley Page Type F. Concern was equally strong in France, where the Blériot Company had begun to suspect structural weaknesses in their own aircraft. So Colonel Seely, the new Secretary of State for War, resolved to take no more risks. Following the Hotchkiss tragedy an edict was issued banning the flying of all monoplanes by the Military Wing of the RFC. The French authorities followed suit.

The effect of this on firms like British & Colonial and Handley Page need hardly be stated. The Bristol firm felt it particularly keenly as, following the military aeroplane competition, they had every reason to believe that considerable orders for their military monoplane would follow: the War Office had already agreed to purchase the two used in the trials,[116] and it was on the first of these that Hotchkiss was killed. Moreover, a licence for foreign production had been negotiated with the Caproni firm of Italy, precipitated by no less a person than Major Giulio Douhet – at that time Chief of Italy's first Aviation Department, but subsequently the renowned 'prophet' of air power.[117] Instead, despite success in overseas markets, the company found its business at home temporarily confined to the production of Farnborough (i.e. Royal Aircraft Factory) sub-contracted BEs. A batch of four of these had been successfully tendered for in June 1912, prior to the military trials.[118] Following the monoplane ban a contract for a further seven was awarded.[119]

Yet with at least some turnover at home, and success abroad, the company

[115] For details of investigations into these accidents see ibid. AIR 1 733/199/7, and AIR 1 2100/207/28/11 (report of departmental committee on accidents to monoplanes). In the Bristol monoplane case there had been an accidental opening of the quick-release catch securing the anchorage of the flying wires. This had been fitted to facilitate the dismantling of the aircraft during the military trials: Bruce, *Aeroplanes of the Royal Flying Corps*, 155. For more background to the accident see P. Wright, *The Royal Flying Corps 1912–1918 in Oxfordshire*, Oxford 1985.

[116] Minutes, 30 Aug. 1912, B&C papers.

[117] Ibid. 8 Nov. 1912. The agreement establishing the Società Italiana Bristol Aeroplani was formally ratified on 31 Dec. 1912. Disputes subsequently arose and Douhet later sought to convert the Italian Bristol machines to biplanes: ibid. 6 Feb. 1914.

[118] Ibid. 3 July 1912.

[119] Ibid. 19 Feb. 1913. Such Farnborough-derived batches became a regular – and

was still able to announce a £150,000 capital increase at the turn of the year.[120] Coanda, however, despite his inclinations, was now compelled to turn his attention to biplane design. This, of course, had already become Gordon England's sphere of influence. Given Coanda's difficult temperament, a clash was perhaps inevitable. The rights and wrongs of what became a bitter personal feud are difficult to establish in retrospect. What is certain is that it continued to poison accounts of the company's history for many years. One event in particular fuelled the dispute, although it occurred after England's resignation in February 1913. Gordon England's brother, Geoffrey, who had also joined British & Colonial as a test pilot, was killed in yet another monoplane accident in March 1913. Like Hotchkiss, he had been flying a Coanda Bristol military monoplane – one of the Rumanian orders – but this time the machine broke up in the air through no fault of the pilot. This had two effects. First, it caused some interruption in delivery of the order, and led eventually to the batch's conversion into biplanes, and second, it tragically confirmed Gordon England in his already jaundiced view of Coanda's work and character. His statements in this regard should consequently be treated with caution.[121]

Before following British & Colonial into the war years it might be useful to stand back for a moment in order to survey the extent of the company's international dealings. These began back in 1910 with an order for eight Boxkites from the Russian government, yet the firm was to be equally successful in Spain, Germany, Italy, Japan, France, Turkey, Bulgaria and Rumania. The practice was for a member of the company's staff, such as Henry White Smith or Captain Dickson, accompanied by one or more of the firm's test pilots, such as Harry Busteed, Howard Pixton or Sydney Sippe, to lead a mission to a particular location, where potential markets would be investigated and reception tests for the firm's products undertaken.[122] It is evident, however, that at least half of the countries involved wanted an aeroplane capability for immediate military use. This was true not only of Bulgaria and Rumania but also, for example, of Italy and Turkey, who went to war in 1911. British & Colonial emissaries travelled widely throughout central and eastern Europe in an effort to awaken interest in their product in precisely those areas of most heightened tension. To this end the volatile region around the Balkans and the disintegrating Ottoman Empire proved particularly fertile.[123]

In some instances licences were negotiated for the overseas production of

---

increasingly controversial – feature of the incipient aviation market from about this time. The B&C minutes amply illustrate this trend.

[120] Capital increased to £250,000, through the creation of 150,000 new £1 shares: minutes, 20 Jan. 1913, B&C papers; George White papers, 35810/GW/T/30 & 32.

[121] See, for example, Penrose, *Pioneer years*, 433–4.

[122] *Flight*, 11 Jan., 30 Aug., 20 Dec. 1913.

[123] See, for example, details of Capt. Dickson's 'Near East' tour, encompassing Turkey, Bulgaria, Rumania and Serbia: minutes, 16 May 1912, B&C papers. Note also M. Paris, 'The first air wars: North Africa and the Balkans, 1911–13', *Journal of Contemporary History* xxvi (1991), 97–109.

Bristol models, not just their export. The company minutes for May 1912 record, for example, the establishment of the Deutsche Bristol-Werke at Halberstadt.[124] (The British military attaché in Berlin had reported, in March 1912, how impressed the 'Prussian military authorities' were by Bristol aircraft, and he indicated that they – the Prussian authorities – were, in fact, behind this development.)[125] An Italian equivalent – as seen – followed, while the directors of British & Colonial themselves turned down a proposal from Russia.[126] All this is to say nothing of the many European nationals who consequently attended the Bristol schools in England, among them Captain Alfredo Kindelán, who subsequently commanded the Nationalist Air Force in the Spanish Civil War.[127] Indeed, such was their success that military aviation schools based on the Bristol example were subsequently established in Spain, Rumania and Germany – this last being attached to the Deutsche Bristol-Werke.[128]

The period immediately preceding the outbreak of war, however, was principally distinguished by the firm's development of various biplane designs. This began as simple expediency; in particular, over the summer of 1913 the structurally suspect Coanda monoplane was modified and converted to a tractor-biplane configuration, known thereafter as the TB8. In this mode it proved moderately successful – it was this design that the French firm Breguet subsequently acquired a licence to manufacture, at the suggestion of the French military authorities, who could officially procure only French-built machines – [129] but by the end of the year any new Bristol design faced competition from both the Avro 504 and the Sopwith Tabloid. Against these something more than an expedient would be needed. This was recognised by Frank Barnwell, who had contributed to the later TB8 revisions. At the instigation of Harry Busteed – the Australian test pilot of the X-3, who had come over to England with Harry Hawker, creator of the Sopwith Tabloid – Barnwell began planning

---

124 Minutes, 16 May 1912, B&C papers. This German licence is further evidence of the nation's initial backwardness in aviation. Many indigenous firms survived only by producing foreign designs: Luftverkehrsgesellschaft (LVG), for example, manufactured Farman's and Nieuport's, while Allgemeine Elektrizitaets Gesellschaft (AEG) produced Breguet's – although Breguet were themselves soon to adopt a Bristol patent: Morrow, *Building German air power*, 41–3; minutes, 6 Feb. 1914, B&C papers.

125 PRO, AIR 1 685/21/13/2243. Shortly before the war Deutsche Bristol-Werke were ordered by the German authorities to discontinue the production of Bristol machines in favour of government-sponsored aircraft: minutes, 12 May 1914, B&C papers. This was prudent. German manufacturers clearly could not afford to remain reliant on overseas component suppliers in the event of war.

126 Ibid. 30 Jan. 1912.

127 C. G. Grey, 'Bristol's', 221.

128 The 'Deutsche Bristol school' was run in association with the General Inspectorate of the German War Ministry. Albatros, Gotha and other manufacturing-derived German flying schools were subsequently remodelled on the Bristol pattern: Morrow, *Building German air power*, 75.

129 Minutes, 6 Feb. 1914, B&C papers.

a purpose-built British & Colonial equivalent.[130] This became the Bristol 'Baby', the prototype single-seater Bristol Scout. Employing staggered, aileron-controlled, single-bay wings of 22 ft span and powered by an 80 h.p. Gnôme engine, it was test flown by Harry Busteed at Larkhill in February 1914. Here it quickly attained a speed of 95 m.p.h.

At a stroke Barnwell had superseded all the previous Bristol designs. Realising this, British & Colonial's directors resolved to make the prototype their central exhibit at the following month's Olympia Aero Show. In other words they could not get it on to the market quick enough.[131] Yet before the type had really had time to establish itself war broke out.

To a greater extent than other private sector designs the Bristol Scout, modified by an increase in span of some 2½ ft in April 1914, was subsequently utilised by the War Office in the early years of the war, and most active RFC squadrons retained at least one or two until the emergence of co-ordinated fighter-aircraft and the full opening of the military market in 1916. As with other aircraft the early models lacked effective armaments, but enterprising pilots skilfully rectified the situation. At this stage armament was still largely a matter of individual initiative. The first two production 80 h.p. Gnôme Scouts – designated Scout Bs – were acquired by the War Office in August 1914, and a fresh batch of twelve – designated Scout Cs – was ordered the following November. Deliveries began in April 1915. It was on one of these, armed with an obliquely aligned Lewis gun, that Captain Lanoe Hawker won the first VC for aerial combat.[132] The War Office ultimately procured a total of eighty-nine B or C class production types, although many were obliged to carry Le Rhône rather than Gnôme engines.

More pointedly, both the Larkhill and Brooklands Bristol schools were requisitioned on the outbreak of war, while orders for a further twenty-two BEs had already been received by the end of July 1914.[133] On 7 August, at the conclusion of British & Colonial's first board meeting since the outbreak of hostilities, Sir George White made a speech to his assembled workforce emphasising the need for everyone to 'do their part' in producing the machines ordered by the government.[134] Having all along aimed at the military market the war brought about no great policy crisis for the company. Difficulties arose, instead, in the maintenance of an effective workforce for the massive expansion to come.

In October 1914 Coanda left the company and returned to Rumania. The

---

[130] Busteed recounted details of this to Penrose, *Pioneer years*, 472–4. Barnwell, in fact, adopted the abandoned fuselage of a proposed Coanda single-seat monoplane (the SB5), receiving permission to convert this into a fast single-seat reconnaissance biplane: Barnes, *Bristol aircraft*, 91.

[131] Minutes, 6 Feb. 1914, B&C papers. This prototype was subsequently purchased by Lord Carbery (1892–1970) as a racing biplane. He installed an 80 h.p. Rhône engine.

[132] Ibid. 15 Sept., 30 Nov. 1914; Bruce, *Aeroplanes of the Royal Flying Corps*, 161–2.

[133] Minutes, 7 Aug. 1914, B&C papers.

[134] Ibid.

obvious choice to succeed him as chief designer was Frank Barnwell, but faced with the prospect of simply adapting government BE contracts he chose instead to join the RFC.[135] Large sections of the workforce followed,[136] and soon the entire shopfloor was disintegrating just when it was most needed. The board decreed that 'immediate measures' be taken to broaden the workforce – the situation being described as 'urgent'. By December, however, the works manager was still reporting that the greatest difficulties were being experienced in 'increasing our staff as desirable'.[137] Ultimately, like A. V. Roe & Co., British & Colonial were compelled to introduce female labour.[138]

Increased demand meant that the firm's actual shopfloor area had soon to be expanded. Here the company was fortunate in having ancillary premises to exploit. As early as September 1914 plans were being laid for a further encroachment on the Bristol Tramways works.[139] British & Colonial were then to request that, in return for supplementary expansion, the War Office guarantee 'forthwith' a contract for 500 aircraft. This stipulation was accepted in July 1915.[140] The idea was successfully extended in May 1916 when the company proposed an additional £88,000 expansion in return for a guarantee of 1,500 orders.[141]

Yet the most pressing problem for British & Colonial, as for the majority of prominent aircraft manufacturers, was the restrictive nature of the War Office's initial procurement programme. Not even the early orders for the Scout were without contention. In December 1914, for example, the Admiralty ordered twenty-four Type Cs from the company. Within the week Colonel Brancker, Henderson's deputy at the Directorate of Military Aeronautics, was on to the firm, insisting that the War Office – who had themselves ordered a batch of twelve the previous month – should have been consulted first and demanding that the order be transferred to the army.[142] This could only have been because the Directorate wanted to keep a tight rein on its principal sources of supply.

A tetchy compromise was reached on this particular occasion, but Admiralty–War Office procurement tensions continued to cause the company difficulties. Thus, although the Admiralty eventually procured a total of seventy-four Type Cs, the overwhelming emphasis – dictated by the DMA's control of all military orders – remained on the private sector adapting its

[135] Ibid. 30 Nov. 1914. Coanda, it records, 'left the company without notice or explanation'. Barnwell's resignation, on the other hand, was looked on as only temporary – and so it proved.
[136] The company's pilots were the first to go; Sippe, Busteed and Dacre had all joined the RNAS on the outbreak of war: ibid. 7 Aug. 1914.
[137] Ibid. 20 Dec. 1914.
[138] Ibid. 29 Mar. 1916.
[139] Ibid. 15 Sept. 1914. See also 10 Feb., 5 May 1915.
[140] Ibid. 16 June, 7 July 1915.
[141] Ibid. 30 May, 9 June 1916.
[142] Ibid. 28 Dec. 1914.

means to the production of Farnborough-designed BEs. As was becoming only too apparent, however, this policy of – effectively – stifling original designs was suffocating the industry, frustrating and driving out potential designers of the future. The Bristol company's minutes for the early months of the war amount to a laconic record of the developing strains in the firm's dealings with the War Office.[143]

It was the success in combat of the German air service's Fokker Eindecker that finally compelled the War Office to re-examine its procurement policy. Under these circumstances, in August 1915, Barnwell was given 'extended leave without pay' to re-enter the company's service as its chief designer.[144] He immediately revised the Scout in light of its combat performance, and a total of 210 further machines – designated Type D class – were constructed between November 1915 and December 1916, 130 for the RFC and eighty for the RNAS. By then, however, the need was for new combat aircraft, as British & Colonial's directors were well aware. As an incentive Barnwell had already been offered a bonus of £5 a machine for each new machine of his design sold. (A similar bonus scheme had earlier been accorded to the workforce.)[145] By the summer of 1916 he had devised the two-seater Bristol F2b fighter. Powered, when available, by Rolls-Royce Falcon engines, these were subsequently produced on a prodigious scale, soon dwarfing previous orders.[146] It did not require much foresight to realise that a crippling slump was bound to follow the war, but for the time being this was a secondary consideration.

In these years the Bristol organisation became one of Europe's largest and most successful manufacturers of aircraft. It achieved this position primarily because it was the first aeroplane company founded specifically and discerningly by an established industrialist, with more than adequate financial backing. Not for this firm the humiliating round of commercial begging letters, such as characterised the emergence of A. V. Roe & Co. Sir George White, on the contrary, mingled in influential circles with the ease of an industrial aristocrat. Such was his natural milieu. British & Colonial's ability to despatch well-connected emissaries to a variety of potential overseas markets only further underlined the advantage of a secure industrial base. In their business methods,

---

[143] Note, in particular, ibid. 30 Nov. 1914. In the course of this meeting, during which the directors discussed the shortcomings of the War Office's procurement programme, the secretary reported that orders for a further 50 BEs had been received. See also ibid. 28 Dec. 1914. B&C were to manufacture some 1,150 BEs of various modification before army aircraft procurement was reformed and Farnborough orders completed.

[144] Ibid. 30 Aug. 1915.

[145] Ibid. 16 June, 30 Aug. 1915.

[146] It is not the function of this study to trace the technical evolution of this aircraft. For this see Barnes, *Bristol aircraft*, 104ff, and Bruce, *Aeroplanes of the Royal Flying Corps*, 169–79. Barnwell was to remain Bristol's chief designer throughout the inter-war period. He was killed in an air accident in August 1938. Thus he failed to see the wartime employment of the Blenheim fighter-bomber, the prototype development of which he had overseen, or learn of the death in combat of his three sons – two while flying Blenheims.

in their ample financial backing of research projects and intelligent lobbying of the authorities, the directors of British & Colonial were innovators in the market.

Vickers

Shortly after the industrialist, Sir George White, decided to diversify into the aviation market, the directors of the Vickers corporation resolved to follow suit. Vickers is too well-known an organisation for any résumé of its origins to be necessary;[147] suffice to say that the firm's directors were no strangers to diversification. They had already demonstrated an admirable flexibility in the face of volatile markets. In the 1890s, for instance, they had moved from a steel manufacturing base into the construction of warships. Operating principally as armament manufacturers, however, there was never any question of aircraft being seen as anything other than pieces of weaponry.

In fact, the company had been commissioned by the Admiralty in 1909 to build the first British naval airship, but this had not proved a success. The appropriately titled 'Mayfly' (512 ft long, with a beam of 48 ft and powered by two 160 h.p. Wolseley engines) was not finally completed until September 1911, and it promptly broke in two while being manoeuvred out of its construction hangar for a preliminary launch.[148] Vickers pressed the Admiralty for further orders in an effort to maintain their recently established aeronautical department, but they must have perceived the limits of this market when Murray Sueter, the Royal Navy's Inspecting Captain of Airships, earlier suggested that they diversify into aeroplane manufacture.[149] At any rate, by the time of this airship reverse Vickers had already adopted what they hoped would be a suitable aeroplane patent to develop.

The directors had been advised in their search by Captain Herbert F. Wood, the pugnacious former cavalry officer who had been one of the first graduates of the Bristol schools. Bertie Wood, as he was commonly, though seldom affectionately, known, was the well-bred son-in-law of Quintin Hogg, founder of the Regent Street Polytechnic. Vickers's managing director, Trevor Dawson, was a close personal friend, and no doubt believing Wood to be more experienced in aviation matters than was the case, was instrumental in securing his services as an advisor. Probably influenced by the Bristol Boxkite, it was Wood who drew the board's attention, in the first instance, to the Sommer biplane, and then, as an alternative, to the REP monoplane. Vickers duly opened negotiations with the designers of both these aircraft.

Of the pair Robert Esnault-Pelterie (REP) had quickly proved the more

---

147 See J. D. Scott, *Vickers: a history*, London 1962.
148 PRO, AIR 1 2306/215/15; AIR 1 2464. For the background to this episode see R. Higham, *The British rigid airship 1908–1931: a study in weapons policy*, London 1961, 34ff.
149 Penrose, *Pioneer years*, 290.

responsive. Indeed, as early as November 1910 Vickers's agent in these negotiations – the much vilified arms dealer, Basil Zaharoff – indicated that he was ready to close on the deal. But it is evident from his correspondence that Zaharoff was working for his friend Esnault-Pelterie as much as for his firm Vickers, so quite how objective his advice was is questionable.[150] Sueter appears to have been contacted for a second opinion partly for this reason; however, he simply endorsed continued negotiations with both parties.[151] The Royal Navy were clearly regarded as the potential market for any Vickers aircraft, so Sueter's views were doubtless canvassed as an expression of Admiralty opinion.[152] In the event, this twofold approach proved unsustainable. Sommer, arguably the more important designer, received Vickers's advances with an indifference bordering on contempt. A company despatch of 9 December 1910, little hiding its exasperation, reported him as 'seeming to make light of the very important position of the Vickers firm, and the great asset he would be getting in obtaining their co-operation'.[153] These negotiations were consequently abandoned.

All Vickers's aviationary ambitions were thus soon centred on REP. In January 1911 they gained that firm's British patent rights.[154] Vickers had already set aside a section of their Erith works, in Kent, for aircraft construction, and a design office was now also incorporated into the company's central office, Vickers House, in Westminster. Without delay a steel-framed monoplane powered by a 60 h.p. REP engine was offered to the Admiralty for £1,500. Citing recent American developments, Vickers suggested that this might be used to make scouting sorties from warships.[155] The Admiralty, which had recently signalled its intention to accept Francis McClean's offer of the loan of two Short-Sommer biplanes, was unmoved. There were, it stated, no specific plans for aeroplane procurement 'at present'.[156] Vickers were evidently hoping to establish themselves in the role of semi-official aircraft contractors to the Admiralty, just as British & Colonial had tried to do with the War Office. Their position as constructors of the Royal Naval Airship No.1 had, perhaps, given grounds for this aspiration. The Admiralty's formal, not to say terse, response on this occasion effectively dashed this unlikely hope. Vickers, like everyone else, would have to compete in open market.

The initial approach to the Admiralty had, in fact, been somewhat precipitate, for it was to be the following July before the Vickers No.1 monoplane, as

---

[150] See B. Zaharoff to F. Barker (Vickers), 16 Nov. 1910, Vickers papers, 284.

[151] F. Barker (Vickers) to M. F. Sueter, 25 Nov., 2 Dec. 1910; M. F. Sueter to Barker, 29 Nov. 1910, ibid. 284; 1006A.

[152] In correspondence on the subject Vickers specifically referred to 'the matter of aeroplanes for the government'. This was before any patent had been acquired: F. Barker (Vickers) to B. Zaharoff, 2 Dec. 1910, ibid. 1006A.

[153] F. Barker (Vickers) to Capt. H. E. Macdonnell, 9 Dec. 1910, ibid.

[154] Copy of Vickers–REP agreement, effective from 1 Jan. 1911, Vickers papers, 285.

[155] Vickers to Admiralty, 3 Jan. 1911, Vickers papers, 284.

[156] Admiralty to Vickers, 17 Jan. 1911, ibid.

it was subsequently designated, made its debut flight. By that time a private test-ground had been established at Joyce Green, near Dartford.[157] Meanwhile, in March 1911, Captain Wood had been officially appointed manager of Vickers's new aviation department. Faced with the task of recruiting an integrated technical staff from scratch he of course cast his net in the direction of former Bristol associates and secured the services of George Challenger, Archibald Low and test pilot Leslie Macdonald. In addition, a French mechanic, named Gabriel Bourcier, was recruited as what was termed the department's 'aviation motor engineer'.[158]

One of the aviation department's first responsibilities, delegated to Archibald Low, aided by a young apprentice named Rex Pierson, was to incorporate tandem-seating into the adopted REP monoplane design and then fit it with a stronger twin-skid undercarriage. Including the revised No.1 monoplane, five such aircraft were ultimately built. Despite extensive lobbying, however, none were taken by the services. The first crashed; the second was purchased by Douglas Mawson for his Australian Antarctic Expedition (Mawson had been Sir Ernest Shackleton's guest at the May 1911 Parliamentary Aerial Defence Committee's Hendon display);[159] and the remainder went to the Vickers flying school, which opened at Brooklands in early 1912. Here three Gnôme-powered Hewlett & Blondeau Farman-derived boxkites were also used with some success, Harold Barnwell, Frank Barnwell's brother, eventually becoming chief instructor.

In June 1912 the sixth Vickers-REP monoplane was modified by George Challenger, in anticipation of that August's military aeroplane competition. In particular, the aircraft's wing-span was reduced, a new simplified undercarriage (with a single pair of wheels and one central skid) fitted, and pilot and observer seats placed side-by-side. This last amendment was for increased reconnaissance potential, and constituted one of the competition's entry requirements. It duly performed adequately – within the limits of its frustratingly inefficient 70 h.p. Viale radial engine – but failed to win any of the available awards.

The Vickers corporation had, in fact, already decided to end the REP agreement and to develop its own indigenous designs.[160] The outcome of the military trials reinforced this decision. A new Vickers-Low (70 h.p. Gnôme) tractor-biplane prototype, deriving from the No.6 monoplane fuselage, was constructed by the close of the year. Before it had had an opportunity to

157 C. F. Andrews and E. B. Morgan, *Vickers aircraft since 1908*, London 1988, 35; Joyce Green aerodrome, Vickers papers, 910.

158 Vickers–G. Bourcier agreement, 14 Oct. 1911, ibid. 284.

159 Lt. H. E. Watkins, Essex Regiment, was appointed to fly the aircraft during the Antarctic expedition, but crashed it on a trial flight in Australia. The machine was consequently taken south as a wingless tractor, to drag sledges: D. Mawson to H. E. Scrope, 16 Mar. 1956, ibid. 570.

160 R. Esnault-Pelterie to Vickers, 10 June 1912, confirming cancellation of contract, ibid. 285. Unfortunately this was far from the end of the matter. Much litigation was to follow.

establish itself, however, the machine crashed into the Thames near Dartford on 13 January 1913, with the loss of the firm's chief test pilot, Leslie Macdonald.[161] As a result, the tractor-biplane configuration was temporarily abandoned.

Two more Gnôme-powered monoplanes were subsequently constructed, one in the pre-military trials mode, and one in the manner of the revised No.6 trials machine; but these were essentially a stop-gap, while Low developed a new indigenous design. This proved to be for a pusher-biplane, with a nacelle, or forward compartment, encasing both the pilot and passenger, and incorporating a swivel-hinged Vickers-Maxim machine-gun in its protruding nose. Such a configuration was, at root, a derivative of the boxkites used as the company's instructional machines, and plans for experimenting with a Vickers-gun fixed to a Sommer biplane had been considered as early as November 1910; however, according to Low, Captain Wood 'hooted with joy' on being presented with the initial drafts of this new design, and 'dashed to the War Office'. There he seems to have received an immediate order for three machines.[162]

The prototype was unveiled at the Olympia Aero Show of February 1913. Powered by an 80 h.p. Wolseley engine it was officially designated the EFB1 (Experimental Fighting Biplane No.1). Given its salient feature, however, the machine was soon known colloquially as the 'Destroyer', or, more generally, the Vickers Gunbus. Its framework and fuselage were nearly all metal, made from steel tubes coated with duralumin, and the design incorporated wing-warp-operated staggered mainplane wings. It constituted what was probably the world's first purpose-built combat aeroplane, as opposed to being simply an adapted military reconnaissance aircraft. The gunner sat in the nose of the machine with an unrestricted view and line of fire. There was, of course, no interrupter device at this time, so this forward-firing position was possible only in aeroplanes of a pusher configuration.

A perplexing variety of models followed, but all were derived from this essential structure. For example, the EFB2 incorporated a semi-circular tailplane, unstaggered wings with a slight top-wing longitudinal overhang and large celluloid windows or panels. Windows were then absent from the aileron-controlled EFB3, of December 1913, six of which were subsequently ordered by the Admiralty.[163] Production was then amended to War Office needs, which included a return to a rectangular style of tailplane and a more blunted nacelle, designed to accommodate a Lewis-gun mounting. It was this 'production' FB, designated the FB5, that came to be officially labelled the Gunbus. All these later variants were powered by 100 h.p. Gnôme Monosoupape engines.

---

[161] H. E. Scrope, 'Golden wings: 50 years of aviation by the Vickers group of companies', unpubl. manuscript 1960, 22.

[162] Penrose, *Pioneer years*, 412; F. Barker (Vickers) to Capt. H. E. Macdonnell, 14 Nov. 1910, Vickers papers, 1006A. Andrews and Morgan indicate that a contract for the construction of an as yet unformulated 'fighting biplane' had been agreed with the Admiralty as early as 19 Nov. 1912: *Vickers aircraft*, 43.

[163] Ibid. 48–9.

The type even kindled interest overseas. The German government, in particular, sought more details, and in February 1914 a Vickers agent was appointed in Berlin. Three twin-engine 'German Fighting Biplanes', each with a five-hour fuel-carrying capacity, were subsequently ordered, but the outbreak of hostilities brought an end to their construction.[164]

For all the success of the Gunbus, however, Vickers – like British & Colonial before them – found the market intensely competitive. There was no lack of turnover as such once the War Office had chosen to place its first BE sub-contracts with the firm. The fact that Vickers were already established arma-ment contractors undoubtedly influenced this decision.[165] None the less, the company soon felt compelled to devise its own equivalent to the Avro 504, the Sopwith Tabloid and the Bristol Scout. Through sharing instructional facilities at Brooklands with the manufacturers of these aircraft their respective features were well-known to Vickers's personnel. Indeed, Harold Barnwell, the firm's chief instructor, was the brother and erstwhile colleague of the designer of the Bristol Scout. Given this evidently successful trend in aeroplane design, development of a fast tractor scout must have seemed the natural progression. By late 1913, however, the personal animosity between Archibald Low and Captain Wood had led to Low's dismissal as Vickers's chief designer, so this new project had to be initiated while there existed something of a vacuum within the organisation's aviation department.

A tandem-seat tractor-biplane, devised by Harold Barnwell and George Challenger, was presently displayed at the 1914 Olympia Aero Show. It did not prove a success.[166] Responsibility for the configuration thereafter fell into abeyance for some time. The initiative for its revival came, again, from Harold Barnwell at Brooklands. Appropriating the Olympia scout, he surreptitiously began to reconstruct it as a single-seat 100 h.p. Gnôme Monosoupape tractor-biplane in the latter months of 1914. The finished machine incorporated a circular-section fuselage of some 20 ft 3 in. length and unstaggered wings of some 24 ft 4 in. span. Christened the 'Barnwell Bullet', it first flew early the following year. The undercarriage, however, collapsed under the impact of landing, and it was given to Rex Pierson to repair and revise. It re-emerged with the official designation ES1 (Experimental Scout No.1). Two further revised models, designated ES2s, followed, both powered by 110 h.p. Clerget rotary engines. One of these was fitted with a gun-cum-propeller interrupter-device, designed by George Challenger, which enabled a fixed machine-gun to be fired directly through the propeller blades, thereby allowing the pilot to take aim by simply directing the aircraft itself.

Despite this breakthrough, the type provided the company with no long-term advantage. Sopwiths were to devise an improved synchronising-device, and the Vickers ES2 never looked like threatening the superiority of, for

[164] Ibid. 57–8.
[165] Orders for Vickers aircraft, Vickers papers, 289.
[166] Penrose, *Pioneer years*, 476, 512.

example, the emerging Sopwith and Bristol scouts and fighters. It was felt to have a poor field of vision and was revised as the confusingly labelled FB19 tractor-biplane. Vickers's time would have been more profitably employed marketing the Gunbus. On the advent of war the existing EFB types found effective employment in the aerial defence of London. This was on account of their perceived anti-Zeppelin capability. Once in service the 'Experimental' prefix was dropped, and the production class became simply the FB5 Vickers Gunbus. Subsequently, in December 1914, Darracq secured the European patent rights.[167] Meanwhile, within Vickers's aviation department, Howard Flanders, A. V. Roe's former assistant from his Lea Marshes days, was recruited to replace Archibald Low.

Flanders had, in fact, been with the company for less than a month when hostilities commenced, and was reportedly 'teeming' with ideas.[168] Despite this, however, the early war years saw Vickers employed chiefly in the manufacture of BEs, and later SE5As, for the government. Factory space for this work was acquired next to the Brooklands circuit, which had itself been requisitioned by the War Office.[169] Aircraft construction was also transferred to Vickers's Crayford works, some three miles from Erith. A substantial reliance on sub-contracted orders naturally curtailed the company's own scope for development, but in the event the crisis engendered by the War Office's restrictive procurement programme was to break open the market again. In 1916 the Vickers FB9 Streamline Gunbus, with rounded wing-tips and an aerodynamically revised nacelle, came into production. Then, towards the end of the war, the company produced its most significant contribution to aviation to date, the twin-engine Vickers Vimy tractor-bomber, devised by Rex Pierson.

Construction of this type of aircraft had been developing for some time. Howard Flanders had designed a twin-engine tractor-biplane, intended to carry a Vickers one-pounder gun (the EFB7), in 1915, and by July 1917 the Air Board was urging the company to carry this idea further. Particular encouragement came from Alec Ogilvie, formerly one of the first Short-Wright pilots, but now the Air Board's controller of technical design. The Air Board initially wanted a bomber built around a surplus stock of 200 h.p. Hispano Suiza engines, but when the time came indigenous Rolls-Royce Eagle VIII engines were used. The finished Vimy was some 43 ft 6½ in. long, with a 68 ft wing-span and an endurance capability of up to eleven hours: this with a three-man crew and 2,476 lb bombload. The first contract, for 150 of the type, was received by Vickers on 26 March 1918. A second contract, for a further 200, followed soon after. The armistice was to prevent the order's completion, but it was patently evident that, as with the later Handley Page products, an important precedent was being established. The type not only exemplified the direction in which military aviation was moving, it also proved amongst the most adaptable to

---

167 Darracq–Vickers agreement (draft), 10 Dec. 1914, Vickers papers, 289.
168 Penrose, *Pioneer years*, 553.
169 PRO, AIR 1 727/160/3.

civilian needs. Its range potential and heavy load capability made it suitable as a transport aircraft. It was on an amended service-type Vimy, indeed, that Alcock and Brown flew the Atlantic in June 1919. The design was subsequently revised as the oval-section Veron, the aircraft which the AOC, No.45 Squadron, Mesopotamia – Arthur Harris, the future C-in-C, Bomber Command – was to adroitly adapt as the RAF's main bomber in its Middle East 'air control' operations of the early 1920s.

The war, then, ultimately brought out the best in Vickers and its aircraft. Initially the firm had struggled to find its niche. Despite the fact that the company started trading with all the financial and industrial advantages of a large, expansive manufacturing concern (one, moreover, with close military ties), it at first failed to match the success of smaller competitors such as Sopwiths, Avro and Shorts. This was because its products had generally lacked the innovative quality of those of its main rivals. Yet, increasingly, financial backing for expensive development programmes was becoming the key to design innovation. In that respect, both Vickers and British & Colonial – the first aviation concerns founded by recognised industrialists, rather than merely by technological entrepreneurs – perhaps more accurately typified the nature of the developing industry.

Unfortunately neither Sir George White nor Major Bertie Wood survived the war years. Sir George died suddenly in November 1916, while Wood, who had seen active service with his regiment in France, was a victim of the 'flu epidemic which swept Europe shortly after the armistice. British & Colonial officially became the Bristol Aeroplane Company in March 1920, when the original firm – like Sopwiths – went into voluntary liquidation as a means of circumventing the government's excess profits duty.[170] Vickers, on the other hand, were to acquire Supermarine as a subsidiary company in 1928.

The publicists: George Holt Thomas and Claude Grahame-White

In the public realm interest in aviation was meanwhile being cultivated and shaped. The real explosion of enthusiasm undoubtedly occurred in 1909, the year of the first Rheims meeting and Blériot's successful cross-channel flight – this last being the result of another Northcliffe *Daily Mail* challenge. These specific events are very well documented.[171] What is of more immediate interest is the way in which public perceptions were influenced and why. The

170 Memo, 8 Mar. 1920, B&C papers. As has been shown, a subsidiary company under that name had been registered in February 1910, with a nominal capital of £100.
171 See Gibbs-Smith, *Aviation*, 144–6; H. S. Villard, *Contact!: the story of the early birds*, Washington, DC 1987, 73ff, and *Blue ribbon of the air*, 21ff. The Blériot cross-channel flight occurred on 25 July 1909. The first Rheims meeting ('La Grande Semaine d'Aviation de la Champagne') ran from 22 to 29 August 1909. Its highlight was the Gordon Bennett Cup, won by Glenn H. Curtiss on his 'Rheims Racer', a 50 h.p. derivative of the Herring-Curtiss 'Golden Flyer' pusher-biplane.

Daily Mail, of course, had sponsored the cross-channel flight, so could perhaps be expected to proclaim aviation mania following that event;[172] what was more surprising was the public's response. Over 120,000 people crammed into Selfridges in Oxford Street to see Blériot's monoplane over the four days it was subsequently displayed[173] and there seemed, for a time, no saturation point of popular interest. This was the year when many of the aviation journals first appeared: *The Aero*, for instance, or more influential still, *Flight*, founded and edited by the experienced aeronautical journalist, Stanley Spooner. 'There is not room for all these periodicals', complained Baden-Powell, by that time co-editor of the *Aeronautics* journal, to C. G. Grey, in May 1909.[174] But this did not prevent Grey – one of the leading contributors to *The Aero* – from successfully adding to the number when he launched his own polemical *Aeroplane* journal some two years later. Through this he was to become the principal critic of the government's Farnborough regime. 1909 also saw the emergence of the Aerial League as one of the new technology's most prominent, if not progressive, agitating bodies. It held a spectacularly successful rally at the Mansion House in April, before subsequently falling foul of the military authorities.[175]

It was within this hothouse atmosphere that the journalist and entrepreneur, George Holt Thomas, consciously adopted the mantle of aeronautical herald, thereby effectively laying the ground for Grahame-White's great pre-war publicity campaigns. Holt Thomas's father had founded the *Graphic* and *Daily Graphic* newspapers and on leaving Oxford Holt Thomas had managed these titles, whilst simultaneously launching his own *Empire Illustrated* journal. Having grown up in a media environment he was imbued with the value of good publicity. As his interests turned towards aeronautics he was particularly impressed by Northcliffe's campaign of aviation promotion in the *Daily Mail*, a feature of which, from November 1906, was the offer of £10,000 for the first flight from London to Manchester. This prize became the focus of his distant ambition.[176] He began by integrating himself within recognised aeronautical circles, travelling, like Brabazon, to France and introducing himself to Henri Farman. He attended the 1909 Rheims meeting and there met Louis Paulhan, the French aviator whom he was subsequently to manage. Using Paulhan, Holt

---

172 *Daily Mail*, 26 July 1909. The paper later had to defend itself against charges of giving too much space to aviation. See *Daily Mail*, 'The Outlook', 28 Aug. 1909. Even Northcliffe realised that the aeroplane could be oversold, witness his letter to T. Marlowe (editor, *Daily Mail*), 15 Mar. 1911, complaining of 'too much aeroplane in this morning's paper': Northcliffe papers, Add. 62198.

173 *Daily Mail*, 30 July 1909.

174 B. F. S. Baden-Powell to C. G. Grey, 18 May 1909, RAeS archive. The periodical *Aeronautics* – not to be confused with the *Aeronautical Journal*, the journal of the Aeronautical Society, with which Baden-Powell was also associated – had itself only been established two years previously, in association with J. H. Ledeboer.

175 The organisation and aims of the League are examined in ch. 5.

176 Reminiscences of G. Holt Thomas (1913), PRO, AIR 1 728/171/1; *DBB, s.v.* Thomas, George Holt.

Thomas made it his mission to bring such flying as he had witnessed at Rheims back to England itself, thereby, he hoped, demonstrating 'the actual possibilities of flying to the general public, and so . . . influencing the Government'.[177] He helped organise the October 1909 Blackpool flying meeting, and then endeavoured to secure a similar venue nearer London, ultimately persuading the Brooklands authorities to adapt their motor track for the purpose. An area of about thirty acres in the centre of the circuit was cleared and levelled and a purpose-built shed constructed to coincide with the closing car meeting of the season. Paulhan then flew on three consecutive days, 28–30 October, on one occasion carrying the track's proprietress, Dame Ethel Locke King, as a passenger. The response of the public was such that further improvements in the circuit's layout were immediately initiated and hangars constructed for rent.[178]

By this time Holt Thomas was actively preparing his protégé for an attempt on the London to Manchester prize.[179] This Paulhan was eventually to win amid enormous publicity – and against the keen competition of Claude Grahame-White – in April 1910. Later that year Thomas reported for the *Daily Mail* on both the French and British army manoeuvres, in which aeroplanes participated for the first time. His articles, carried under banner headlines proclaiming 'The New Arm Totally Neglected' and similar assertions, were clearly designed to embarrass the authorities.[180] Less well publicised was the fact that Holt Thomas was by now either acquiring or seeking to acquire so substantial an interest in the developing aviation industry that his objectivity must have been compromised.[181] One need only note his representation of Paulhan's and Farman's interests in the UK, and his subsequent acquisition of British patent rights for the Farman aircraft and the Gnôme engine. Indeed, he was to facilitate the sale of both a Farman and a Paulhan biplane to the Royal Engineers' Balloon School within weeks of the 1910 manoeuvres.

After successfully securing the Farman rights, Holt Thomas registered the formation of his own Aircraft Manufacturing Company, better known as Airco, in June 1912. From this root would eventually emerge both the Gloster and de Havilland companies. It would be wrong to assume, however, that Holt Thomas was uniquely compromised as an advocate of military aviation, or, more particularly, as a campaigner for increased spending on military aviation. A similar charge might very well be made against so celebrated a figure as Claude Grahame-White. The difference lay in the fact that, far from displaying any equivocation on the subject, Grahame-White, by 1910–11 a name as

[177] Reminiscences of G. Holt Thomas, 8.
[178] Ibid. 6–7; *The Aero*, 16 Nov. 1909.
[179] Reminiscences of G. Holt Thomas, 8.
[180] Ibid. 14. This 'neglect' was in respect of British efforts: the French manoeuvres had been a notable success. See ch. 6.
[181] The information Holt Thomas received as a prominent air advocate he certainly exploited for personal gain: Penrose, *Pioneer years*, 497.

familiar to the general public as that of any sports star or Antarctic explorer, positively rejoiced in the association.

Born in 1879 into a wealthy Hampshire family from Bursledon, on the river Hamble, and educated at Bedford School, Claude Grahame-White's early life exemplifies the usual accomplishments of the archetypal mechanical pioneer. An engineering apprenticeship at sixteen, coupled with a privately cultivated passion for cycling, led inevitably to an interest in, and devotion to, the automobile. In 1897 Grahame-White became a founder member of Frederick Simms's Automobile Club of Great Britain and Ireland (later the Royal Automobile Club) and, not unlike his contemporary, C. S. Rolls, soon established himself as a successful automobile dealer. He also became a member of the sport-oriented aero clubs of Great Britain and France. There may have been an element of social climbing behind this, something to which Grahame-White, the Edwardian dandy, was always susceptible. He also had before him the example of his elder brother, Montague, who had obtained a position with the Daimler company of Coventry in 1897 and who went on to win renown as a pioneering sportsman, automobilist and yachtsman.

The switch to aviation came in the aftermath of Blériot's July 1909 cross-channel flight. Inspired by this event and the resultant publicity Grahame-White determined to attend the following month's Rheims meeting, where he aimed to make Blériot's acquaintance prior to procuring one of his celebrated monoplanes. (He had already had dealings with Blériot's automobile accessories firm.)[182] Once in France, however, his eye was caught by the rather more unconventional two-seater Blériot XII tractor-monoplane, which the indefatigable Frenchman had constructed in a somewhat rash effort to gain more speed. In marked contrast to the standard Blériot – a rudimentary tractor-monoplane, incorporating a long thin fuselage and seat-level propeller and wings – in this revised configuration the deep fuselage-frame entirely encapsulated the pilot and any passenger, who were situated immediately behind the 50 h.p. ENV engine, directly under the aircraft's wings. The engine powered a single propeller, itself positioned above the pilot, on a level with the raised mainplane, i.e. the aircraft's wings. Within days of arriving at the display ground a deal was struck whereby Grahame-White would purchase this machine at the end of the meeting. But as if by way of a warning the aircraft burst into flames after a forced landing before the transaction could be completed.

This was not Blériot's first accident and he energetically persuaded the distraught Englishman that another such aircraft could be constructed with the minimum of difficulty. Grahame-White assisted at the Blériot workshops

---

[182] 'Papers from H. Harper and *Times* re. Grahame-White's efforts in early aviation', PRO, AIR 1 7/8/1. Harry Harper, 1880–1960, was chosen by Northcliffe as his 'air correspondent' in 1906. He became a well-known figure in the pre-war aeronautical world, collaborating with Grahame-White on a number of early aviation books, but his writings now appear painfully banal. 'I think', wrote Handley Page, in 1958, 'Harper puts a rather higher value on his work . . . than most people would accord to him': Handley Page to Lord Brabazon, 20 Feb. 1958, Brabazon papers, box 25(2).

in Paris during the period of its construction. The new model was completed in November 1909, and fitted with a 60 h.p. ENV engine. Grahame-White somewhat dramatically labelled the finished machine the 'White Eagle'. In Blériot's absence, he then impulsively commandeered it for a preliminary trial at Issy-les-Moulineaux, in the course of which he succeeded in both flying and crash-landing what to even an experienced aviator would have been an unorthodox and awkward aeroplane.[183] Between them Blériot and Grahame-White unconvincingly ascribed this accident to the restricted nature of the Issy plain, and the decision was taken to transfer the aircraft to Blériot's more commodious aerodrome at Pau. No sooner had the 'White Eagle' been redeployed, however, than another smash resulted from a poor, forced landing.

This time terrain could not be blamed. In fact, as was becoming increasingly apparent, the Type XII configuration was disproportionately front heavy, and in the glide gave a dangerously steep angle of descent. This third consecutive mishap clearly illustrated the resulting dangers, so Blériot reluctantly withdrew the model from service. To mollify Grahame-White he furnished him instead with a pair of conventional Type XI monoplanes. It was on one of these that Grahame-White secured his pilot's licence from the Aero Club de France (certificate No.30, dated 4 January 1910). He thereby became the first Briton, aside from Henri Farman, a naturalised Frenchman, to gain an internationally recognised pilot's certificate.[184] Blériot, meanwhile, placed the unstable Type XII in storage. It would re-emerge the following year in rather different circumstances. Its loss does not seem to have greatly concerned Grahame-White. Emboldened by gaining his licence he promptly acquired a further half dozen Blériot Type XI monoplanes and established a British flying school near Blériot's Pau aerodrome. Amongst its first pupils were the rich American sportsman, J. Armstrong Drexel (who, with W. E. McArdle, later opened his own Blériot flying school at Beaulieu, in the New Forest, and who himself became a celebrated competition pilot), Lieutenant F. L. M. Boothby RN (later one of Commander Oliver Schwann's colleagues, attached to the Naval Airship Tender, HMS *Hermione*), and a Miss Spencer Kavanagh. This curious lady evidently operated as a professional parachutist under the name 'Viola Spencer', an activity which cost her her life in July 1910, following which she was revealed to have been a Miss Edith Maud Cook.

The school's success unleashed within Grahame-White a tidal wave of ambition. No sooner was it established than he turned to a still greater challenge. Disregarding his own lack of cross-country experience, he returned to Britain the following spring intent, like Holt Thomas, on securing the *Daily Mail's* long dormant £10,000 London to Manchester prize. As he possessed a pilot's certificate he qualified to participate in public displays and competitions. By now he was sufficiently familiar with the mechanics of flight to appreciate

183 PRO, AIR 1 7/8/1; Bruce, *Aeroplanes of the Royal Flying Corps*, 133.
184 Lord Brabazon to M. Grahame-White, 14 June 1955, Brabazon papers, box 25(2); G. Wallace, *Claude Grahame-White: a biography*, London 1959, 45ff.

the limitations of the Anzani-powered Blériot XI monoplanes in his possession, so he sought for this extended passage a Farman boxkite biplane with a Gnôme rotary engine. Indeed, he personally visited Farman at his Mourmelon workshop, some fourteen miles from Rheims, where he helped supervise the construction of a new machine. This eventually cost him some £1,500. It was completed in March 1910 and shipped to England the following month, Grahame-White having taken the lease of a hangar at Brooklands. In the early hours of Saturday 23 April, dressed in riding breeches and puttees and wrapped in a fur-lined gabardine jacket, Grahame-White set off for Manchester from the Royal Agricultural Society's former showground at Park Royal, near Ealing. In accordance with the competition's rules he had twenty-four hours to achieve the flight, allowing for two refuelling stops. He got as far as Lichfield, where he landed for the second time, only for his Farman biplane to be blown over and smashed by a gust of wind.[185]

It was while Grahame-White was preparing for his second attempt on the Manchester prize that Paulhan, backed by Holt Thomas, officially entered the contest. This brought press speculation on both sides of the channel to fever pitch.[186] In an effort to steal a march on his rival, Paulhan, in another Farman, set off from Hendon at 5.30 p.m. on 27 April 1910. (The *Daily Mail* stipulated that competitors had to take off from – or at least circle – a point within five miles of its London office, and similarly alight within five miles of its Manchester office.) On learning the news, Grahame-White, amid near hysterical crowd scenes at Wormwood Scrubs, set off in pursuit; but by then he had lost nearly an hour's daylight.[187] By dusk Paulhan was 117 miles on, at Lichfield; Grahame-White a mere sixty miles on, at Roade. Grahame-White realised that he could only win the competition by flying on into the night – a dangerous and unprecedented expedient. Consequently he set off again at 2.50 a.m., groping his way through the darkness, following the headlamps of accompanying automobiles, the most important of which was the tender vehicle, carrying mechanics from the Gnôme engine factory. In the event turbulent conditions forced him down still ten miles short of his opponent. Thousands of spectators had, meanwhile, gathered at what had become the designated finishing post – a large field in Manchester's southern suburb of Didsbury. Among them was the young Roy Chadwick, then barely seventeen-years-old, and with the thought of becoming A. V. Roe's personal assistant beyond his wildest dreams. Loudspeakers kept the crowd in touch with events until, at 5.30 a.m., Paulhan's victorious aircraft came into view.[188]

Despite losing the race to his French rival, Grahame-White's daring

[185] PRO, AIR 1 7/8/1; manuscript, London–Manchester air race (1935), Grahame-White papers, B796.

[186] Witness the enormous scrapbooks on the event in Grahame-White papers, B705–9, 767.

[187] Manuscript, London–Manchester race, 5–6; transcript, BBC radio interview (1950), ibid. B799.

[188] PRO, AIR 1 7/8/1; transcript of lecture given during 1910 American tour, Grahame-

exploits, and the sporting manner in which he accepted defeat, meant that his name was made. He subsequently went on to win many other competitions and systematically set about exploiting his fame for commercial and propagandist purposes. After toying with the idea of opening a flying school on the site of his initial launching ground at Park Royal,[189] he began drawing up long term plans for the development of a London aerodrome on some 207 acres of land between Colindale and Hendon – the site of Paulhan's recent launching ground.[190] He was not alone in perceiving the potential of the location. In October 1910, while he was furthering his plans, Blériot launched a flying school on the site, having built there eight hangars, three of which were leased to Horatio Barber's Aeronautical Syndicate Ltd, previously located at Larkhill. When this last organisation subsequently folded, in April 1912, its sheds and storage space were acquired by Holt Thomas's emergent company, Airco.[191]

In a very short time Grahame-White became probably the most famous English aviator of his day. In the autumn of 1910 he toured America with A. V. Roe. While there he met President Taft and later, in Washington, landed his Farman biplane on Executive Avenue, prior to lunching with some of the capital's service chiefs.[192] This was just a year after introducing himself to Blériot at Rheims. At the Belmont Park meeting, on Long Island, he piloted a 100 h.p. (Gnôme) Blériot monoplane to victory in both the second Gordon Bennett contest (the first, held at Rheims the previous year, had been won by the American, Glenn H. Curtiss) and the FAI's highly contentious $10,000 round Statue of Liberty race. In England he received gold medals from both the Royal Aero Club and the Aerial League, Lord Roberts making the latter presentation. His every move became tabloid news. His likeness went on display at Madame Tussaud's, and his liaison with the leading West End actress, Pauline Chase, became a *cause célèbre*.

With such a public reputation behind him Grahame-White sought to push

---

White papers, B768; Penrose, *Architect of wings*, 8. John Alcock was also among the spectators. He was then eighteen-years-old.

[189] *The Winning Post*, 21 May 1910 (Butler collection, xi. 60).

[190] PRO, AIR 1 727/160/3. This site, adjacent to Colindale Avenue, just off the Edgware Road, was being used by Messrs Everett & Edgecombe (or Edgcumbe), who had recently constructed a Blériot-derived monoplane with the help of the young Richard Fairey: Penrose, *Pioneer years*, 231. Grahame-White himself had inspected the location, with Kenelm Edgecombe, at the beginning of the year, and *The Aero* was reporting on 18 Jan. 1910 that the Grahame-White flying school had by that time already acquired flying rights over the locality. The site should be 'quite ready by April', Grahame-White is reported as saying. In the event, April proved to be the month of the Manchester race. Wallace states that as early as January Grahame-White had secured an option to purchase the site: *Claude Grahame-White*, 46–7.

[191] The rest of the Aeronautical Syndicate's stock was acquired by Handley Page Ltd: Penrose, *Pioneer years*, 354. The organisation's original holdings at Larkhill (hangar, lease and flying rights) had been acquired by British & Colonial: minutes, 24 May 1911, B&C papers.

[192] PRO, AIR 1 7/8/1; Wallace, *Claude Grahame-White*, 106, 110–11.

his fledgeling organisation in the direction of manufacturing. Early in 1911 he moved his operation from Brooklands to Hendon and endeavoured to float a limited company in conjunction with Sir Hiram Maxim and Louis Blériot: but even names like these were not enough to attract investors to so speculative an industry. Clearly the populace had to be convinced that aviation was something more than merely an exciting spectacle if sufficient investment was to be forthcoming. Consequently, when Grahame-White initiated his national 'Wake Up, England!' publicity campaign the following year, his interests ran deeper than its persuasively altruistic façade. Meanwhile, of course, the immediate Maxim-Blériot proposition had collapsed. Grahame-White Aviation was consequently re-established as a private concern, tided over by means of a £10,000 loan from Arthur du Cros, MP, chairman of the Dunlop Rubber Company and the founder, in 1909, of the Parliamentary Aerial Defence Committee – of which he subsequently became honorary secretary.[193]

The great publicity drive began at Hendon, where an initial ten year lease of the entire site was taken out. The aim was to make what was now called the London Aerodrome one of the capital's most prominent public attractions. With the aid of a newly-appointed aerodrome general manager, Richard Gates, and under the legal auspices of the Royal Aero Club, Grahame-White launched his opening season of international flying events there in April 1912.[194] Grandstands, pavilions and other amenities were built on the site, while posters appeared all over the metropolis, advertising tournaments, exhibitions and competitions, all within easy reach of central London. The whole spectacle was inaugurated – in less than auspicious weather conditions – with a three-day meeting over the Easter Bank Holiday weekend. The object was to promote what was called the 'Hendon habit', and make it as familiar to people as a day out at Lords or at the Zoo.[195]

The attempt was outstandingly successful, despite press coverage being initially somewhat overshadowed by the sinking of the Titanic liner.[196] But of course Grahame-White's diversion of the populace was not an end in itself – profitable though it soon became. He was principally aiming to demonstrate aviation's utility to potential investors. To this end, as well as aerial races and other contests, war displays were prominent on the Hendon programme. Aviation's military potential was consistently and dramatically enacted throughout the summer months of 1912. Dummy battleships, for example, were routinely pounded from the air, a set-piece which was particularly popular at night, when searchlights and explosions created a dazzling effect.[197] Over

---

[193] Ibid. 132.

[194] RAeC Committee minutes, 26 Mar. 1912, RAeC papers.

[195] PRO, AIR 1 727/160/3; Wallace, *Claude Grahame-White*, 164.

[196] PRO, AIR 1 7/8/1.

[197] Ibid. Some indication of the intensity of these air-to-surface 'attacks' can be gauged from the official reprimand the aerodrome received as a result of its sanctioning of low flying over the heads of spectators: RAeC Committee minutes, 1 Oct. 1912, RAeC papers. This was not a new antic. As a publicity stunt Grahame-White had dive-bombed the real fleet,

this first Hendon season Grahame-White Aviation grossed upward of £10,000 in gate money alone. This they confidently expected to double the following year.[198] For major events, such as the annual Aerial Derby (a London-encompassing air race, inaugurated in June 1912 as aviation's answer to the Cup Final or the Wimbledon Tennis Championships), as many as 50,000 spectators would pass through the turnstiles.[199] 'These contests and exhibitions', commented Grahame-White in 1913, 'have done more than anything else . . . to educate the British public to the vast potentialities of the modern aeroplane.'[200] For that reason, above all, the success of the venture had been 'most gratifying'.

As if to distinguish the purpose of all this activity from the inevitable hyperbole, Grahame-White had meanwhile begun cultivating the military authorities directly; and he was to continue this more subtle method of military ingratiation under the very shadow of his Hendon displays. In fact he had proposed a scheme for the attachment of an 'aeroplane corps' to the 1st London Territorial Division as early as February 1911, and the indications are that the War Office took this, unlike Butler's proposed 'balloon corps' some years before, very seriously.[201] For his scheme Grahame-White offered the use of three aeroplanes, three pilots and six mechanics, the accumulated cost of the project to be borne entirely by Grahame-White Aviation. In return, if the experimental 'corps' proved successful, Grahame-White, on behalf of his company, wanted the task of assisting 'in the formation of an aeroplane corps for the whole . . . army'.[202] In other words, like British & Colonial and Vickers before it, the Grahame-White Aviation Company was angling for a special relationship with the military authorities. Grahame-White may not have helped his own cause in this respect. Despite the confidential nature of the negotiations it is evident that he injudiciously communicated the details to other potential lobbyists. Thus we find Northcliffe writing to the editor of the Daily Mail, Thomas Marlowe, in May 1911, advocating support for the idea.[203] The Parliamentary Aerial Defence Committee also appears to have been consulted.[204] This, it should be remembered, was a self-appointed parliamentary pressure group, with no government mandate.

In the event, the idea was not adopted. This did nothing, however, to curtail

---

off Torbay, as early as July 1910 – shortly before joining the second Blackpool meeting: Daily Mail, 28 July 1910.

[198] G. Mansfield (sec., Grahame-White Aviation) to P. Hill, 2 Nov. 1912, Grahame-White papers, B771.

[199] PRO, AIR 1 7/8/1; AIR 1 727/160/3; Wallace, Claude Grahame-White, 168, 194.

[200] Draft of speech commemorating foundation of Hendon aerodrome, undated, but probably late 1913, Grahame-White papers, B775. The document includes lists of meetings/gates over the 1912 and 1913 seasons.

[201] War Office to Grahame-White, 4 Feb. 1911, Grahame-White papers, B769.

[202] Memo to War Office re. details of proposed 'aeroplane corps' attachment, undated, ibid.

[203] Northcliffe to T. Marlowe, 26 May 1911, Northcliffe papers, Add. 62198.

[204] PRO, AIR 1 7/8/1.

Grahame-White's lobbying of the authorities. Indeed, under the auspices of the Parliamentary Aerial Defence Committee, with whose founder and secretary he was, of course, financially associated, he pre-emptively hosted a formal military flying display at Hendon on 12 May. The programme on this occasion included such events as bomb throwing, despatch carrying, gun transporting, a mock dirigible attack and reconnaissance tests – with serving officers as observers: and among the aviators contributing were S. F. Cody, Gustav Hamel, Pierre Prier, Robert Loraine and A. V. Roe.[205] It proved a notable success, if only in the enormous publicity drawn to the London Aerodrome. Guests included Asquith, Haldane, Northcliffe, Churchill, Balfour, Lloyd George, and in all over 200 MPs, as well as contemporary celebrities like the Antarctic explorer, Sir Ernest Shackleton, and military and naval dignitaries such as Lord Roberts, Lord Fisher and Lord Charles Beresford. Grahame-White subsequently offered to take any of them up in his Farman biplane, and McKenna and Balfour – at that time the First Lord of the Admiralty and the Leader of the Opposition respectively – were among those who responded.[206]

By this time Grahame-White Aviation was beginning to market aircraft of Grahame-White's own design. These machines were not startlingly original. The Grahame-White 'Baby', for example, the prototype of which had been developed by Starling Burgess as a result of Grahame-White's US tour and which formed the centrepiece of the firm's 1911 Olympia display, was simply a scaled-down Farman, while the Grahame-White Boxkite proved equally derivative.[207] This diversification consequently offered no serious challenge to more intrinsically pioneering manufacturing organisations, such as Sopwiths, Avro or Shorts. The company, acknowledging this, subsequently engaged as designer a talented young engineer named John North, who had previously served an apprenticeship with Horatio Barber's short-lived Aeronautical Syndicate (a Hendon neighbour of Grahame-White Aviation). Displaying noticeably more competence as an aircraft designer than his nominal employer, North devised a military pusher-biplane, with a quick-firing Colt gun mounted within a forward-projecting nacelle – rather in the manner of the recently conceived Vickers EFB Gunbus. This particular design may have derived, at least in part, from an earlier, abortive, Aeronautical Syndicate blueprint,[208] but whatever its antecedents the prototype was publicly unveiled at the 1913 Olympia Aero Show with the unequivocal designation 'Grahame-White Type VI'. Unfortunately, however, neither this machine nor a variety of derived nacelle-Farman synthesised pusher-biplanes significantly advanced the

---

[205] Programme of events, RAeS archive.
[206] PRO, AIR 1 7/8/1.
[207] Wallace, *Claude Grahame-White*, 134, 167. The company were quick to bring this diversification to the attention of the War Office, with an offer to accommodate a 'military aviation depot' within their Hendon works: Grahame-White Aviation Co. to War Office, 2 Dec. 1911, PRO, CAB 16/16, appendix x, 194.
[208] Penrose, *Pioneer years*, 422.

company's commercial prospects, although there were sporadic service orders.[209] This seems to have been accepted with practical equanimity. Lacking design ingenuity the company's best hope of significant profits lay in tendering for sub-contracts. The government's subsequent patronage of the private sector through the sub-contracting of Farnborough-designed BEs soon made this an extremely lucrative corner of the market. Grahame-White thus still had everything to gain from exhorting the military to increased aeroplane procurement.

By the time of North's recruitment Grahame-White had already resolved to take his 'Wake Up, England!' publicity campaign to the provinces, having acquired for the purpose the sponsorship of Lord Northcliffe and the Daily Mail. An intrepid corps of aviators, led by Grahame-White himself on a newly acquired vivid blue and suitably emblazoned interchangeable wheel-float Farman biplane, duly stormed the country's major towns and coastal resorts over the summer months of 1912, in what was effectively the precursor of Alan Cobham's publicity-seeking inter-war flying circuses. Yet for all the Bank Holiday gaiety there was never any pretence that the 'Wake Up, England!' tour, as it was labelled, was anything other than a campaign to whip up popular agitation for increased military aircraft procurement. 'The object', affirmed the Daily Mail, echoing Grahame-White's own previous statements, 'is to educate the people of this country as to the qualities and potentialities of the "new arm" and to stimulate the Government and the War Office to make good the deficiency caused by past neglect.'[210] In addition to Grahame-White, aviators as celebrated as B. C. Hucks, a Grahame-White protégé who had become the company's leading test pilot, and Gustav Hamel, the undisputed hero of the Hendon crowds, participated in these theatrical displays.

One might have thought Grahame-White's behaviour during this time a little too brazen for government officials, but again they really do seem to have taken his views seriously. Early the following year, when he forwarded to Whitehall another proposal for an aerial service, it was discussed in considerable depth by a CID sub-committee. This latest scheme involved the establishment of a chain of aircraft stations around the country. Each station would accommodate twelve aeroplanes, eight of which would remain on stand-by. For coastal stations there would be five seaplanes, four being maintained on stand-by. A total of 450 pilots, including 100 airship pilots, and an additional 300 mechanics, would then be trained annually, in conjunction with the Grahame-White flying school.[211] Grahame-White indicated that he would be both willing and able to raise sufficient funds to establish such a programme, provided either the service departments or the Treasury would then agree to subsidise it. He had, in fact, come to an understanding of some kind with the

---

[209] In particular, numbers of Type XVs were later supplied to the RNAS as training aircraft. For service acquisitions see Bruce, Aeroplanes of the Royal Flying Corps, 254–61.
[210] Daily Mail, 25 June 1912, quoted in Wallace, Claude Grahame-White, 171.
[211] PRO, AIR 1 2317/223/24/4.

international financier Sir Edgar Speyer, of the London banking house of Speyer brothers. Speyer proposed putting up the capital if the government would pledge a percentage investment. Once established, commercial potential in a passenger and freight service based on routes between the various stations was foreseen. In other words what Grahame-White was after was government subsidy for an aerial communications system, which would become the structural basis of a United Kingdom aerial defence organisation in the event of war. Indeed, in its geographical divisions the proposal may be said to have been prophetic.

The scheme was essentially planned as a Grahame-White Aviation Company project, with, it was hoped, military representatives joining the firm's directorate to help select the various aircraft to be employed. To subdue fears of too close a commercial association, the precedent of government subsidy of the merchant shipping industry was invoked. Grahame-White also candidly admitted, in confidence, that his object was as much to stimulate the incipient aviation industry as to defend the country:[212] not that the two aims could easily be distinguished in his mind. Sub-committee members accepted this as a perfectly valid, indeed laudable motive. The main disadvantage, as they saw it, was that such aircraft as were chosen would need constant, and perhaps prohibitively expensive, upgrading. The sub-committee was thus ultimately divided over whether or not to recommend that the scheme be adopted. (They did, however, warn Grahame-White against communicating with the press this time.)[213] In the end they deferred any decision by proposing that a technical sub-committee be appointed to reconsider the matter. There negotiations effectively ended.[214]

The failure of this scheme left Grahame-White free to criticise government policy again, and this he did with renewed vigour. He formulated a mock appeal, under the banner 'Wanted: £1,000,000 for flying', as another means of bringing pressure on the authorities. 'Does our Treasury . . . intend to allocate a sufficient sum for aviation?', he asked, in a draft leaflet publicising the new campaign. 'This matter can no longer be trifled with. Either we are to have our place among nations in aerial armament [sic], or we are not; and, if the Government still refuses to take adequate action, a national movement should compel it to do so.'[215]

Aviation is presented here as nothing but an extension of the armaments industry. In other words Grahame-White's publicity campaigns were primarily designed to stimulate what was essentially a military–industrial complex. This was not, of course, explicitly spelt out to the general public (unlike civil servants); they were simply presented with the supposed need to 'protect our

---

[212] Ibid. point 27.
[213] Ibid. point 43.
[214] Wallace's rather anecdotal account of this episode should be viewed with considerable caution: *Claude Grahame-White*, 178–9, 182–3.
[215] 'Wanted: £1,000,000 for flying', c. 1913, Grahame-White papers, B774.

shores from the menace of aerial invasion',[216] with little intimation of the vested interest behind that alleged imperative. This was not, perhaps, the best way to proceed. Clearly, the country would benefit from its own indigenous industry, and obviously this needed support; would not, then, both private enterprise and the government profit from some measure of co-operation? Grahame-White had greatly impressed the CID sub-committee when arguing from this rational premise. Careful petitioning against government scepticism on the one hand, and against military suspicion of the private sector on the other, might have yielded more gratifying results – although the 1911 manufacturers' deputation to Colonel Seely had not proved conspicuously successful.

Such, however, was not the temper of the times. Instead, Grahame-White, exploiting his enormous fame, chose to play on popular war scares and, in increasingly intemperate tones, to denigrate governmental 'indifference' to the 'air threat'. These sentiments were then eagerly amplified and echoed throughout the Northcliffe Press. As a result, public perceptions of aviation were increasingly couched in terms of an arms race.[217] Indeed, insofar as Grahame-White was representative of the industry, public perceptions had become part of the military–industrial complex.

Somewhat ironically, the sudden outbreak of hostilities in the summer of 1914 left Grahame-White behind. For all his previous agitation he played no significant part in the prosecution of the air war. He sought an interview with Churchill and talked him round to the idea of a private air-strike against the Kiel Canal, but, not surprisingly, Commodore Sueter vetoed the idea.[218] He was already preparing aircraft for offensive operations. Meanwhile Hendon itself was requisitioned by the Admiralty, as an incidental result of which Richard Gates was tragically killed after he crash-landed a Farman biplane upon returning from an unofficial RNAS anti-Zeppelin patrol over London.

The Grahame-White Aviation Company compliantly accepted its subservient position as, predominantly, government BE manufacturers. The directors did not really face the problems of their more successful competitors, British & Colonial for example, who were effectively denied a market for their own designs by the War Office's increasingly restrictive procurement policy. This is not to suggest that the company did not suffer from enormous supply difficulties or that it did not experience any number of technical and financial disagreements with the relevant departments of the Admiralty and the War Office; simply that its own design ingenuity was not thwarted in quite the same manner as that of Avro and British & Colonial.

Grahame-White himself had only an intermittent influence on his company's wartime affairs. He initially accepted a commission in the RNAS, and

---

216 Ibid. 8.
217 Grahame-White actually talked in terms of 'the race for aerial supremacy': ibid. 4.
218 Grahame-White papers, B1433; PRO, Adm. 116/1278; Wallace, *Claude Grahame-White*, 196.

served in the main as a flying instructor at Hendon. Returning to full-time management in June 1915, he then helped co-ordinate the numerous sub-contracted orders being awarded to his firm. Grahame-White Aviation still employed its own design-staff, but with limited success. 1916 saw the company constructing primarily Avro 504s, and 1917 saw it producing large numbers of DH6s. After the war the newly-created Air Ministry made a compulsory purchase of the Hendon aerodrome. Grahame-White subsequently severed all connections with aviation. He died at Nice in 1959.[219]

Aviation's military–industrial complex

Despite the prominence given the subject by Grahame-White's activities, aviation's military–industrial complex was not simply a product of the few years before the Great War. On the contrary; although the necessity of military investment in what was by then an embryonic industry did certainly distinguish the growth of aviation in those pivotal years, an examination of aeronautics from its inception establishes clearly that the military dimension was there from the beginning. This can be traced right back to the ballooning exploits of the Montgolfier brothers. Beyond its reconnaissance potential, indeed, the balloon was seen to have little practical value. Aeroplanes thus followed – and were seen to follow – a long-established precedent, whether explicitly or implicitly. Of the early twentieth-century pioneers, for example, S. P. Langley was actually financed by the US Army Board of Ordnance and Fortifications, while the Wright brothers seem never to have imagined the commercial exploitation of their invention outside the arms market. By their time, of course, the international environment made that market seem particularly ripe for exploitation. When negotiations with the representatives of a particular government faltered, the brothers played up the fear that a rival power might pre-emptively acquire an air capability, cynically pushing their sales drives during periods of heightened tensions.[220] These negotiations related to individual machines, rather than to any investment in industrial growth as such, but when the next wave of technological entrepreneurs came forward it is little wonder that they too were drawn toward the military market. It was effectively the only regular market.[221]

It is important to differentiate between general use of the term 'military–

---

[219] For these subsequent dealings see A. D. George, 'Aviation and the state: the Grahame-White Aviation Company, 1912–23', *Journal of Transport History* ix (1988), 211–13.

[220] Gollin, *No longer an island*, 122, 140.

[221] There is even evidence that John Stringfellow, who prophetically investigated the possibility of heavier-than-air powered flight in the mid-nineteenth century, thought along these lines, for all his non-conformist Liberal sympathies. Indeed, he also developed and patented an 'armoured cart' for military purposes at the time of the Franco-Prussian War, thus inviting comparison with later pioneers like Simms and Maxim: and it is no coincidence that the latter's work developed out of a career in the arms industry. See H. Penrose,

industrial complex' and its more specific connotation. Specific use, usually taken to derive from C. Wright Mills's *The power elite*, published in 1956, i.e. in the midst of the Cold War, pertains to the view that in free market economies there exist or existed conspiring elites with a vested interest in maintaining high defence expenditure. Studies arguing from this premise, popularised by President Eisenhower's use of the expression in his celebrated 1961 farewell address, are largely confined to the post-World War II defence market, and are thus for the most part outside our field of inquiry. But more recently the term has been applied to a wider historical framework.[222] It is in this wider sense, denoting a requisite coincidence of interests between private enterprise and military bureaucracies, that the expression is applied here.

With recently evolved businesses it was, of course, the manufacturer's own responsibility to persuade the military that their interests coincided. This is the most consistent theme of the various stages of the aviation industry's development. As each new company evolved it vied, by divers means, for the military's favour. The main disadvantage of such an episodic approach, however, is that one can fail to see the wood for the trees. Ideally, some notion of the collective needs and demands of the numerous new companies clamouring for attention at this time should be fixed, before moving on to examine the other side of the equation, the military's own response to the astonishingly rapid growth of aviation and, more particularly, a competitive aviation industry in these years.

Fortunately, in the middle of this period there occurred an event which bound together the many enterprises in just this way. Dissatisfied with official responses to their needs a deputation of British aircraft manufacturers assembled in December 1911 to petition the then Under-Secretary of State for War, Colonel J. E. B. Seely. This deputation – the precursor of the Society of British Aircraft Constructors (SBAC), founded in March 1916 – neatly crystallised the major arguments of the new industrialists. Its submissions are therefore of some significance.

The British aircraft manufacturers' deputation

The deputation, which met Seely at the House of Commons on 5 December 1911, contained representatives from all the major indigenous aviation firms, as well as spokesmen from the Royal Aero Club and the Aeronautical Society. Among those present were A. V. Roe, Handley Page, Horace Short, Grahame-White, J. W. Dunne (the former British army Balloon Factory pioneer, now representing his own Blair Atholl Syndicate), Captain H. F. Wood (manager of Vickers's aviation department) and Captain Bertram Dickson (British &

An ancient air: a biography of John Stringfellow of Chard, the Victorian aeronautical pioneer, Shrewsbury 1988, 123–4, 129–30, 135.

[222] For example, by W. H. McNeill, in *The pursuit of power: technology, armed force, and society since A.D. 1000,* Oxford 1983, 269.

Colonial's London and Continental representative).[223] J. W. Dunne opened the proceedings, declaring that the delegation's over-riding need was for practical government assistance. He called, in particular, for consistent military orders and a measure of industrial protectionism. As an extension of the armaments industry, he maintained, aviation should not be subject to the play of international free trade forces. The government should treat it, rather, as it would any other element of the indigenous armaments industry, making the establishment of an internal source of supply a matter of priority.[224] The fact that the War Office had recently announced that it was to open its long-awaited military aeroplane competition to all nations gave added impetus to this demand.

Howard Flanders then came forward to place the emphasis specifically on conditional contracts as a means of supporting research and development. Stipulate your requirements, he urged Seely; all the manufacturers asked was that should they be fulfilled, purchases would be guaranteed. New businesses, he added, could not be expected to 'live on hope'.[225] The War Office was advised, however, not to be too dogmatic in its stated requirements: technology was advancing so rapidly that the best designs could be outmoded within months. They must allow a measure of flexibility.[226]

Seely was less than encouraging in reply to all this. It would not do, he commented, to depend too much upon military orders: these could not 'possibly be sufficient' for the purpose of establishing an industry on the scale now envisaged. This was not what the deputation wanted to hear, and indeed, in his response, Seely may have been unnecessarily dismissive. He refused to be drawn as to the desirability of conditional contracts, and undoubtedly alienated several members of the delegation, to the extent that he became to many, like C. G. Grey, the personification of what appeared government wrongheadedness.

This perception was sharpened significantly by an ill-advised Commons exchange which occurred some seven months after the controversial competitive trials known as the military aeroplane competition. During the introduction of the 1913 Army Estimates Seely (by then Secretary of State for War) was pressed on the expenditure allotted to aviation. Without going into details, he rashly claimed on 19 March 1913 that the RFC, Military Wing, had 101 serviceable aeroplanes. This statement, variously amended, was challenged by opposition members Arthur Lee (since 1909 chairman of the Parliamentary Aerial Defence Committee) and William Joynson-Hicks. A lively controversy ensued. The War Office, having had its bluff called, endeavoured to disguise

---

[223] 'List of deputation to Col. Seely, 5 Dec. 1911', Mottistone (i.e. Seely) papers, box 19.
[224] 'Deputation to Col. Seely: J. W. Dunne on general policy', ibid.
[225] Howard Flanders, ibid.
[226] J. H. Ledeboer, ibid. The deputation's arguments have been abridged. The speakers made their submissions in the following order: J. W. Dunne, J. H. Ledeboer (Aeronautical Society), R. W. Wallace (RAeC), Capt. H. F. Wood, R. L. Howard Flanders, F. May (Green Engine Co.), H. G. Burford (Humber Ltd).

the fact that Seely's impulsive claim was, at best, a gross exaggeration, and, indeed, attempted a clumsy cover-up. War Office officials trawled the country in an effort to gather together as many aeroplanes as could be immediately acquired – irrespective of their quality. These machines, including a number only purchased, in some desperation, after Seely's *faux pas*, were then farmed out to Farnborough and Larkhill, to be brazenly presented to Joynson-Hicks's parliamentary investigatory commission – consisting of Joynson-Hicks himself and Captain George Sandys, MP, Arthur du Cros having been refused permission of the War Office to join the party – as proof of the Secretary of State's claim.

One need only examine the batch of at least half-a-dozen Grahame-White aircraft procured as a part of this attempted cover-up to appreciate the effrontery masking this (in some quarters, reluctant) defence of Seely's ill-considered assertions. Found among them were Grahame-White's own much-travelled 'Wake Up, England!' Farman biplane and a battered and largely expended school boxkite.[227] These acquisitions were wholly cosmetic, and are unlikely to have impressed anyone, least of all the service's pilots. The cynicism thus displayed in concealing the truth behind a needless parliamentary gaffe fatally undermined Jack Seely's credibility in the eyes of aviation's civil–industrial representatives. His foremost critics, notably C. G. Grey in the press and Joynson-Hicks in parliament, mercilessly exploited the government's ill-disguised discomfort, transforming the affair into a highly publicised scandal. Joynson-Hicks was particularly effective in maximising the Secretary of State's embarrassment.[228]

In marked contrast to certain of his more vociferous associates – prominent air advocates like Holt Thomas, Grahame-White and, a little later, the infamously vitriolic Pemberton Billing – William Joynson-Hicks (a future cabinet minister) was careful to retain his integrity as chief parliamentary critic of the government's air programme. When H. V. Roe, inspired by his philanthropic work in Manchester, as well as his support for the aviation industry in parliament, at some point in about 1912 invited him to become chairman of the Avro board, Joynson-Hicks, a devout evangelical, declined, lest his advocacy of aviation in the Commons be compromised.[229] Such scrupulosity was far from universal.[230]

These, then, were the sorts of demands and hostilities the War Office faced

---

[227] Bruce, *Aeroplanes of the Royal Flying Corps*, 252. In fact most of Grahame-White's early War Office orders probably resulted from this incident.
[228] See, in particular, William Joynson-Hicks, *The command of the air or prophecies fulfilled: being speeches delivered in the House of Commons*, London 1916, passim. Note also R. D. Brett, *The history of British aviation 1908–1914*, London 1933, 221ff; and A. Gollin, *The impact of air power on the British people and their government, 1909–1914*, Basingstoke 1989, 246–50.
[229] H. V. Roe, 'Pioneers'.
[230] Pemberton Billing reportedly disposed of his shares in Supermarine on entering parliament as an Independent 'Air Member' in 1916: *P-B, the story of his life*, London 1917, 147.

from the newly evolved aviation industry at the time it began drawing up plans for the formation and maintenance of a Royal Flying Corps. How it dealt with such pressure and what it hoped to achieve through the creation of, effectively, the nation's first air force is the next subject for consideration.

4

Military origins

The military aeronautical tradition

In tracing the development of heavier-than-air powered flight from its inception up to the First World War it has become clear that it was primarily a mechanical and entrepreneurial progression, the military dimension being cultivated principally as the new industry's potential market. The response of the military has been thus viewed only from the industrialists' perspective, the military appearing in the role of market consumers. Aeronautics and the aeronautical industry must now be examined from the other side, from the perspective of the military itself, and the way in which they sought to deploy the new technology in the field must be analysed.

Whereas aviation's engineering-based manufacturers and entrepreneurs – from the Wright brothers on – developed with and through the general technological advances of the period, the approach of the military and of military aeronautical establishments throughout Europe and America sprang from an entirely different source. It derived primarily from a gradual acceptance of the spherical balloon's utility on the battlefield. These two elements were ultimately to converge, as manufacturers sought a market and military authorities sought to procure navigable aeronautical devices, but this was in the long term. Initially, in Britain at least, the two sectors operated independently. Even after the emergence of an embryonic competitive industry relations remained strained. From the beginning the War Office sought to internalise any potential military development of heavier-than-air machines, and it was only the technocratic difficulties which resulted that by degrees compelled them to open their doors to the private sector. Thus when private enterprise was at last ready to negotiate the sale of practical aeroplanes to the British army it was met by an already well-entrenched aeronautical establishment.

This is the essential background to any consideration of the military's response to the developing industry. It was not a case of having to create a new establishment to accommodate the emergence of an unprecedented technology. Rather, it was a matter of assimilating experimental advances within an already established framework. In other words the military, particularly in Britain, were cautiously adapting to a dynamic technological environment; they were not, as most manufacturers would have wished, seeking to build a new arm from scratch specifically to meet the advent of heavier-than-air powered flight.

Beginnings: the US Civil War

Military potential was an intrinsic part of the balloon's earliest development. By 1782, a year before the first successful manned ascent, Joseph Montgolfier was already speculating as to the practicability of using the device as a means of transporting troops into the British garrison at Gibraltar. The first passenger in the Pilatre de Rozier-piloted captive Montgolfier balloon, André Giraud de Vilette, similarly remarked immediately upon their reconnaissance potential, recounting his experiences in a letter to the *Journal de Paris* on 20 October 1783.[1] Events following the French Revolution simply added momentum to early speculations. In 1793 the Committee of Public Safety was urged by a member of its Scientific Commission, Guyton de Morveau, to employ gas balloons as an aid to the revolutionary armies in the field. At this time, however, there was a practical problem in that all the necessary sulphur supplies were going directly into the making of gunpowder: consequently sulphuric acid was no longer available as a lifting agent. De Morveau therefore consulted the distinguished chemist, Antoine Laurent Lavoisier, who recommended the use of hydrogen, which could be produced by water decomposition, i.e. by passing steam over red hot iron filings in a brick kiln. Aided by two physicists-cum-aeronauts, Jean Coutelle and Nicolas Conté, de Morveau successfully produced and stored a quantity of this gas, and then despatched Coutelle to General Jean-Baptiste Jourdan, Commander of the Army of the North, in an effort to persuade him to adopt a balloon adjunct.[2] But although Jourdan eventually accepted in principle the idea of an aerial arm, he thought experiments still too rudimentary for use on campaign. They should rather, he suggested, be continued back in Paris, at least until such time as a fully evolved balloon corps was available. Thus in early 1794 the former arms depot at Château-Meudon became the world's first military aeronautical establishment.[3] An aerostatic corps of the artillery service was formed in April of that year: initially just one company – a captain, a lieutenant, four NCOs and twenty men. A second company was sanctioned the following June. The newly-commissioned Captain Coutelle thereafter commanded the corps, while Conté directed the Meudon foundation. A national military balloon school was established there the following October.

These *aérostiers*, as they were called, achieved their finest hour at the battle of Fleurus on 26 June 1794. On that day Coutelle twice ascended with General Morlot, Jourdan's acting Adjutant-General, in the 1st Company's 'Entrepren-

---

1   Extracts from the letter are quoted in the scrupulously documented introductory chapter to F. Stansbury Haydon, *Aeronautics in the Union and Confederate armies*, Baltimore 1941. See also Rolt, *The aeronauts*, 159.

2   A. L. H. Hildebrandt, *Airships past and present*, London 1908, 128; and – on this method of hydrogen production – Lt. G. E. Grover, 'On the uses of balloons in military operations', *Professional Papers of the Royal Engineers* xii (1863), 80.

3   D. Lacroix, *Les aérostiers militaires du Château de Meudon 1794–1884*, Paris 1885.

ant' balloon, and together they witnessed virtually the entire engagement – of some nine hours – from aloft. Throughout the battle communications with ground forces were maintained by means of both flag-signalling and despatches, these being dropped from the balloon via a weighted-bag attached to the mooring cable. The tactical advantage gained was certainly useful, if not, as some have claimed, decisive.[4] Unfortunately, however, the corps could scarcely yet be considered fully mobile, or its equipment sustainable in a prolonged campaign of movement, so they were unable to maintain this level of achievement. Indeed, the 1st Company and all its equipment were captured by the Austrians following Jourdan's defeat at Wurzburg in 1796. The end came in Egypt in 1798 when the revived unit's equipment was destroyed by the British during the naval battle of Aboukir Bay. Napoleon, who had never accepted the utility of balloons, took the opportunity to disband the corps. The Meudon foundation was closed shortly afterwards.

England, in the meantime, had her aeronauts – most notably in the person of James Sadler of Oxford – but no effort was made to exploit their work militarily. In 1803 one of Sadler's better-known contemporaries, Major-General John Money, did publish *A short treatise on the use of balloons and field observators [sic] in military operations*; however nothing resulted from it, and there, effectively, ended the first phase of military aeronautics. Little progress occurred over the next half century. In a practical sense the spherical balloon had already reached its apogee, being merely a non-navigable gas-bag with obvious inherent limitations. The scientific community soon lost interest in it, as did the military – this being, generally, a time of peace. Consequently, by the mid-nineteenth century the balloon had changed little from the 1780s or 90s. It had come to be seen, indeed, as little more than a vulgar fairground attraction, although even there, as the novelty wore off, increasingly sensational feats had to be devised to keep the populace interested. As a result, ballooning became an occupation of ill-repute, the preserve of charlatans and showmen – the original 'gas-bags' and 'hot air merchants' – hardly likely to attract influential members of the armed forces.

There is intermittent evidence of the balloon's use on the battlefield in the years between 1815 and 1861, but to no notable effect. Not until the American Civil War was this negative trend reversed; and then it was due above all to one man, Thaddeus Lowe. Lowe was a native of New Hampshire, who had begun his aeronautical career in search of backing for a proposed transatlantic balloon crossing. Most of his early ascents were undertaken with this object ultimately in view. In the event war intervened before his plans came to fruition. On the outbreak of hostilities Lowe immediately offered his services

---

[4] Stansbury Haydon, *Aeronautics*, 10; Rolt, *The aeronauts*, 161–3; Maj. R. S. Waters, 'Ballooning in the French army during the revolutionary wars', *Army Quarterly* xxiii (1932), 332.

to the Federal government in Washington.[5] That they were accepted resulted more from his methods than his innate qualities as an aeronaut, of which he was but one of a number available. In particular, he had devised a crude system of portable generators, which made rapid inflation in the field possible. 'Portable' here is a relative term. It still took up to seven wagons to haul the necessary materials for maintaining three balloons – of either 13,000 or 26,000 cu. ft capacity – in transit.[6] Even then, however, Lowe was given no official status, beyond that of civilian adjunct to the Federal forces. (He adopted the title 'Professor', although he had no right to it.) Given a complement of some fifty men he was beset from the start with logistic difficulties, deriving directly from both his own and his corps' administratively ambiguous position. The corps had no established place in the Federal army's command structure: the men, ostensibly regulated as Topographical Engineers, held no rank. For supplies the unit was dependent upon a reluctant and grudging Quartermaster Corps. By early 1862 five newly-constructed observation balloons were on the strength, but with limited resources to maintain them. That Lowe was so consistently employed in the early stages of the war was due principally to General George B. McClellan, of the Army of the Potomac, who took him under his wing and sanctioned the accompaniment of a Balloon Corps (comprising the 'Intrepid', 'Washington' and 'Constitution' balloons) during his Peninsula campaign of March–August 1862. McClellan was by training an army engineer, and his staff officers and Lowe would each work from maps marked with identical grid references, facilitating the communication of reconnaissance information.[7] How far the Federal forces under McClellan's command benefited from this information, however, is a matter of some dispute.

Lowe found himself employed soon after McClellan had disembarked his troops at Fort Monroe on the Virginian Peninsula. In April 1862 his balloons were used to reconnoitre the Confederate defences and dispositions at Yorktown. The enemy were clearly concerned – as they might well have been, given that under General John B. 'Prince' Magruder they were effectively practising mass deception, to disguise the initial weakness of their forces – and Confederate artillery unsuccessfully attempted to range on the balloons. However, aerial observations did nothing to alter McClellan's misconception of the enemy's strength. Even had useful information been forwarded McClellan was disinclined to exploit the situation. His inherent hesitancy and indecision as a soldier was reinforced by the realisation that the gap between his own political objectives and those of the progressively more abolitionist Republican

[5] J. D. Squires, 'Aeronautics in the Civil War', *American Historical Review* xlii (1936), 655–6; Stansbury Haydon, *Aeronautics*, 154ff.

[6] Squires, 'Aeronautics in the Civil War', 660; Capt. F. Beaumont, 'On balloon reconnaissances as practised by the American army', *Professional Papers of the Royal Engineers* xii (1863), 95–6.

[7] B. F. S. Baden-Powell, 'Military ballooning', *Journal of the Royal United Services Institution* xxvii (1884), 739.

government in Washington was steadily widening. Moreover, to be in any sense effective, balloon observations required placid meteorological conditions, and these were often unavailable – even in spring and summer. Thus information derived from balloon reconnaissances was for the most part patchy and not to be taken as definitive.

The dangers of ascending in even comparatively mild winds were dramatically illustrated on 11 April when McClellan's closest corps commander, Brigadier-General Fitz John Porter, made a solo ascent in the 'Intrepid' balloon, only for it to break free of its moorings and drift directly towards enemy lines. A wind-shift saved the craft at the last moment, but both Porter and McClellan were shaken by the incident, and thereafter senior officer volunteers to accompany Lowe were scarce.

In an effort to match this Federal activity the Confederate forces at Yorktown secured a balloon of their own; but it achieved little. Indeed, Lowe's reaction was nothing short of contemptuous. The improvised aerostat was dismissed as no more than a fire (i.e. a hot air) balloon, with an aerial buoyancy of little more than half an hour. However, not even Lowe perceived how uncomfortable that duration was for his opposite number. The balloon was secured to the ground with a single mooring-rope, the strands of which had a tendency to unwind – gradually spinning the solitary basket around like a top. The effect of this on the unfortunate Confederate aeronaut, Captain John Bryan, 21-year-old ADC to General Magruder, can be imagined. At the start of only his third ascent the balloon's coil of rope caught the leg of a bystander – of whom there were always too many – and the line had to be cut with an axe. Bryan was left to experience the same free-flight helplessness so recently endured by General Porter. The ordeal was made worse by the fact that it was still dark. He survived, but abandoned ballooning.

On the evening of 3 May 1862 the Confederate forces, now under the command of General Joseph E. Johnston, threw up a wide artillery barrage, behind which they began their staged withdrawal to the outskirts of Richmond. All the Federal Balloon Corps could do was confirm, the following morning, that the Yorktown defences had been abandoned. By then it was too late to do anything about it. In the subsequent fighting around Richmond, however, balloons played their most prominent part in the war, notably at the battle of Fair Oaks/Seven Pines on 31 May 1862.

The Federal army named this battle Fair Oaks to commemorate the fact that they had fought best around the Fair Oaks station of the Richmond & York River Railroad; the Confederates called it Seven Pines to memorialise the fighting around the crossroads of that name. In fact the greater number of casualties occurred at Seven Pines, some seven miles east of Richmond. The armies clashed here because McClellan had divided his forces on either side of the Chickahominy river. General Johnston sought to isolate and engage those divisions which had crossed to the south of the river, relying on the effects of recent storm-flooding to prevent reinforcements coming to their aid. It was a cleverly conceived plan and, despite delays and considerable confusion in the co-ordination of the advancing Confederate army, the Federal forces were caught to a dangerous degree unawares. No aerial reconnaissance was operating on the morning of 31 May, although Lowe had two balloons stationed on the north bank of the Chickahominy, whence good views could be obtained of Richmond and its suburbs. Indeed, even when alerted to the fact that their forces on the south side of the river were under concerted attack, it was 2.00 p.m. before wind conditions permitted exploratory balloon observations. By then the fighting was well advanced, and what useful information could be gleaned from the air was limited by the heavily wooded terrain in which most of the battle was occurring. It would thus have been a near practical impossibility to discern the particular manoeuvres the Confederates were attempting to employ – even supposing their confused movements could be precisely interpreted, which is doubtful. However, in an effort to promote his corps Lowe was subsequently to imply that his observations had indeed shaped the Federal army's successful recovery of the situation. In fact he retrospectively amended his original despatch to give this impression, claiming that he had actually ascended at noon on 31 May. Whatever the precise time, there is no evidence that he either did or could have decisively influenced the issue.[8] In reality, the Confederate attack had become desperately confused and unco-ordinated, and, against expectations, the Federals, led by John Sedgwick's division, successfully crossed the storm-risen Chickahominy river to reinforce the endangered southern positions. But an aeronautical myth had been born, and, irrespective of the truth behind it, acted as a powerful spur to military-balloon advocates in the following decades.

That this episode should have become the most renowned example of balloon-support in the war gives some indication of the limited role balloons played in subsequent events. Lowe participated in the battles of The Seven Days, which followed Fair Oaks, but was no more successful than before in countering McClellan's complete miscalculation of Confederate numbers and dispositions. As at Yorktown, General Magruder's deceptions proved effective, particularly during the Gaines's Mill engagement on 27 June. More notable were the Confederates' renewed attempts to improvise a rival balloon unit. Captain Langdon Cheves, an engineer with the South Carolina Ordnance, famously amassed dress silk from the cities of Savannah and Charleston, from which he constructed a new balloon envelope. This was transported to Richmond, to be filled with 'illuminating gas' from the city's municipal gasworks, after which it was conveyed to the front roped to a York River Railroad boxcar. Piloted by Major Porter Alexander, one of General Johnston's staff officers, it

<hr />

8   S. W. Sears, *To the gates of Richmond: the Peninsula campaign*, New York 1992, 54–5, 125–6, 412 n. 12. This study concerns itself with the campaign as a whole and persuasively counterbalances the less discriminating accounts of Lowe's achievements evident in most aeronautical-centred histories (for example in T. D. Crouch's *The eagle aloft: two centuries of the balloon in America*, Washington, DC 1983, 390).

was then successfully employed on 27 June, reporting the northward crossing of the First Division of General Franklin's Sixth Corps across the Chickahominy river – towards Gaines's Mill. Ironically, the Federals believed that the enemy balloon was overseeing preparations for a – non-existent – assault south of the river. This in some ways typifies the closing stages of the Peninsula campaign. As events unfolded, General McClellan's assumptions and actions became increasingly divorced from reality. Almost regardless of the battles being fought he continued an eastward withdrawal, until encamping at Harrison's Landing, on the James river, where his army came within support of the Federal gunboats. Lowe himself was by this time on sick leave, as a result of the fever sweeping through the ranks of the long-encamped Union divisions.

The Army of the Potomac eventually evacuated the Virginian Peninsula in August 1862, having achieved nothing. Shortly before, on 4 July, Porter Alexander had affixed his silk balloon to an armed tug (the 'Teaser'), in order to steam down the James river in an effort to observe the Federal army's dispositions; but the tug ran aground and was eventually captured with its accompanying aeronautical equipment. The Confederate balloon unit was never revived. (A second silk balloon was captured at Charleston the following spring.)

Lowe and the Union Balloon Corps did not survive much longer than their improvised counterpart. McClellan was finally removed from command in November 1862, and his removal effectively saw the plug pulled on co-ordinated balloon-support, for when obliged to liaise with other officials, intent on rather more movement in the field than McClellan, Lowe found them considerably less sympathetic to his needs. One of Lowe's balloons supported General Burnside's crossing of the Rappahannock river on 11–12 December 1862, but this merely preceded the disaster at Fredericksburg, so hardly enhanced the corps' reputation. By the time of Chancellorsville, at the turn of May 1863, many viewed the use of balloons with a jaundiced eye. The administrative reforms which followed led to Lowe's expected resignation – he had promised it the month before. Shortly afterwards the Union's Balloon Corps disbanded.

As this suggests, it would be wrong to assume that all reports of Lowe's activities were uncritical. Lieutenant-Colonel H. C. Fletcher, of the Scots Guards, who was attached to McClellan's HQ as a military observer, noted the degree of portability that had been achieved (three balloons carried by six wagons, as he recorded it; an inflation taking approximately three hours and lasting up to four weeks), but for limited ends. Balloons had proved useful in spotting for artillery, but for general reconnaissance, he felt, their use had been disappointing. This, as Fletcher conceded, may have been due to the nature of the terrain.[9] Yet ultimately, notwithstanding Lowe's improvements, difficulties of transportation, the time engaged in inflation, and meteorological and topographical limitations were still formidable obstacles to the balloon's utility

[9] Report dated 3 June 1862, PRO, AIR 1 2404/303/1.

on the battlefield. Nevertheless, the achievements of this small irregular band of aeronauts directly inspired the first serious agitation for an air arm in the British army.

European developments

By the time of the Franco-Prussian War there was a wider appreciation of the balloon's military potential among most European armies, and the Prussians were to call upon the services of the English aeronaut, Henry Coxwell, to help them establish a balloon corps for that campaign. As the British army was also to consult Coxwell about its own early aeronautical investigations a brief profile might be useful.

Coxwell was born in March 1819, at Wouldham, on the Medway in Kent, the son of a Royal Navy commander. After leaving school he ended up, rather incongruously, as a dentist. As a boy, however, he had witnessed the ascents of many of the great aeronauts of the day – men like Charles Green and John Hampton – and from an early age he was drawn to aeronautics. This was not the sort of calling to which young men of respectable pedigree, let alone dentists, were expected to admit: some tact was therefore required in pursuit of the object. Furtively, Coxwell made the acquaintance of Hampton. By 1844 he was assisting him in the construction of a new balloon. In August of that year, under the pseudonym of Henry Wells – and with Hampton as pilot – he made his first ascent.[10]

Intoxicated by the experience Coxwell never looked back. In 1845, still under the pseudonym of Wells, he attempted to found the world's first aeronautics periodical – *The Balloon or Aerostatic Magazine* (*sic*). Unfortunately it was discontinued after just four editions, but this did nothing to check Coxwell's enthusiasm. Within a very short time he was one of the country's most proficient aeronauts. With growing confidence he crossed to the continent in 1848, and for the next three years toured Europe, and in particular Germany, demonstrating, amongst other things, what he perceived to be the balloon's military potential. This included the dropping of explosives from the air. Coxwell was, therefore, well-known to the Prussians as an advocate of war ballooning long before the conflict of 1870. He became, indeed, the subject's most prominent authority, publishing in 1854 a pamphlet entitled *Balloons for warfare*, which criticised those military personnel unable to grasp the significance of the new medium.[11]

By this time Coxwell had thrown off the cloak of pseudonymity and become a self-professed professional aeronaut. He made regular public ascents from

---

[10] Henry Coxwell, *My life and balloon experiences, with a supplementary chapter on military ballooning*, London 1887, 48ff; J. E. Hodgson, *The history of aeronautics in Great Britain*, Oxford 1924, 264–5.

[11] *Balloons for warfare: a dialogue between an aeronaut and a general*, Tottenham 1854.

both London and the provinces, countering, so he hoped, the sensationalist showground impression created by his more disreputable colleagues. In association with the distinguished meteorologist, James Glaisher, for example, he undertook, from 1862, a celebrated series of investigatory ascents into the upper atmosphere. This was in a specially constructed 90,000 cu. ft balloon labelled, not inappropriately, the 'Mammoth'.[12] Coxwell's reputation seemed to grow in proportion to his balloons, for in this same year his persistent lobbying of the War Office also began to pay dividends. When, through the agitation of a Royal Engineers officer, Captain Frederick Beaumont, and of others, the Ordnance Select Committee was ordered to report on the whole question of military ballooning, it recommended not only that preliminary experiments be initiated at Aldershot but that Coxwell be called in to undertake them.[13]

Little wonder, then, that when the Prussians began preparing an aeronautical contingent for the war of 1870 they too should have turned to Coxwell for advice. It was in Germany, after all, that he had first publicly demonstrated many of his ideas. The fruit of this union, the first German Balloon Corps, was composed initially of two companies of twenty men each, commanded by a colonel and a lieutenant respectively. The actual balloons Coxwell supplied, however, seem to have been somewhat makeshift, consisting, according to the Prussian aeronaut, Captain Alfred Hildebrandt, of a 40,000 cu. ft and an uneven 23,500 cu. ft capacity.[14] But after preliminary trials it was decided that these twenty-men detachments were insufficient for the task of managing both balloons anyway, so the force was amalgamated and sent to the front with just the smaller model. It was subsequently employed in operations around Strasburg. Here initial ascents were promising enough, but major difficulties arose when rapid transportation and reinflation were required. In fact, these problems effectively precluded the corps from playing any further part in the campaign. Even when the detachment at last reached Paris it was unable to reinflate, so the unit was disbanded, with little opposition, on 10 October 1870.[15]

This failure to accomplish much of any tactical value on the battlefield, largely – as certainly the British were to realise – through the lack of adequate accompanying apparatus, was soon overshadowed by the communications role foisted on the balloon by the siege of Paris. The French authorities had been

---

[12] Henry Coxwell, My life and balloon experiences, London 1889 (second ser.), ii, 92ff; James Glaisher, Travels in the air, London 1871, passim.

[13] Coxwell was always adamant that this recognition by the authorities came only after many years of sustained and 'vigorous advocacy' of military ballooning on his part: H. Coxwell to B. F. S. Baden-Powell, dated simply 1897, RAeS archive.

[14] Hildebrandt, Airships, 141. See also Maj. W. L. Möedebeck, 'The development of aerial navigation in Germany', RAeS Journal vi (1902), 25, and Coxwell, My life and balloon experiences (second ser.), ii, 255–6.

[15] Hildebrandt, Airships, 141–2; Möedebeck, 'Development of aerial navigation in Germany', 25.

essentially negative about the use of aerostats prior to the war, but once Paris was surrounded these same aerostats became the capital's principal means of communicating with the outside world. Consequently, balloon workshops and an aeronautical training centre for volunteer pilots were hastily established at the by then redundant Parisian railway stations, much of the organisation being supervised by the veteran balloonist, Eugene Godard. Ultimately, some sixty-five coal-gas balloons were to leave the city during its investment, including one carrying the Minister of the Interior, Léon Gambetta, who aimed to continue the war. In undertaking such a voyage, however, one was literally throwing caution to the wind, for, as this episode illustrated, spherical balloons were non-navigable. One could drift out of Paris, but that was all. Even so, fifty-nine of these outward and, for the most part, varnished cotton balloons did somehow manage to land safely in friendly territory – although usually following alarming journeys. (Coal-gas gave rise to erratic movements in the air and was highly combustible.)[16]

Limited though the effect of all this activity may appear now there is evidence that the possibilities of what might have happened – and more important, what could in future happen – were not lost on the European nations. This is apparent when much of the correspondence relating to the Hague Peace Conference of 1899 is examined. This was the first occasion on which the idea of prohibiting the discharge of projectiles from the air was seriously mooted. In discussing the 1870 precedent the British delegation privately surmised that, while it was true that had the besiegers of thirty years ago had 'the artillery of today' (i.e. 1899) Paris would have fallen in half the time, 'if balloons, dropping heavy charges of . . . explosive on the camps, barracks, arsenals and c. within the investing lines had been used as an adjunct, it is improbable that the place would have held out much longer than a week'.[17]

The trouble is that whoever drafted this statement clearly had no experience of balloons, for their principal feature – certainly in 1870 – was their essential non-navigability. It would have been impossible to direct such aerostats as then existed into the sort of precision bombing suggested here, or indeed into any useful bombing at all. It was simply anachronistic to suggest otherwise. On the other hand, the thirty years since the siege had been dominated by the search for aerial control, and the Hague Conference coincided with the apparent achievement of this through, for example, the emergence of the Santos-Dumont and Zeppelin airships. That was what was beginning to frighten people, and was making projected reconstructions of what supposedly could have happened in 1870 appear worse than the case in reality warranted. Given the same set of circumstances in 1899 then the mayhem the writer goes on to portray as possible in 1870 might very well come to pass – or so it seemed. Recall the exasperation occasioned by the bombardment of Paris, the

16 Hildebrandt, Airships, 143; M. Howard, The Franco-Prussian War, London 1961, 325–6; A. Horne, 'By balloon from Paris', History Today xiii (1963), 441ff.
17 PRO, FO 412/65, 101.

document urged; then imagine a population subjected to the perpetual discharge of explosives from the air. The city would be laid bare and lost from the moment of its investment.

This context, the search for aerial control and the emergence of the airship, should be seen as the backdrop against which all military ballooning developed after 1870. That said, however, military aeronautical establishments over this period remained primarily ballooning establishments. This was because balloons played specific ground-support roles on the battlefield. They either spotted for artillery units or undertook tactical reconnaissance duties. This can be seen in the emergence of a British air arm in the years following the US Civil War and the Franco-Prussian War, and it was in order to facilitate this practical work that the Central Establishment for Military Aerostation was instituted by the French army's Corps of Engineers in 1877. This establishment readopted the Château-Meudon site, or more accurately the park of Chalais-Meudon, which had been military land for some time, and which contained workshops and barracks, which were suitably refurbished. A German balloon corps – modelled to some degree on its British counterpart, with its method of transporting compressed gas in steel cylinders – was similarly readopted some seven years later. It was administered by, and wore the uniform of, the Railway Regiment, which had diverse duties not dissimilar to those of the Royal Engineers. It grew steadily throughout the remainder of the century, from four officers and ten men in 1884, to five officers and fifty men in 1886, six officers and 140 men in 1893, finally becoming a battalion in 1901.[18]

By the late nineteenth century, then, experienced aeronautical establishments would exist within most of the major European armies. More so, ironically, than in America, where a balloon company had to be hurriedly extemporised for the Spanish-American War of 1898. It would be the task of these organisations to adapt to new technological developments as they occurred.

British developments

It was the use of captive balloons in the American Civil War that initially brought the subject to the attention of the British military authorities. Captain Frederick Beaumont RE, who, like Lieutenant-Colonel Fletcher of the Scots Guards, had been a military observer with Lowe's Federal Balloon Corps, and a fellow sapper named Lieutenant George Grover, began a persistent campaign of agitation from 1862 for the introduction of a similar organisation into the British army. Partly, one suspects, in an effort to mollify these particular officers, the War Office instructed the Ordnance Select Committee, of which both

---

[18] Hildebrandt, *Airships*, 158–60; Möedebeck, 'Development of aerial navigation in Germany', 25. Note also Committee on Ballooning report, 4 Jan. 1904, PRO, WO 32/6930, and Lt. H. B. Jones, 'Military ballooning', *Journal of the Royal United Services Institution* xxxvi (1892), 276, for some consideration of German developments.

Grover and Beaumont were members, to look into the matter.[19] To this committee Lieutenant Grover then presented a paper on 'The employment of balloons in warfare', reiterating instances of their use in recent campaigns and drawing predictably positive conclusions. The other members seem to have been duly impressed: 'The Committee are of [the] opinion that there are many critical occasions when a captive balloon at a moderate elevation might render very valuable services to an army', they concluded, allowing in some instances for topographical limitations. Should a commander in the field feel that his operations would be facilitated by the use of such a resource, 'it should be furnished as [a] matter of course'.[20]

The committee had to concede, however, that although the subject was 'no longer an experimental one', extensive trials would still be necessary to determine the best means of managing balloons in the field. They therefore asked to be allowed to communicate with 'some experienced aeronaut' on the matter. This was the genesis of the 1863 Coxwell trials at Aldershot and Woolwich, the first official military aeronautical trials in the UK. Coxwell, with no small sense of his own importance, supplied the balloons and, on both occasions, Beaumont and Grover ascended with him.

Any enthusiasm stimulated by these trials was soon tempered by an understandable and, given the constraints the War Office was working under, justifiable sense of official caution. Costs had to be calculated against likely efficiency in the field, which had certainly not been proven. This was very far from the blinkered conservatism sometimes portrayed. Indeed, although the War Office were as yet unaware of the fact, the war-strained American military authorities were themselves on the point of abandoning their balloon corps. It was not simply the ascents that were the problem, but the means of generating, storing and transporting hydrogen, and the lack of a really impermeable balloon material.[21] The War Office's chemist, Frederick Abel, when ordered to consider these matters, reported that such means could not be 'extemporised', so, for financial reasons, it was decided not to pursue the subject, this being considered unnecessary in a time of 'profound peace'.[22]

In the event, profound peace in Europe was not to last very long. In 1870, following the outbreak of the Franco-Prussian War, the Royal Engineers' Committee was reconvened to consider recent technological improvements in field equipment and a permanent Balloon Sub-Committee developed from

[19] Maj.-Gen. W. Porter, History of the Corps of Royal Engineers, ii, London 1889, 190.

[20] PRO, AIR 1 2404/303/1. Grover's paper, retitled 'On the uses of balloons in military operations', was subsequently published in the Professional Papers of the Royal Engineers xii (1863), 71–86.

[21] Beaumont gave an interim account of the Coxwell trials in an appendix to his paper 'On balloon reconnaissances as practised by the American army', but their modest aim, as he now saw it, was merely to establish that ascents 'really [do] convey the advantages . . . represented'. Maintenance in the field was another question. It was, however, this latter difficulty that the trials were originally appointed to investigate.

[22] Broke-Smith, 'Early British military aeronautics', 2.

it. This consisted initially of the original Woolwich balloon committee, comprising Grover, Beaumont and Abel, but Grover was almost immediately succeeded by Major P. H. Scratchley, and in 1873 Lieutenant C. M. Watson replaced Beaumont.[23] From the sub-committee then emerged, in 1878, the Balloon Equipment Store at Woolwich, which was made responsible for the experimental production of balloons and accompanying field equipment. Over these years it was Watson who took up the mantle of the Royal Engineers' leading balloon advocate, but his name was associated, from 1875, with that of a militia captain, J. L. B. Templer, whom both Watson and Abel were now instrumental in getting attached to the new Equipment Store. With that establishment's initial grant of £150 practical aeronautics in the British army may be said to have truly begun.

The Templer era

James Templer may not have emerged from the Victorian army's enormously resourceful Corps of Royal Engineers but he was to become the driving force of British military aeronautics for the next quarter of a century. He had, in fact, been born into a wealthy, predominantly naval, family at Greenwich in May 1846, and his first inclinations were towards a naval career. His father, however, determined that the boy should follow him in reading for the Bar, so he was despatched, with little enthusiasm on his part, to Harrow and then Cambridge. Only then did Templer manage to break the shackles of his imposed vocation. Turning his back on the Inns of Court he plumped for the army. Presumably because the usual channels of entry were closed to him, he obtained a commission in the militia (the 2nd Middlesex Militia, subsequently incorporated into the prestigious King's Royal Rifle Corps – the 'Green Jackets').[24]

According to his daughter, Templer's interest in aeronautics stretched back as far as his schooldays at Harrow, whence he would abscond to witness the London ascents of such figures as Charles Green and Henry Coxwell.[25] Suitably inspired, he then seems to have taken up ballooning privately soon after leaving school. By the time he joined the Equipment Store he was both an enthusiastic and skilled aeronaut. Pending construction of the Store's first balloon, indeed, it was on Templer's own 25,000 cu. ft coal-gas 'Crusader' balloon that the experimental team practised.

The first British army balloon, the 'Pioneer', was to be of a 10,000 cu. ft capacity and made from a fine varnished cambric linen. Presumably as a result

---

[23] For details of the Royal Engineers' Committee Balloon Sub-Committee see Watson's own paper, 'Military ballooning in the British army', *Professional Papers of the Royal Engineers* xxviii (1902), 40, and Broke-Smith, 'Early British military aeronautics', 2. Note also PRO, AIR 1 686/21/13/2245, and Stanley Lane-Poole, *Watson Pasha: a record of the life-work of Sir Charles Moore Watson*, London 1919, 36, 85ff.

[24] Foreword, Blakeney reminiscences, Templer papers, 7404/59.

[25] Statement by Mrs U. Goold, Sept. 1962, ibid. 7404/59/3, 35ff.

of the delay in its construction, Templer, accompanied by Captain Henry Elsdale RE, initiated practical inter-service training by taking his antiquated 'Crusader' balloon to the Volunteer Review at Dover in 1879 and, similarly, to Brighton in 1880. Little record remains of its performance on these occasions, but Coxwell seems to have been at Brighton and he reported it as 'a great success'.[26] That same year a balloon detachment participated in the Aldershot manoeuvres.

The Balloon Equipment Store was transferred in 1882 to a more commodious site at Chatham, where it came under the direction of the School of Military Engineering. Here the work was formally divided into two distinct channels, research and development on the one hand, and practical field work on the other. Templer was given charge of the first, through control of a small workshop; the latter was left to regular Royal Engineer officers. Unlike Templer, however, these officers were still obliged to undertake tours of foreign service, so they were regularly rotated. Thus the command alternated initially between Majors Henry Lee, Henry Elsdale and the recently promoted Charles Watson.[27]

The first balloon made at Chatham, the 'Sapper' (of 5,600 cu. ft capacity), was constructed of silk soaked in linseed oil, but this material was still felt to be insufficiently impermeable.[28] This problem became the bane of Templer's existence until he happened upon a Jewish-Alsatian family, the Weinlings, in the East End of London, who were in the habit of constructing toy balloons by a process of envelopment known as goldbeater's skin. This terminology derived from the method used to apply gold leaf, but the material was actually taken from the lower intestine of an ox. When grafted together in small layers it proved considerably more impervious to hydrogen than any other fabric, in addition to being uncommonly light and strong. Templer promptly employed the entire family and shifted their operation piecemeal to Chatham, where work was begun on a large balloon envelope. This eventually emerged as the 10,000 cu. ft 'Heron' balloon. Tested alongside the 'Sapper', it proved in every respect superior. The new method of construction was formally adopted without delay.

The goldbeater's skin manufacturing process was cloaked in secrecy, the formula being known only to Templer and the Weinling family. It was not the material that was the mystery so much as the method of superposing the skins. The Weinlings were fiercely jealous of this skill. Only with the greatest difficulty could they be persuaded to initiate new hands into the ritual, and at

26 'The day was just made for the experiment', he reported to B. F. S. Baden-Powell: 30 Apr. 1880, RAeS archive. Baden-Powell later recorded, more dispassionately, that poor early conditions and the limiting effects of gun-smoke gave way to a clear afternoon, leading to useful observations: 'Military ballooning', 742. This view is corroborated by a report in the *Argus* newspaper of 30 Mar. 1880.
27 Broke-Smith, 'Early British military aeronautics', 4–5; Watson, 'Military ballooning', 45; PRO, AIR 1 728/176/3; AIR 1 686/21/13/2245.
28 Ibid. AIR 1 686/21/13/2245.

the slightest provocation they would down tools. On one occasion, indeed, work on the 'Heron' balloon was seriously disrupted because the principal hand was jailed for three months for an assault on the police. Templer, grasping the inevitable nettle, was compelled to ask the president of the Engineers' Committee in October 1883 to assign two sappers to the task instead.[29] This was duly sanctioned, but only in the face of determined resistance from the formidable Weinlings.[30]

Unfortunately, this family paranoia was eradicated only to be replaced by a War Office version and, incredibly, it was Templer himself who next fell a victim of it. Some five years after the 'Heron' incident rumours emerged that a foreign power was seeking to emulate the goldbeater's method of manufac-ture, and through, it seems, the imputations of a brother officer, Templer was suspected of having leaked the information. Worse than that, he was accused of selling balloon secrets to a foreign government. As a result, to his lasting anguish, he underwent trial by court martial in April 1888. The fact that he was honourably acquitted did nothing to lessen the pain of this occasion. The affair should never have got to the trial stage; it seems to have been a complete false alarm. In reality, the secret of the goldbeater's method of construction was kept until it became irrelevant.[31]

Three sizes of balloon were standardised under the process: 10,000, 7,000 and, though seldom employed, 5,600 cu. ft capacity. It remained, however, an enormously expensive method of construction. A 10,000 cu. ft balloon was estimated in 1900 to cost in the region of £1,000.[32] There also remained the difficulty of how to form mobile field units from these balloons. Lack of portability – not just of the aerostats, but of hydrogen supplies and generating plant – was still the most compelling objection to any attempts to maintain aeronautical units in the field. Mobility would be of particular importance in those colonies where the British army was most frequently employed. So Templer next bent his mind to overcoming this problem.

The possibility of transporting compressed hydrogen in steel cylinders had been under direct investigation at Chatham since 1882. When, in February 1884, this method was finally adopted, Templer could at last set about designing

[29] Memo dated 22 Oct. 1883, ibid. AIR 1 728/176/3; Watson, 'Military ballooning', 45–6.
[30] Col. H. B. Jones, 'Reminiscences of early balloons', PRO, AIR 1 2310/220/1. Weinlings continued to be employed as balloon-construction supervisors as late as 1906 (ibid. AIR 1 728/176/3/31), and the last of the family was not to retire from the Royal Aircraft Establishment until 1959: *RAE News*, May 1959.
[31] Blakeney reminiscences; statement on court martial, Chatham, 5 Apr. 1888, Templer papers, 7404/59/3. See also Templer's obituary, *Royal Engineers Journal* xxxviii (1924), 144. Interestingly, in his 1896–7 glider patent submission Pilcher speculated on the possibility of employing goldbeater's skin as wing covering material, but his use of the term 'may be used' suggests only limited knowledge and perhaps a vague hope of official help: P. Jarrett, *Another Icarus: Percy Pilcher and the quest for flight*, Washington, DC 1987, 47, 163.
[32] 'Scouts in the sky', *Daily Mail*, 9 Mar. 1900. See also PRO, AIR 1 686/21/13/2245, and AIR 1 2310/220/1.

transport units round it.[33] By this means he was able to circumvent the need for a portable generator. Whether the authorities were initially obstructionist over sanctioning these innovations is difficult to say, but certainly some of Templer's subordinates thought so. Robert Blakeney, who was subsequently to command the 3rd Balloon Section in the Boer War, indicated as much in a private memoir of his former chief written in 1925.[34] Judging from Blakeney's invectives, however, it seems that this suspicion was exacerbated by inter-service rivalry over the number of horses to be assigned to this new method of carriage. It was partly to alleviate this problem that Templer began developing the use of steam traction engines.

The practical adoption of the latest innovations was further handicapped by the ambiguous administrative position of the Store's actual field unit. Templer, of course, had no influence here. It was left to the School of Military Engineering to detail working parties, and their trials had to be confined within its limited resources. In January 1883 the idea of a semi-autonomous Balloon Section was expressly rejected.[35] A little over a year later Major Elsdale urged, at the very least, the permanent retention of two or three officers, but this too was rejected. Elsdale was worried by the extent to which general corps duties were drawing designated personnel away from essential aeronautical training. His own temporary command of the unit was itself subsumed within wider corps responsibilities. Templer, at the Store, was more fortunate; being a militiaman rather than an Engineer he was not subject to the same adminis-trative round. In June 1884 his continued retention was approved.[36]

Soon the embryonic field unit was experiencing its first taste of active service, further emphasising the need for cohesion within the army's collective aeronautical organisation. In fact a balloon unit of some sort had been tentatively included in the lists of potential imperial expeditionary forces for some time. As far back as 1873 plans were drawn up for a detachment to accompany Sir Garnet Wolseley on his Ashanti expedition, but upon calcula-tion of costs as against likely efficiency, the idea was abandoned.[37] This was, of course, a full decade before the development of either goldbeater's skin or portable gas cylinders. Then, in 1882, the year the Equipment Store was transferred to Chatham and the celebrated soldier-adventurer Frederick

---

33 Memo dated 25 Feb. 1884, ibid. AIR 1 728/176/3. See also Broke-Smith, 'Early British military aeronautics', 5, and Watson, 'Military ballooning', 42. The latter recalled discus-sions on this subject as early as 1875, but no action was taken until the following decade.
34 Blakeney reminiscences.
35 Memo dated 15 Jan. 1883, PRO, AIR 1 728/176/3.
36 Memo dated 14 June 1884, ibid. He was subsequently, in 1887, gazetted 'Instructor in Ballooning to the Royal Engineers'.
37 Watson, 'Military ballooning', 42; Broke-Smith, 'Early British military aeronautics', 3; Lane-Poole, Watson Pasha, 36–7. Watson had asked Coxwell to supply two balloons for the expedition, but Abel subsequently vetoed the idea. He was right to do so. The environment would not have been conducive to cumbersome balloon support. See Sir Evelyn Wood, From midshipman to field marshal, i, London 1906, 256ff.

Burnaby of the Blues succeeded in crossing the channel by balloon, a detachment was placed on stand-by for service in Egypt. On 11 September the order was given for Captains Lee and Templer and Lieutenants Trollope and Hawker to proceed. Four days later, however, the instruction was revoked. Wolseley's destruction of the rebel Egyptian army at Tel-el-Kebir, on 13 September, had obviated the need for reinforcements. The aspirant aeronauts could only look on as their best chance of vindication to date evaporated before them.[38] Undeterred, they continued to scan distant horizons for signs of unrest; but when, at the close of 1884, readiness did finally lead to embarkation, two campaigns erupted virtually simultaneously, severely straining the Store's meagre resources. These campaigns were centred on Bechuanaland, in South Africa, and Suakin, in the Sudan. No sooner had Major Elsdale and Lieutenant F. C. Trollope been despatched to the former, with a ten-man detachment and the Store's best equipment, than Major Templer and Lieutenant R. J. H. L. Mackenzie were ordered to hold themselves in readiness for the latter. In the event, only eight NCOs and Sappers could be found for the Sudanese expedition.[39]

For all the first detachment's preparedness, however, no fighting actually occurred in Bechuanaland – Rhodes's crucial 'Suez canal to the interior' – where just three years after the First Boer War and simultaneously with the Convention of London tensions had arisen between the indigenous population and Boer freebooters from the Transvaal, leading to the illegal establishment of two new Boer republics, Stellaland and Goshen, in the territory and well-founded government fears of a concerted German infringement from South West Africa. The unit's influence was consequently limited. None the less, using compressed hydrogen from portable steel cylinders a number of ascents were undertaken, including on one notable occasion near Mafeking, when the commander of the field force, Major-General Sir Charles Warren, and members of his staff viewed the surrounding terrain. Lieutenant Trollope reported that at an altitude of 1,000 ft a range of some ten to twelve miles was possible. (Warren had previously been Chief Instructor in Surveying at the School of Military Engineering, so he will have been familiar with the work of both the Balloon Store and its field counterpart. On this occasion, which was more in the manner of a demonstration than a serious reconnaissance, he ascended in the 10,000 cu. ft 'Heron' balloon, constructed at Chatham only the year before.)[40]

The balloon contingent of the Sudanese expedition, on the other hand, was

---

[38] Memos dated 21 Aug., 11, 15 Sept. 1882, PRO, AIR 1 728/176/3. See also Watson, 'Military ballooning', 44, where he remarks that this lost opportunity was 'much to be regretted'.

[39] Memo dated 18 Nov. 1884 (for Bechuanaland), 9 Feb. 1885 (for Suakin), PRO, AIR 1 728/176/3; Watson, 'Military ballooning', 47–8. Note also B. Robson, 'Mounting an expedition: Sir Gerald Graham's 1885 expedition to Suakin', Small Wars and Insurgencies ii (1991), 232–9.

[40] For Trollope reference see H. B. Jones, 'Military ballooning', 279; for Warren's ascent

hampered from the start by substandard equipment and inadequate transport. Nevertheless, operating from the Red Sea littoral port at Suakin, the detachment formed a more immediately integrated part of the expeditionary force it was accompanying, the Eastern Sudan Expeditionary Force. Commanded by Lieutenant-General Sir Gerald Graham, who had led a smaller expedition to the area the year before, the aim of this latest campaign was threefold: to protect the Nile flank, where Wolseley's Khartoum relief force was still operating, despite the death of Gordon; to crush the forces of the local Dervish leader, Osman Digna; and to begin construction of a Suakin–Berber (Red Sea to the Nile) railway.

Graham's revamped force numbered some 13,600 fighting men and included colonial contingents from India and, for the first time, from Australia. Having disembarked at Suakin on 12 March (the balloon detachment had left England on 19 February 1885 and reached Suakin on 7 March), Graham confronted an enemy force on 20 March, at Hashin, some seven miles west of Suakin, hoping to follow this up with an advance on Osman Digna's main camp at Tamai, fourteen miles south-west of Suakin. As a preliminary, however, zeribas, that is enclosed fortified camps constructed from the indigenous mimosa thorn-bush, which had the consistency of barbed-wire, were to be maintained *en route*.

On 22 March the first zeriba, at Tofrek, seven miles south-west of Suakin, was subjected to a ferocious attack from approximately 2,000 Dervish warriors. It survived, but was badly mauled. Some 150 men had been killed and 300 wounded. As a result, a series of relief and resupply convoys were despatched from Suakin. The earliest of these, on 23 and 24 March, were themselves harassed by the enemy, so a tighter square formation – incorporating captive balloon support – was adopted to escort a fresh convoy of some 2,250 men, 500 camels and seven carts on the 25th. During this advance Lieutenant Mackenzie reported from the detachment's raised balloon-basket, which was maintained at an altitude of between 300 and 400 ft, while Templer operated the ground-support. Several Dervish patrols were identified, but the square traversed both legs of the journey unmolested.[41] An additional unsupported column the following day was not so fortunate. (Given that General Graham attributed the previous convoy's success to the inflated balloon frightening the natives, it seems odd that he should not have re-employed it on the next day. Perhaps, as elsewhere, winds proved prohibitive.)[42]

The disputed causes of the near disaster at Tofrek led to much bitterness between the local commander, Major-General M'Neill, and General Graham. Graham, in his official despatch of 28 March 1885, commented with regard to

see *Graphic*, 6 June 1885. Note also Maj. H. Elsdale, 'Military ballooning', *Minutes of Proceedings of the Royal Artillery Institution* xvii (1890), 47; and PRO, AIR 1 2310/220/1.
[41] *Royal Engineers Journal* xv (1885), 114.
[42] Col. R. H. Vetch, *Life, letters and diaries of Lieutenant-General Sir Gerald Graham*, Edinburgh 1901, 459: final despatch, 30 May 1885.

the area's terrain that the 'density of bush [was] so great that even mounted men [were] unable to see beyond a very short distance'. This being the case, and considering the inherent slowness of any telegraph cable-laying zeriba force, the Suakin balloon detachment should perhaps have been employed from the start. Scouts must have reported the problems of ground-level visibility. The Assistant Adjutant-General, Intelligence Department, Suakin, was none other than Major George E. Grover RE – the military balloon advocate of the 1860s. It can be safely assumed that he would have pressed the case for balloon support whenever the opportunity arose.

On 2 April 1885 General Graham marched a strong square – eventually amounting to some 8,600 men, 1,639 camels and 930 mules – from Suakin to Tamai, in the hope of at last encountering the main force of Osman Digna's army. This time the balloon detachment was included from the beginning. In the event its performance demonstrated both the strengths and weaknesses of captive balloon support. Despite tearings from the local vegetation effective reconnaissances were made during the approach, but they were prematurely curtailed by increased winds. It made little difference. By the time Graham reached Tamai the following morning the site was deserted, and he was not to get another chance. General Wolseley arrived at Suakin on 2 May to begin the withdrawal of the expeditionary force. General Graham himself had left the area by 17 May. It cannot be said that the campaign had enhanced his reputation, although, as a Royal Engineer, he had at least been willing to actively employ the balloon detachment. The detachment's most notable achievement, however, proved to be almost incidental to active operations. On 25 April, at Tambuk, some twenty-eight miles west of Suakin, a little six-stone Arab, known as Ali Kerar, who was employed as a light aeronaut, made a 2,000 ft ascent in the modest 5,600 cu. ft 'Fly' balloon and reported in the process verifiable observations of the whole of the distance between Tambuk and Suakin. This event became something of a legend within the unit.[43]

Despite the limited nature of these colonial campaigns, some degree of official satisfaction at the balloon's performance can be deduced from Templer's subsequent mention in despatches.[44] But if the Balloon Store's recent introductions of goldbeater's skin and gas cylinders had proved for the most part viable under service conditions, then the damaging absence of permanently constituted field units was equally well demonstrated – at least to those prepared to consider the subject *in toto*. Critics sceptical of the balloon's military utility were apt to see in the two detachments' extemporised nature

---

[43] Broke-Smith, 'Early British military aeronautics', 7; PRO, AIR 1 728/176/3/1. For the detachment's performance in the campaign generally see RE Committee extracts, report of Maj. Templer, Balloon Detachment, Suakin Field Force, Apr. 1885, Templer papers, 7404/59/3. See also *Royal Engineers Journal* xv (1885), 119; Porter, *History of Royal Engineers*, ii. 193–4; and B. Robson, *Fuzzy-wuzzy: the campaigns in the Eastern Sudan 1884–85*, Tunbridge Wells 1993, 143–4, 148, 152.
[44] Templer's obituary notice, *Times*, 4 Jan. 1924.

proof of their essential impracticability on campaign, whereas in fact their expedient nature stemmed directly from the administratively ambiguous position of the original peacetime field unit. Renewed calls for the unit to be placed on a more autonomous footing met with little immediate success,[45] but were not entirely without effect.

In February 1886, when balloon manufacture and primary training were confirmed at Chatham, it was decided to send two detachments to Aldershot annually, for practice 'as nearly as possible' under service conditions.[46] This merely underlined the inadequacy of normal training. It was clear that what was needed was not so much extracurricular excursions from Chatham as a permanent aeronautical training ground. Discussions were soon begun as to its possible location. Imagining the authorities to be dithering on the matter, however, Templer characteristically attempted to force the issue. Relying on what he took to be verbal assurances of remuneration, he acquired, entirely by his own means, a farm at Lidsing, near Chatham, and here he established a balloon camp. (The site had evidently been used intermittently for training purposes.) If he really imagined this ploy was going to work he was soon disillusioned; not only did the *fait accompli* fail, but he was left in a serious financial predicament. Meanwhile the campaign for a more autonomous field unit continued.[47]

It was the reunified balloon detachment's performance in the 1889 Aldershot manoeuvres that finally tilted the balance in the aeronauts' favour. The GOC, Aldershot Division, General Sir Evelyn Wood, was so supportive of their claims in his subsequent report that the decision was taken to go ahead with the establishment of a permanent Balloon Section – within the Corps of Royal Engineers – the following year. According to numerous sources, General Wood was of the opinion that the detachment required much more integrated tactical training. This was probably the motive behind the decision to finally establish the new organisation at Aldershot, rather than in the Chatham area.[48] Difficulties arose, however, when it was decided to re-establish the Balloon Store there too. On this point Aldershot Command objected, maintaining that the district should be reserved exclusively for military training. The Balloon Store at this time hardly constituted a major manufactory, but this did not seem to matter. Sensitive noses detected the fumes of industry and

---

45 Memo dated 4 May 1886, PRO, AIR 1 728/176/3; H. B. Jones, 'Reminiscences of early balloons'; Broke-Smith, 'Early British military aeronautics', 9.
46 Memo dated 27 Feb. 1886, PRO, AIR 1 728/176/3. Similarly, a detachment was to be sent annually to the siege artillery range at Lydd: memo, 11 June 1888, ibid.; H. B. Jones, 'Reminiscences of early balloons'.
47 Foreword, Blakeney reminiscences. Watson records that the War Office did actually 'hire' this site from Templer for 'two or three seasons', but is vague about how formal this arrangement was supposed to be: 'Military ballooning', 50. Note also Lane-Poole, *Watson Pasha*, 93; and H. B. Jones, 'Military ballooning', 266.
48 H. B. Jones, 'Reminiscences of early balloons'; idem, 'Military ballooning', 269–70, 278; PRO, AIR 1 686/21/13/2245; Watson, 'Military ballooning', 52.

the Depot, as it now came to be called, was given its first intimation of the hostility that was to plague it throughout its time at Aldershot.[49]

Ironically, Templer himself (by then Lieutenant-Colonel) also objected to the move at first, preferring his own Lidsing site. Neither of these objections was sufficient to alter the decision, but they delayed its implementation. The move to Aldershot was not finally completed until 1892.[50] At the time of its removal the Chatham establishment had seventeen employees: eight men, one boy and eight women – most, presumably, members of the Weinling family.[51] Additionally, two new classes of balloon had been added to the lists: of 13,000 and 11,500 cu. ft capacity.[52] The newly-established Section and Depot consisted – excluding civilian staff – of three officers and thirty-three other ranks – all initially under the command of Lieutenant, soon Captain, H. B. Jones.[53] Facilities on the new site included a large balloon shed or hangar, of sufficient size to house several inflated balloons; a machine shop; a foundry; a carpenter's shop; and several gasometers. Horses, on the other hand, had to be borrowed from adjacent units – a 'highly unpopular arrangement', as Broke-Smith recalled.[54] Captain Jones was to command the Section from its inception to 1895, when he was succeeded initially by Captain B. R. Ward, and soon after by Major Gerard Heath. The Depot, of course, remained Templer's responsibility. At some point during this period the Store or Depot acquired what was to become its familiar title, the Balloon Factory – or Aircraft Factory, as it became – but it is difficult to determine precisely when. By 1894 we find Templer signing himself 'Superintendent, BF', but it was 1897 before this nomenclature was officially recognised.[55] It may have been essentially an institutionalised colloquialism.

The new Section enthusiastically participated in the 1893 Aldershot manoeuvres. However it failed to make any significant contribution to events. This was largely as a result of what Blakeney called the 'absurd regulation' forbidding ascents in winds of over 20 m.p.h.[56] Yet it was one of the captive balloon's inherent limitations that it could only be employed in comparatively placid conditions. This, again, underlined the need to develop alternative

---

[49] Memo dated 22 May 1890, PRO, AIR 1 728/76/3. Templer himself was said to lack tact, and was not, perhaps, the best person to calm troubled waters in this regard: see Capper's obituary of Templer, *Royal Engineers Journal* xxxviii (1924), 142.

[50] Broke-Smith, 'Early British military aeronautics', 11.

[51] Memo dated 22 May 1890, PRO, AIR 1 728/176/3.

[52] PRO, AIR 1 686/21/13/2245.

[53] Broke-Smith, 'Early British military aeronautics', 11. Writing in 1892 Lt. Jones gave the initial formation as: *Depot* – 1 instructor, 1 military mechanic, 1 clerk, 6 other ranks; *Section* – 1 Capt., 2 Lts, 1 Coy Sgt Maj., 1 Sgt, 23 other ranks, and Section transport comprising 1 balloon wagon, 4 tube wagons and 1 general service wagon: 'Military ballooning', 266.

[54] Broke-Smith, 'Early British military aeronautics', 11. For a record of Depot/Section facilities see Watson, 'Military ballooning', 52–3; H. B. Jones, 'Reminiscences of early balloons'; and AIR 1 728/176/3/35.

[55] Ibid. AIR 1 728/176/3; AIR 1 686/21/13/2245.

[56] Blakeney reminiscences, 8.

navigable devices; and it further explains why the authorities were initially so reluctant to sanction a permanent air unit. As it was, Section members were still required to train as ordinary engineers for when balloons were not in use (a stipulation which was subsequently to become a serious encumbrance to the development of an autonomous aviation unit). The alternative to captive ascents – free runs – were sometimes undertaken as part of the Section's training, but only when a sufficient quantity of gas had so deteriorated that it needed dissipating anyway. Such excursions were expensive and, given the lack of control, of limited tactical value.[57]

By the early 1890s then, a plateau of sorts had been reached. A regular Balloon Section was an established part of the British army's strength and would be included in any future campaign. As it happened, the subsequent decade proved disconcertingly peaceful. Inactivity made the airmen nervous. What had been acknowledged achievements began to look less significant as the years passed by. The 1893 manoeuvres had all too clearly identified the unit's limitations. The Section felt itself vulnerable again. Then at the close of the decade, in October 1899, the Boer War broke out in South Africa and Section members found themselves at the forefront of activity.

The Hague Peace Conference, 1899

Before examining the Section's performance in the Boer War, the views on military aeronautics expressed at the Hague Peace Conference in 1899 ought briefly to be noted. At the conference the whole question of technological advances in warfare was considered, with a view to deciding whether limitations should, or could be imposed. Actually, it is doubtful how serious many of the delegates were in their ostensible aim of achieving restraints, but all were anxious to appear to be so.[58] The British, certainly, felt they had everything to lose by any such limitations. 'Under a stationary regime of weapons and armament[s]', wrote Major-General Sir John Ardagh, the Director of Military Intelligence, 'numerical superiority in the field – in which we cannot now compete with the great military Powers – would . . . have [an even] greater influence.'[59] This was scarcely in their interests. He recommended that no restrictions be agreed relating to either the 'employment of further developments in destructive agencies', or 'the methods of employing them'.[60]

The British delegation undoubtedly aimed to follow this advice, but, in the event, the declaration specifically prohibiting 'the shooting or dropping [of]

---

[57] H. B. Jones, 'Reminiscences of early balloons'. Jones estimated that, at most, six free runs were undertaken each season.

[58] G. Best, *Humanity in warfare: the modern history of the international law of armed conflicts*, London 1983, 139–40.

[59] Correspondence, Hague Peace Conference, PRO, FO 412/65, 83.

[60] Ibid. 87.

projectiles or explosives from balloons' (or 'other novel analogous contrivances') occasioned some gentle diplomatic manoeuvrings. Initially, not only Britain, but France and Germany were against such a restriction. So, in an effort to avoid a confrontation, the principal US delegate introduced a compromise whereby prohibition would be limited to five years. Sir John Ardagh opposed even this, but eventually agreed to endorse the amendment in return for American opposition to the proposed prohibition of Dum-Dum bullets. A memo on the subject was forwarded to the Commander-in-Chief, Lord Wolseley, for comment. This suggested that the proposed restriction was in reality a penalisation of the country's scientific and manufacturing potential;[61] that being understood, the question of whether opposition ought to be temporarily withdrawn in exchange for US support of Dum-Dum bullets was probably best left with the Foreign Office – which seemed particularly anxious to cultivate friendly relations with those former colonies.

Wolseley too also initially opposed acceding to any of the three 'declarations' under consideration – prohibiting the employment of asphyxiating gases, Dum-Dum bullets and air-launched projectiles – again on the grounds that they would 'seriously hamper us as a small military power', but seemingly concurred that a final decision on the last point be left to the Foreign Office. At any rate, this is what happened, and the five year prohibition was duly ratified as part of the conference's Final Act.[62]

In theory then, the Balloon Section may have been under some restriction as to its activities in South Africa. However, little of this appears to have been conveyed to the unit. This may have been because it was essentially only a captive balloon reconnaissance formation, with little bombing potential:[63] but it was probably also because no-one accepted the Hague Declaration as rigidly binding anyway. It would, in any case, only properly apply in the event of war with a cosignatory.[64] At all events, there is no evidence that the soon-to-be-expanded sections in the field gave the matter the slightest consideration.

---

[61] Ibid. 496–7. See also Lady Susan Ardagh, countess of Malmesbury, *The life of Sir John Ardagh*, London 1909, 317–18.

[62] Correspondence, Hague Peace Conference, PRO, FO 412/65, 497 (communication dated 7 Oct. 1899); 'Final Act' of conference, in *The Hague conventions and declarations of 1899 and 1907*, ed. J. B. Scott, New York 1915, 1–31 (specifically point IV, declaration I), signed and ratified by Great Britain.

[63] The *Daily Mail* (9 Mar. 1900) reported that Templer's own 'explosive throwing experiments' were curtailed as a result of the Hague conference, but these trajectory investigations probably never amounted to much more than Templer occasionally dropping an explosive from a captive balloon. No mention is made of the field section(s), which seem to have had little or no experience in this regard.

[64] 'Pseudo-legal' was how F. W. Lanchester later dismissed the Hague decree: *Aircraft in warfare: the dawn of the fourth arm*, London 1916, 194. See also D. C. Watt, 'Restraints on war in the air before 1945', in M. Howard (ed.), *Restraints on war: studies in the limitation of armed conflict*, Oxford 1979, 60.

The Boer War

On the outbreak of war in South Africa it was immediately clear that the Balloon Section would not be able to meet all the demands now to be made on it. In September 1899 the entire balloon organisation (Section and Depot) had a strength of just four officers and forty other ranks, with no indigenous mounted formation[65] – and its services were required in more than one area. So the single section was split, like a genetic cell, into two units, which were reconstituted as the 1st and 2nd Balloon Sections and brought back up to strength again.[66] Transport, however, remained a major problem. Each balloon wagon was by this stage supposed to trail an equipment carriage and not less than six tube wagons, each of which carried nine 9 ft gas cylinders (six wagons supplied sufficient gas for two balloons of up to 13,000 cu. ft capacity – i.e. three wagons provided one fill): yet horses for the unenviable task of hauling all this were never scarcer. Ultimately, the sections became reliant on teams of mules and oxen, beasts which proved noticeably less amenable for the purpose. Oxen had the advantage of requiring only grazing for subsistence, but travelled at a crawling pace, necessitating either a slow column or the detachment of a strong escort. Mules, on the other hand, needed forage and were scarcer. Even oxen, however, could be, and often were, requisitioned at a moment's notice.

Balloon depots, incorporating field hydrogen factories, were quickly established at base camps in Cape Town and Durban; but unfortunately the more efficient electrolysis system of hydrogen production, perfected by Templer some three years before as an alternative to the lengthy and highly dangerous zinc-acid process, lacked sufficient portability for reassemblage on site.[67] Working under these constraints the restructured sections quickly found themselves in the van of the British army's campaign to subdue the Boer republics.

The 2nd Section was – paradoxically – the first despatched, being on its way, indeed, before the actual declaration of war. Commanded by Major Heath, supported by Captain W. A. Tilney of the 17th Lancers, and 2nd Lieutenant C. Mellor RE, it was one of the last units to reach Ladysmith before the town's investment, taking part in and contributing effectively to the preliminary battle of Lombard's Kop, known also as the battle of Farquhar's farm, on 30 October 1899. On this occasion it came under heavy fire.[68] During the subsequent siege intermittently useful observations were made, reconnoitring

---

65 Broke-Smith, 'Early British military aeronautics', 15.
66 H. B. Jones, 'Reminiscences of early balloons'; Watson, 'Military ballooning', 56.
67 Superintendent's comments, 21 Oct. 1901, PRO, AIR 1 728/176/3/3. For the background to the electrolysis system of hydrogen production see Watson, 'Military ballooning', 55, and Broke-Smith, 'Early British military aeronautics', 12. For the evolution of wagon transport see Watson, 'Military ballooning', 56ff; Broke-Smith, 'Early British military aeronautics', 13–14; PRO, AIR 1 728/176/3/2; and WO 108/246.
68 Col. S. Waller, 'History of the Royal Engineer operations in South Africa 1899–1902', unpubl. manuscript RE Library, 1904, 14.

enemy forces and, until the Naval Brigade objected, assisting in the directing of artillery fire from the heavy naval guns smuggled into the town just prior to its encirclement. Certain of the enemy's own field guns were neutralised in this way. Hostile positions, Heath later recalled, could be seen at anything up to ten or twelve miles distance.[69] Communication with the outside world and, more specifically, with the advance guard of General Redvers Buller's relief force was also maintained from the air by heliograph (a compact mirror-encasing signalling apparatus, used to transmit morse-coded messages). Unfortunately, however, much of the unit's tail had been lost in the initial rush to get into the town, and as there was no longer any means of replenishing hydrogen stocks from the depot established at Durban, the situation soon grew critical. In a little under a month supplies were exhausted and ballooning ceased.

Section personnel, being engineers, were subsequently employed in the maintenance of field defence works. This was initially simply an expedient, but as it transpired the 2nd Section was never to be revived as a balloon unit in South Africa. After the break-up of the Boer armies the campaign reverted to a style of guerilla warfare unconducive to captive balloon support, and the Section was ultimately converted to a troop of mounted Sappers – the 3rd Field Troop – attached to Lord Dundonald's Cavalry Brigade. Out of those remnants ditched by the 2nd Section on its dash to Ladysmith, however, another Royal Engineers officer, Captain G. E. Phillips, who had served with the old field unit back in 1886,[70] was able to extemporise an additional detachment for use with General Buller's relief force. He drew supplementary stores and the necessary skilled personnel from the depot at Durban, and had the luxury of horse transport for his balloon wagon. Thus, coming into service at the end of December 1899, a Royal Engineers balloon unit was able to provide some limited support at those hard-fought engagements which preceded the lifting of the siege, notably in those operations centring on the battle of Spion Kop on 24 January 1900. Captain Phillips was, indeed, wounded during an ascent

---

[69] PRO, AIR 1 728/176/3/3. There seems to have been no little antipathy between the Balloon Section and the Naval Brigade, evident in the terse memos placed before the War Office Committee on Ballooning in 1903. Indeed, Rear-Admiral (formerly Capt.) Lambton and his officers' reports verged on the insulting, eliciting a rejoinder from Lt.-Col. (formerly Maj.) Heath, communicated through Maj.-Gen. Fitzroy Hart, president of the RE Committee: WO 32/6930. Gen. Sir George White, GOC, Ladysmith garrison, mentioned the work of Maj. Heath and the Section in his official despatch of 23 Mar. 1900: WO 105/6. See also Col. E. H. Pickwood RFA, memo to War Office Committee on Ballooning, 17 Sept. 1903, WO 32/6930; Broke-Smith, 'Early British military aeronautics', 17–18; Col. Templer, 'British war balloon operations in South Africa', *RAeS Journal* v (1901), 54–6; 'Memorandum concerning the use of the captive war balloon during the siege of Ladysmith', ibid. vi (1902), 15. This last provides a slightly more critical account of the Section's performance, written, it states, 'by one of the Imperial Light Horse' who was present at the siege. The author emphasised, in particular, that observations must needs be frequent, otherwise subsequent enemy troop movements were liable to invalidate them – as at Lombard's Kop.

[70] H. B. Jones, 'Reminiscences of early balloons'.

undertaken in the 'Thames' balloon at Potgieter's Drift, on the Tugela river, on 20 January, shortly before the battle of Spion Kop. (Sir Neville Lyttelton, GOC, 4th Brigade, had effected a river crossing at this point – a few miles east of the Spion Kop position and in front of the Brakfontein track to Ladysmith – on the night of 16 January. The following day a howitzer battery took up positions along a ridge known as One Tree Hill above the river. From there they could support Lyttelton's 'demonstrations' intended to distract attention from movements further to the west, and help range on the Boer positions on the high ground above Trichardt's Drift, constituting the launching point for the attempted flanking assault on the Tabanyama plateau on 19–20th, and following its stalling, Spion Kop on the evening of the 23rd. From their own chief vantage points, the garrison at Ladysmith, some dozen miles north-east, both watched the work of, and communicated with, the improvised Balloon Section on this occasion.)[71]

Buller's force had been joined by Lieutenant-General Sir Charles Warren's 5th Division by the time of the Spion Kop debacle. Warren – who was to play a leading and controversial role in the battle – was himself an experienced officer of the Royal Engineers and had been the first British army commander to employ a balloon unit on active service. How much use he made of the resource on this later occasion is unclear, but in general his experience was not happy. Rather like McClellan, nearly forty years before, he was severely criticised for his lack of initiative. The improvised Section afterwards assisted in the abortive Vaal Krantz offensive of 5–7 February, before its period of active service concluded with the relief of Ladysmith on 28 February 1900 – after Buller had finally broken through the Boer defences on the Tugela, east of Colenso. Captain Phillips, like the members of the original 2nd Section, subsequently joined Lord Dundonald's staff.

The 1st Balloon Section, under the command of Captain Jones, supported by Lieutenants A. H. W. Grubb and R. G. Earle, had in the meantime been despatched to the Orange Free State, where, drawing its supplies from the depot established at Fort Knokke, Cape Town, it joined Lord Methuen's 1st Division on the Modder river on 9 December 1899.[72] Unfortunately Methuen came to perceive the balloon's value only as a result of his own miscalculations. In an effort to break through to Kimberley he ordered the Highland Brigade under Major-General Wauchope to advance frontally, on 11 December, over the open veld situated before the Magersfontein kopje. The aim was for a dawn attack on the southern end of the ridge, but the division as a whole was lacking

---

71 D. Macdonald, *How we kept the flag flying: the story of the siege of Ladysmith*, London 1900, 203–5; H. W. Nevinson, *Ladysmith: the diary of a siege*, London 1900, 245.
72 'No.1 Balloon Section, Royal Engineers, in the Boer War', contemporary diary of 1st Balloon Section, ed. A. Cormack, *Journal of the Society for Army Historical Research* lxviii, lxix (1990–1), lxviii. 257. The Section, it is recorded, embarked for Cape Town on 4 November with three officers and 34 NCOs and other ranks. See also report on organisation and equipment of Engineer arm, South Africa, appendix F (Capt. Jones's report, 1 July 1900), WO 108/246.

adequate reconnaissance information regarding either the terrain or the enemy's dispositions. There had previously been opportunities to carry out reconnaissances of the area, following the 1st Division's securing of the Modder river position on 28 November, and if Methuen had waited but one more day he would have had the benefit of captive balloon support: the Balloon Section was included in the order of battle and the 10,000 cu. ft 'Titania' balloon arrived in position just hours after the attack had gone in. The meteorological conditions were ideal, although Jones recorded in his diary that by the time ascents were made the enemy positions were extremely difficult to delineate. In the event, using telephonic and cable communications, the Balloon Section provided effective support for the artillery, under cover of whose fire the Highland Brigade was withdrawn on the afternoon of 11 December.

Telephonic communication was also maintained with Methuen's Field HQ,[73] but by this time there was little to be done beyond confirming that the whole attack should be broken off. From the air it was evident that the Boers were reinforcing weak points in their line of defences – Methuen acknowledged that it was from the Section that he learned of enemy reinforcements arriving from Abutsdam and Spytfontein – and on the morning of 13 December Captain Jones reported that, unlike at Modder river, the Boers had held their positions throughout the intervening nights and clearly had no intention of withdrawing. Thus, although providing active support by the close of 1899, little was actually accomplished in any strategic sense until the new Commander-in-Chief, Lord Roberts, assumed overall control and, in February 1900, began his indirect march on the Free State capital at Bloemfontein.[74]

Despite the constant problem of sickness within the ranks information obtained from balloon ascents played a crucial part in this advance. Receiving permission to join Lord Roberts's army on the evening of 19 February the Section reached Paardeberg by the morning of the 22nd. Here the retreating Boers had been trapped and encircled. The 12,500 cu. ft 'Duchess of Connaught' balloon was filled and placed in action the following day. On the 24th a 1,100 ft ascent was made and Section observers produced detailed sketches of the enemy positions. After some communication difficulties with the appropriate field batteries verifiably accurate 5 in. and 6 in. howitzer fire was then redirected onto them. This process was repeated on the 26th, when the 'Duchess of Connaught' was punctured by enemy rifle fire. Its hydrogen was transferred to the 11,500 cu. ft 'Bristol' balloon on 6 March.[75] Unfortunately

---

[73] During the battle the Balloon Section was actually situated at Methuen's HQ, and Britain's premier marquess, Maj. Lord Winchester (Coldstream Guards), was shot dead when he turned his back on the enemy to watch the service detachment at work: T. Pakenham, The Boer War, London 1979, 205–6.

[74] See PRO, WO 32/7870, WO 32/7966 for Lord Methuen's despatches. Note also memo to War Office Committee on Ballooning, 18 Sept. 1903, WO 32/6930.

[75] Cormack, 'No.1 Balloon Section, Boer War', lxviii. 261, lxix. 33; Lt. A. H. W. Grubb to Maj. F. C. Trollope, 20 Mar. 1902, PRO, AIR 1 728/176/3/9. See also Sir F. Maurice, History of the war in South Africa, ii, London 1907, 168, 173. Communication was by means

these events did not occur before General Kitchener, Roberts's Chief of Staff, had ordered a series of inexpedient assaults on the Boers' strongly entrenched laager on the north bank of the Modder. None the less, the investment resulted in the total disarming of General Piet Cronje's Boer army on 27 February.[76] The Section, attached to General French's Cavalry Division, was subsequently included in plans for the engagements centred around Poplar Grove, some fifteen miles east of Paardeberg and fifty miles west of Bloemfontein, on 7 March 1900; but once initiated, the Boers declined to give battle. The Free State capital was captured shortly afterwards.

During the following May and June the Section participated in Roberts's converging advance on Pretoria, accompanying the 11th Division under Major-General Pole-Carew. Among those who witnessed its work on this occasion was Winston Churchill, who, as war correspondent for the *Morning Post*, was attached to Ian Hamilton's column. His celebrated comparison of the ascendant captive balloon with 'the pillar of cloud that led the hosts of Israel' generally finds its way into even the most cursory books on the war, and won for the balloon sections in general a modicum of fame. The option of balloon support was provided at the engagement at Vet river on 5–6 May, and ascents were made at Zand river on 9–10 May, before the Section eventually crossed into the Transvaal at Vereeniging, after halting at Kroonstad. Following the provision of support using the 10,000 cu. ft 'Task' and 'Torpedo' balloons during the capture of Pretoria on 4–5 June, the 1st Section assisted the 11th Division – from 19 July – in the eastern Transvaal, until reaching Brugspruit, when on 3 August its transport equipment was transferred to the heavy artillery unit and Naval Brigade.[77] For his part in the campaign Captain Jones subsequently received a brevet Majority, while Lieutenant Grubb – who had been responsible for identifying the Boer positions around Paardeberg – was awarded the DSO.

Three additional balloon sections had, in the meantime, been raised at Aldershot to cover other possible contingencies. If one includes the extemporised detachment accompanying Buller's Ladysmith relief force, this meant that within the first year of the Boer War the British army was maintaining six balloon sections. Of the new batch, the 3rd Section, as it was officially designated, although it became the fourth section in service, was also despatched to South Africa, while the 4th was sent to China, for service with the British contingent of the international peacekeeping force occupying Peking

of flag signals. On 26 February Lt. Grubb – who had little signalling experience – was compelled to communicate directly with batteries on the north side of Cronje's river laager, some three miles away. This sufficed, but, Grubb noted, communicating to ground and thence to battery by field cable would have been more reliable.

[76] Pakenham, *Boer War*, 331–42; Robert's despatch, PRO, WO 105/6.

[77] Capt. H. B. Jones, 'Note on a performance of the "Bristol" war balloon during the South African campaign', *RAeS Journal* vi (1902), 65; Cormack, 'No.1 Balloon Section, Boer War', lxix. 39–43.

in the wake of the Boxer rebellion.[78] A 5th Section, an amalgamation of two nucleus sections, was then released for the Australian federation inauguration in January 1901, where it was commanded by Lieutenant T. H. L. Spaight.

The 3rd Section was placed under the command of Lieutenant (temporary Major) R. B. D. Blakeney, but his immediate subordinate, Captain Bertram Arthur Warry, was drawn from outside the engineering corps, being an officer of the Essex Regiment. The Section disembarked at Cape Town on 30 March 1900, and presently joined Lieutenant-General Sir Archibald Hunter's 10th Division, operating on the left flank of Roberts's advance on Pretoria (covering, on the way, Mahon's relief of Mafeking). As a preliminary to this advance it saw service at the battles of Rooidam on 5 May and Fourteen Streams on 6–7 May; both actions being situated at the point marking the Cape railway's crossing of the river Vaal on the border of the Transvaal Republic, where strong Boer entrenchments were in place.

Fourteen Streams had long been a British–Boer border-point, and had been the site of a celebrated meeting between Sir Charles Warren, Cecil Rhodes and President Kruger in January 1885. Warrenton, on the south side of the river, was being held by Major-General Paget's 20th Brigade, from Methuen's 1st Division, and a deadlock had ensued until 10th Division's arrival. Hunter then crossed the river further south, at Windsorten, moving up to threaten the Boer's right flank. Back at Warrenton a long-range duel was continuing between the Boers and 20th Brigade. The 3rd Section's directing of the artillery fire from 20th Brigade, south of the river, made this engagement – with Paardeberg – the balloon sections' most notable achievement. A camouflaged 6 in. gun was brought up on a railway mounting from Kimberley, and this, under the direction of the 'Thrush' balloon positioned approximately 1,200 ft above it, pulverised some particularly troublesome laagers and the enemy Commandant's HQ hidden up to 7,000 yards away. A frequent observer on this occasion was the Royal Artillery officer, Captain G. F. MacMunn, later Lieutenant-General Sir George MacMunn, as well as Captain Warry. The accuracy of the fire, commented Blakeney in his subsequent report, was a 'distinct surprise' to the Boers, who had imagined these positions to be beyond the range of the British guns. The success also surprised the Section itself. On average the sixth shot at any identified target was effective. Major-General Paget was afterwards effusive in his praise.[79] However, Blakeney's supposition, here illustrated, that balloons observing for artillery should ideally be placed directly above the guns

---

[78] The 4th Section, in the event, arrived too late to be of much practical assistance in China. Difficult terrain had already compelled the British force to devise its own portable field observation platforms. See Col. F. T. N. Spratt Bowring, 'The work of the RE in the China or "Boxer" War of 1900–1901', *Royal Engineers Journal* xiii (1911), 173. Nevertheless, the experimental Bengal Sappers' Balloon Section was drawn from this unit in 1901.

[79] PRO, AIR 1 728/176/3/3, Section I; Maj.-Gen. A. H. Paget, memo to War Office Committee on Ballooning, 24 Sept. 1903, WO 32/6930. See also WO 108/290; Broke-Smith, 'Early British military aeronautics', 18; Waller, *History of Royal Engineer operations in South Africa*, 180–4.

was challenged by both Major Gerard Heath, who commanded the 2nd Section, and the Engineer-in-Chief, South Africa, Major-General Sir Elliott Wood. Heath suggested that a position well to the flank of the guns – a mile or more if possible – would provide a better perspective, while General Wood pointed out that, wherever else they might be placed, balloons should not be so near the guns as to mark their position to the enemy.[80] It is worth noting that Wood had long taken an interest in the military use of balloons. He was a close friend of the former Balloon Section CO, Henry Elsdale, and had acted as guide to Major-General John M'Neill's zariba force in the eastern Sudan in March 1885, thereby participating in the battle of Tofrek on 22 March and witnessing the work of the balloon detachment accompanying the relief and resupply convoy of 25 March.

After the battle at Fourteen Streams several Boer prisoners were to remark on the adverse effect on morale of the British balloon reconnaissances. They often evoked, within the ranks, feelings of vulnerability out of all proportion to the practical intelligence likely to be gleaned from isolated ascents.[81] On the other hand, this last engagement had precipitated arguably the most successful balloon reconnaissance of the war, when enemy reinforcements were seen coming in from Christiana, some twenty miles away. Aware that their flank had been turned and their movements compromised, the enemy on this occasion simply left the British in possession of the field. The siege of Mafeking was lifted ten days later. Far from being complacent about his section's success in the campaign, however, Blakeney was at pains to emphasise the need for crews, in future, to continue observing activity deep in the enemy's rear – ascertaining reserves and monitoring flank movements and counter-strokes. The temptation was to reconnoitre merely those front lines and forward movements within immediate range of ground forces.[82] This was not to be the last such admonishment. One might, incidentally, imagine captive reconnaissance balloons, in these circumstances, at risk from enemy fire; but, in fact, they were immensely difficult to range on, and punctures – when they occurred – were seldom critical, particularly if on the underside, for gas rises.[83] Indeed,

---

[80] PRO, AIR 1 728/176/3/3, Section I. Capt. (formerly Lt.) Grubb, in his memo to the War Office Committee on Ballooning, was, however, adamant that balloons must be kept 'well in the advance, taking such risks as are thereby incurred', and he was dismissive of the value of ascents not undertaken in this manner: memo, 4 Sept. 1903, WO 32/6930.

[81] PRO, AIR 1 728/176/3/3. For similar evidence note also 'The role of English military balloons in the South African War' by 'Colonel Arthur Lynch of the Boer Army', WO 32/6062. For Lynch's background see Walker, *Early aviation at Farnborough: the history of the Royal Aircraft Establishment*, i, London 1971, 34–6, and Gollin, *No longer an island*, 79–81.

[82] PRO, AIR 1 728/176/3/3, Section III.

[83] Ibid. AIR 1 728/176/3/22; Cormack, 'No.1 Balloon Section, Boer War', lxviii. 261, lxix. 40; Lt. A. H. W. Grubb to Maj. F. C. Trollope, 20 Mar. 1902, PRO, AIR 1 728/176/3/9. The Boer forces besieging Ladysmith discovered this for themselves. The balloons employed there were struck two or three times, to no effect. Efforts were afterwards made to hit them during their ascent and descent: Macdonald, *How we kept the flag flying*, 49; 'Memorandum concerning the use of the captive war balloon during the siege of Ladysmith'.

far from being worried by this prospect, when questioned on the matter some years before Templer had retorted that he would be 'very much obliged' if the enemy would fire at reconnaissance balloons, for then their guns could be pinpointed.[84]

At the close of 10th Division's regular campaign the 3rd Section found itself at Potchefstroom, in the Transvaal, where Major Blakeney – who had some experience in such matters – volunteered its services for the working and general superintending of the branch line of the Netherlands railway to Johannesburg. About thirty men from various corps were attached to the Section with a few drivers and the line was worked in this way for seventeen days, with some success. The Section then proceeded to Johannesburg, from where its spare equipment and transport was transferred to the 1st Section at Pretoria. Captain Warry was also attached to the 1st Section in place of Lieutenant Grubb, who had joined the Mounted Infantry. The remainder of the 3rd Section was transferred to the Imperial Military Railways depot.

Colonel Templer, while all this was going on, was in South Africa, but as Director of Steam Road Transport. He embarked in December 1899.[85] As far as the army's aeronautical organisation is concerned he all but fades from the narrative at its most important moment to date. Given the rapid expansion of the sections it was obviously a period of frenetic activity back at the Balloon Factory, and they could ill-afford to lose him – however temporarily. It fell in the end to the noted African surveyor, Lieutenant-Colonel J. R. L. Macdonald, as Acting-Superintendent, to introduce most of the new sections; but he in turn left in August 1900, to command the 4th Section in China.[86] The vacuum was then apparently filled by Templer's old associate, Major F. C. Trollope. However, this cannot have been for long, as Templer himself was certainly back at the Factory, and very much in control again, by early 1901.[87] Templer was thus at least in a position to consolidate the expansion which had resulted from the war, although ironically his personal inclinations were by now drawn more towards the idea of developing an army airship. In recognition of the unit's performance in South Africa the 1901 Estimates confirmed the retention of six balloon sections; five full and one cadre. The idea was originally to allocate one section to each of the six army corps.[88] A mounted establishment was also, belatedly, sanctioned.

---

[84] Capt. J. L. B. Templer, 'Military balloons', *Journal of the Royal United Services Institution* xxiii (1879), 183.

[85] For his activities in South Africa see Templer papers, 7405/58/1.

[86] On arrival in China he, in fact, handed over command of the Section to Capt. A. H. B. Hume RE, who had been a subaltern in the original Balloon Section upon its formation in 1890.

[87] F. C. Trollope to E. S. Bruce, 21 Dec. 1900, RAeS archive; PRO, AIR 1 728/176/3; Broke-Smith, 'Early British military aeronautics', 20.

[88] PRO, AIR 1 2310/220/2.

Post-Boer War developments

By the turn of the century then, the sections had found what appeared to be a permanent niche in the army's lists. In reality, however, the six sections were never to be fully maintained after the war. Just how extensively the original six sections concept was allowed to lapse can be discerned from reports of their subsequent field training. Over the 1904 season work was continued with only the first three sections, the 4th and 5th simply making up the strength of these primary units as required.[89] That there was no great outcry over this is but further evidence of the growing recognition that the spherical balloon, as against other forms of aerostat, had inherent limitations. The emphasis in Europe generally was increasingly on the development of navigable airships: Zeppelin in Germany, and Santos-Dumont and Lebaudy in France, had all recently illustrated this. The British ballooning establishment could hardly ignore the trend.

In fact, there is evidence that Templer had already begun pulling the Balloon Factory in this direction. Asked by the War Office, in July 1901, what progress had been made with 'elongated balloons', Templer replied that one had actually been completed and tested on Salisbury Plain towards the end of 1899 – although he had to concede that this trial had not proved a success. (Indeed, the War Office had apparently forgotten that it had even occurred.)[90] Templer was referring here to what was known as a 'kite balloon', such as had recently been developed in Germany by Major von Parseval and Captain von Sigsfeld. This device consisted of a non-rigid elongated cylindrical balloon, which was set at an angle and which trailed a multi-kite tail. This provided increased stability, particularly in winds prohibitive to captive spherical balloons, but they were much heavier, had a lower altitude limit, and were extremely cumbersome to an army on campaign. Whitehall, however, was primarily concerned with future developments, not past disappointments. The War Office, in other words, was enquiring into the feasibility of initiating a new programme. With the war in South Africa now virtually over, resources were again available for such a project. In November 1901 Templer announced that the Factory was ready to resume experiments in this area. Having been diverted by the exigencies of war for much of the previous two years, however, he requested that he first be allowed to visit Paris, in order to ascertain the precise extent of progress on the continent. This was duly sanctioned.[91]

In December 1901 Templer met his French opposite number, Colonel Charles Renard. But on this occasion Renard denied Templer access to the actual workshops of the French aeronautical establishment at Chalais-

---

89 Ibid. AIR 1 1608/204/85/36. Note also WO 32/6930 para. 46, where it is reported that over the 1903 season the 1st and 2nd Sections had had to be maintained with personnel drawn from the 3rd, 4th and 5th Sections.
90 Memo, 9 July 1901, ibid. AIR 1 728/176/3.
91 Memo, 13 Nov. 1901, ibid. AIR 1 728/176/3.

Meudon, a fact which provides an interesting sidelight onto the state of Anglo-French relations.[92] It made little difference; general exchanges and an encounter with the Franco-Brazilian pioneer, Santos-Dumont, were enough to confirm Templer in his growing advocacy of airship development.[93] Yet no sooner had the programme been initiated than the War Office was insisting that total Balloon Factory costs for the forthcoming year would have to be halved. Templer was understandably distressed by this decision, although it does not seem to have seriously hampered his research work.[94] At this stage he was dealing essentially with abstract plans, which might or might not materialise. In any event, it would be years before any finished article became available for service. Funding would be revised many times before then.

Little is to be gained from tracing the convoluted development of the Factory's first practical airship and its successors. These aircraft were of little intrinsic aeronautical significance. Development of them, however, did mean that the internal combustion engine was to find its way into the Balloon Factory for the first time. The War Office, noting this extension of the organisation's responsibilities, promptly enquired whether the production of spherical balloons need continue at all. 'It is evidently desirable that this pattern should not be perpetuated after a satisfactory pattern of elongated Balloon has been secured', a May 1902 memo reminded Templer.[95] Such reasoning – however rational in theory – revealed in practice an enormous ignorance of the difficulties involved in devising and operating such devices. Indeed, there must have been some level of misunderstanding here, with the War Office imagining that a satisfactory pattern of kite-balloon was being developed, for it would be years before an airship of any utility was produced. Until then reports of the spherical balloon's supersession were an exaggeration, and this would remain true whether or not a pattern of kite-balloon was introduced to the service. The Royal Artillery were particularly anxious that balloon support should not be precipitantly dispensed with, as they valued the direct telephonic communication that had been established in joint RE–RA field gun observations, something neither the dirigible nor eventually the aeroplane appeared likely to provide;[96] and Colonel J. E. Capper certainly did not intend to fade spherical balloons out when he assumed overall command of the field sections in April 1903. Indeed, the question of whether some measure of balloon support should be retained arose when the Commandant of the RE's Air Battalion, Major Sir Alexander Bannerman, gave evidence before the CID flying corps technical sub-committee in February 1912. He concluded that it should be retained, at least as an option. The first thing the

---

[92] He was eventually permitted to visit them in October 1903. See Walker, *Early aviation at Farnborough*, i. 74. The file relating to this visit has since been destroyed by the PRO.
[93] Memo, 2 Jan. 1902, PRO, AIR 1 728/176/3.
[94] Memo, 14 Jan. 1902, ibid. AIR 1 728/176/3; AIR 1 686/21/13/2245.
[95] Memo, 29 May 1902, ibid. AIR 1 728/176/3.
[96] Sir J. Headlam, *History of the Royal Artillery 1860–1914*, ii, Woolwich 1937, 167.

Italian landing force did on arriving ashore at Tripoli, he noted, was to put up two captive balloons to spot for the naval bombardment.[97]

Lieutenant-Colonel John Edward Capper, the field sections' new commander, was to replace Templer as the dominant figure in British military aeronautics and to retain that position for most of the next decade. Born in 1861, the second son of a distinguished Indian civil servant, his progress through life, like that of his brothers, seems to have been untainted by doubts of vocation. He was educated at Wellington College and the Royal Military Academy, Woolwich, from where he joined the School of Military Engineering at Chatham in September 1880. Upon completion of this course, he then spent the years from 1883 to 1899 in India, where he served in conjunction with the Military Works Department on road, railway and bridge construction. At the close of this period, in 1898, he saw active service on the North-West Frontier, in the course of which he supervised construction of the first road for wheeled transport through the Khyber Pass. He was promoted Major soon after, in April 1899.

On the outbreak of the Boer War Capper was sent to the Cape, where he became Assistant Director of Railway Transport. He was responsible for raising the 1st Battalion, The Railway Pioneer Regiment, whose task it was to reconstruct those bridges and junctions previously destroyed by the enemy. Three further battalions were subsequently raised, to ensure the railway services' continued protection. Following the break-up of the Boer armies, and in recognition of his success in defending the army's transport infrastructure from Boer insurgents, brevet Lieutenant-Colonel Capper then became Chief Staff Officer to the Rand Rifles, an organisation described by his official obituarist in 1955 as 'a somewhat unruly force of some 14,000 men, collected for the defence of Johannesburg and the Rand mines'.[98]

Capper appears to have long held an interest in the military utility of balloons, but at the time of his appointment as commander of the field sections in April 1903 his official experience in this area was limited to a brief period of work with Templer at Chatham some twenty years before, when, as a mathematician then serving in the 11th Field Company, he had assisted in the construction of 'The Sapper' balloon. The only other relevant pre-sections reference to Capper in the official records relates to his having attended a short course of ballooning on Salisbury Plain in 1899, in between his Indian and African service.[99] This would seem to confirm that before the Boer War Capper had little familiarity with the practicalities of the subject – to counterbalance his evident enthusiasm and considerable administrative experience. Doubts as to his suitability for the command,[100] however, were soon dispelled. No one

---

97   PRO, CAB 16/16, 5 Feb. 1912, 102; and ibid. appendix xxxi, 'The retention of captive balloons in the British service', 245–6. See also AIR 1 823/204/5/66.
98   *Royal Engineers Journal* lxix (1955), 306.
99   PRO, AIR 1 1608/204/85/36, 38; AIR 1 686/21/13/2245.
100 Ibid. AIR 1 1608/204/85/36, 39.

Plate 1. A. V. Roe (second from right) constructing the Avro 'Friswell' triplane, Lea Marshes railway arches 1909. To the left stands the Roe 1 Triplane.

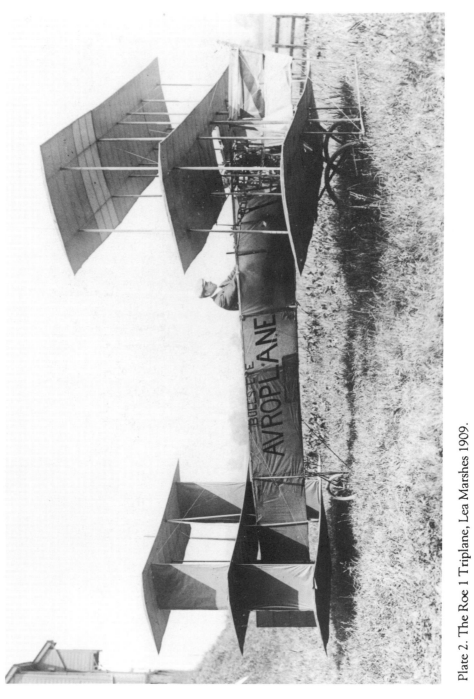

Plate 2. The Roe 1 Triplane, Lea Marshes 1909.

Plate 3. J. T. C. Moore-Brabazon at the controls of Shorts Biplane No. 2, 4 November 1909: left-hand rudder, right-hand elevator, stirrups attached to side control flaps.

Plate 4. The Wright brothers' visit to Mussel Manor, Leysdown, Isle of Sheppey, 4 May 1909. Back row, standing left to right: J. D. F. Andrews, Oswald Short, Horace Short, Eustace Short, Francis McClean, Griffith Brewer, Frank Hedges Butler, W. J. S. Lockyer, Warwick Wright. Front row, seated left to right: J. T. C. Moore-Brabazon, Wilbur Wright, Orville Wright, C. S. Rolls.

Plate 5. Shorts S27 pusher-biplane 1910, incorporating Sommer-derived mono-tailplane section.

Plate 6. Shorts S36, Short Brothers first tractor-biplane, built in 1912 for Francis McClean and loaned by him to the naval flying school at Eastchurch.

Plate 7. Frederick Handley Page in the cockpit of the Handley Page Type A ('The Bluebird'), April 1910.

Plate 8. 80 h.p. (Gnôme) Sopwith 'Tabloid'.

Plate 9. Captain Bertram Dickson on Bristol Boxkite No. 9 during British army manoeuvres, Salisbury Plain, September 1910.

Plate 10a. George H. Challenger, designer of the Bristol Boxkite, photographed in February 1911.

Plate 10b. Archibald Reith Low, designer of the Vickers EFB1, photographed in November 1910.

Plate 10c. Captain Herbert F. Wood, manager of Vickers's aviation department, photographed in November 1910, at the time he gained his aviator's certificate at British & Colonial's Brooklands school.

Plate 11. Vickers EFB2 with Vickers-Maxim machine-gun mounting and side panels.

Plate 12a. Winston Churchill, at that time Home Secretary, talking to Lord Northcliffe outside the Grahame-White sheds during the Parliamentary Aerial Defence Committee's Hendon display, 12 May 1911.

Plate 12b. VIPs at the Parliamentary Aerial Defence Committee's Hendon display: left to right, J. E. B. Seely, Under-Secretary of State for War, Lord Northcliffe, R. D. Isaacs (subsequently Lord Reading), Attorney-General, Lloyd-George, Chancellor of the Exchequer (looking skyward).

Plate 13. The close of the *Daily Mail* £10,000 Circuit of Britain race, Brooklands, 27 July 1911: left to right, Lord Northcliffe, George Holt Thomas, Claude Grahame-White, Lady Northcliffe.

Plate 14. The Balloon Section, Royal Engineers, formed May 1890, photographed at Dungeness artillery range, September 1890. Lt. H. B. Jones is seated in the centre, above lying dog: 2nd Lt. A. H. B. Hume on his right.

Plate 15. Balloon transport *c.* 1889, incorporating balloon and equipment wagons, and gas cylinder wagons: Major C. M. Watson in command.

Plate 16. Colonel J. E. Capper directs the Balloon Sections' training, c. 1904.

Plate 17a. John Duncan Bertie Fulton
RFA, photographed in November 1910.

Plate 17b. George Bertram Cockburn,
photographed in April 1910.

Plate 17c. Lt. Reginald Archibald
Cammell RE.

Plate 18. Geoffrey de
Havilland seated in the cockpit
of 70 h.p. (Renault) Royal
Aircraft Factory BE2.

Plate 19. Mervyn O'Gorman,
Superintendent of the Royal
Aircraft Factory, Farnborough,
1909–16.

could have demonstrated a greater awareness of the dynamic state of aeronautics at this time. Nor, indeed, was his perceptiveness confined to aeronautics; Capper was subsequently to become an early tank pioneer. He became Commandant of the embryonic Tank Corps' training depot, the Machine Gun Corps Training Centre, in May 1917, and was subsequently appointed chairman of the Tanks Committee and Director General of the Tank Corps at the War Office. In between, from the period October 1915 to May 1917, he successfully commanded the 24th Division in France. His conspicuously able younger brother, Major-General Sir Thompson Capper, had been killed in September 1915, whilst commanding the 7th Division at the battle of Loos, and it was in recognition of his abilities that the C-in-C, Sir John French, selected John Capper as commander of one of the recently recruited 'New Army' divisions. Capper was not psc and had never commanded a brigade. It was thus a dedicated soldier of multifarious talents who now came to the fore.

Capper's estimation of the work of the field sections tended to confirm many of the criticisms that came out of the Boer War. In particular, he felt, during reconnaissance ascents 'too much attention was [being] paid to the fighting lines instead of . . . the troops behind them'. He saw remedying this defect as his immediate priority.[101] He also reminded his superiors that Europe was prone to more volatile meteorological conditions than were experienced in South Africa. Thus, as far as practical captive balloon support was concerned, any European campaign was likely to prove considerably more problematical than recent events might have suggested. Hence, again, the need to explore other forms of aeronautical technology.

It was partly to look into this question that a Committee on Military Ballooning was formed by the War Office in June 1903, with Capper and the then Lieutenant-Colonel Henry Wilson among its members.[102] The convening of this committee has sometimes been presented as evidence of a positive development in War Office thinking on the air question. In actual fact, however, it was just one element in a whole series of post-Boer War Royal Engineers investigations (there were analogous committees on the use of railways and telegraph systems), themselves symptomatic of a much wider post-war reassessment. In the event, the Balloon Committee had virtually no impact, merely re-emphasising, as its most important recommendation, the need for the Balloon Factory to develop an airship capability – which it was already doing.[103] Its main effect was almost incidental. As a corollary to the emphasis on airship development it recommended that a more commodious

---

[101] Ibid. AIR 1 1608/204/85/36.
[102] The committee comprised Col. P. T. Buston RE (president); Lt.-Col. H. H. Wilson AAG; Maj. G. M. Harper DAQMG; C. Harris, War Office finance branch, and Lt.-Col. J. E. Capper (sec.).
[103] Walker, *Early aviation at Farnborough*, i, discusses the committee's findings at some length. See PRO, WO 32/6930 for the final report, dated 4 Jan. 1904. For general background see Brig.-Gen. W. Baker Brown, *History of the Corps of Royal Engineers*, iv, Chatham 1952, 38.

test-site be found as a matter of urgency. This led to the Balloon Factory's removal to nearby Farnborough Common.

Events abroad were soon to place these parochial deliberations in a clearer perspective. The Balloon Committee issued its final report on 4 January 1904. Just weeks before that, on 17 December 1903, the Wright brothers had made the world's first heavier-than-air powered flight. Already unsubstantiated reports of this event, which was effectively to negate many of the committee's conclusions, were filtering through. By the autumn Capper had left for America.

The ostensible purpose of Capper's trip was to report on the St Louis World's Fair of September–October 1904. Templer had recommended that an appropriate officer be sent to review the aeronautical section of the festival, and the commander of the Balloon Sections was the obvious choice.[104] Capper, however, using an introduction provided by the aviation-inspired philanthropist, P. Y. Alexander, took the opportunity to contact the Wright brothers at their home in Dayton, Ohio. The protracted and ultimately abortive negotiations that this visit heralded have been so extensively discussed elsewhere as to need little comment here.[105] The meeting immediately alerted Capper to the need to reorganise Britain's aeronautical base. 'We may shortly have as accessories of warfare', he reported, 'scouting machines which will go at great pace, and be independent of obstacles of ground, [sic] whilst offering from their elevated position unrivalled opportunities of ascertaining what is occurring in the heart of an enemy's country.' The authorities should appreciate just how backward the Balloon Factory and field sections were in this respect: 'America is leading the way, whilst in England practically nothing is being done.'[106] What was needed was 'a proper experimental school' where, free from other duties, work could begin in earnest on the study of this new form of flight. At present, he advised his superiors, 'we are entirely out of touch'. What little the sections did know derived from chance acquaintances, such as Patrick Alexander, or more often from civilian societies, such as the Aeronautical Society and Aero Club. Contacts with these bodies should be formalised and a technical library established at Aldershot.

For the moment, the work of the field sections continued as before, but on his return to England Capper took every opportunity to re-emphasise his new message. In his report on the field sections' training over the 1904 season (dated January 1905 – that is directly after his US trip) he acknowledged that as regards captive ballooning 'we are now in a very satisfactory position', yet felt compelled to add that 'we are making but little progress in what may prove a far more important branch of aeronautics', viz flying or scouting machines. The sections, as constituted, were simply not appropriate for the experimental work which was becoming increasingly important. He referred his superiors to

104 PRO, AIR 1 728/176/3/17.
105 See Walker, *Early aviation at Farnborough*, ii; and Gollin, *No longer an island*.
106 Memo, 15 Dec. 1904, PRO, AIR 1 1608/204/85/36.

the Balloon Committee's advocacy of a more compact aeronautical unit, to be known as the Balloon School. This would consist essentially of just two full sections and ample reserves and would therefore greatly facilitate diversification by releasing surplus officers and men for practical research work.[107] As matters stood, the army was in danger of becoming dependent upon the research of private individuals, while her own balloon officers consolidated the increasingly antiquated achievements of the previous decade.

Only four years earlier the balloon sections, through their work in the Boer War, had seemed to have entirely vindicated their existence, and the field organisation was permanently re-established in its expanded form. Now these units were beginning to seem outmoded, if not yet obsolete. With Capper's visit to the United States in October 1904 the dynamic state of the science of aeronautics was suddenly brought home. It was only the beginning.

Explorations of new forms of aeronautical technology

Reforms followed, but with little immediate success. In April 1905 the field unit was restructured into companies, each allotted special subjects in addition to their regular duties. Among the designated subjects were the investigation of motors, fans (meaning propellers) and gliding machines. Based on what had been learned from Chanute and the Wrights in America, model biplane-gliders were constructed to one-sixth scale. The plan was that these would then be rebuilt as full-sized aircraft, incorporating motors and propellers, as and when tests proved satisfactory.[108] In the event, tests never did prove satisfactory and Capper had to look elsewhere for the basis of his first British army aeroplane. Interestingly, however, his obsession with the need for inherent stability was already evident in these inchoate trials. This emphasis contrasted sharply with the Wright brothers' wing-warping method of flight-control, which Capper probably only imperfectly understood. Inherent stability was to be the principal feature of Lieutenant Dunne's experiments.

Capper's position as, by now, the dominant figure in British army aeronautics was confirmed in May 1906, when on Templer's removal from the post he also became Superintendent of the Balloon Factory. Henceforth he was to hold both balloon commands – field and factory – simultaneously. Yet Capper was no jealous rival: Templer's enforced retirement, the culmination, it seems, of Aldershot Command's resentment at the Factory being established in its

---

[107] Ibid. AIR 1 1608/204/85/36. The original recommendation was for a reduction to three complete sections with the means for three more, which was almost the *de facto* situation: WO 32/6930.

[108] Ibid. AIR 1 728/176/3/26. As early as August 1903 Capt. Blakeney, the former OC, 3rd Balloon Section, had recommended to the War Office Committee on Military Ballooning that personnel be trained in motor car maintenance in readiness for 'the inevitable introduction of [dirigibles] or flying machines'. This was before the Wright brothers' first flight had even occurred: PRO, WO 32/6930.

domain,[109] was as big a shock for him as it had doubtless been a disappointment for Templer himself. Templer alone had been supervising the army's 'Nulli Secundus' airship project, and Capper was now expected to pick up the threads. Under the circumstances the War Office was wise to offer Templer a position as part-time adviser at the Factory for the following year – a position which, at Capper's request, was renewed for a further year in March 1907.[110] In addition, the new post of Assistant Superintendent was created, although it was not to be filled on a permanent basis until the appointment of Captain A. D. Carden RE in May 1907.

To confuse matters further, it was at this point, in April 1906, that the field companies were finally reclassified as the Balloon School, with Capper as Commandant and Captain W. A. de C. King RE, as Chief Instructor and Acting Adjutant. As seen, the aim was that the school should be in a position to maintain two permanent field detachments and a third cadre unit, capable of rapid mobilisation in the event of war. But in a report of the organisation's progress over the 1907 season Capper had to admit that the nucleus of this third unit had remained 'practically non-existent'.[111] He attributed this to the fact that, as Royal Engineers, his soldiers were continually drawn away to other duties. A small number of officers and certain of the ranks were classed as balloonists and retained more or less permanently, he acknowledged, but this was scarcely sufficient. The War Office, however, saw no need for change. It was after all the nature of a cadre unit to exist, for the most part, in the abstract. A regular turnover of personnel would, moreover, provide a surplus of trained men from which additional units could be formed in the event of hostilities. The more pertinent question was for how much longer would there be a need for such a reserve? The whole argument was in fact looking increasingly redundant. Much had changed since the six sections of 1900, and few retained much enthusiasm for the old field establishment, whatever its title. There is the impression, from the documents, that the protagonists were merely going through the motions. Two detachments, three detachments – what did it matter? It was the retention of men for research into new technology that was important now. This was Capper's major concern.

Research into other forms of aeronautical technology, operable in more volatile meteorological conditions, had already led to the adoption of

---

[109] See Walker, *Early aviation at Farnborough*, i. 78ff. Templer, as SBF, had been directly responsible to the DFW, not to Aldershot Command: Watson, 'Military ballooning', 56. His replacement by Capper – who as OC, Balloon Sections, was under the latter – removed this anomaly.

[110] Walker, *Early aviation at Farnborough*, i. 84–5; War Office to SBF (Capper), 12 Mar. 1907, approving retention of Templer in advisory role for the financial year 1907–8, PRO, AIR 1 1613/204/88/16.

[111] Ibid. AIR 1 812/204/4/1254. Each company was by now recommended to hold ten 11,500 and two 4,500 cu. ft balloons in readiness for war. Two large and one small balloon would then become first line equipment; four large and one small balloon would become the first reserve at an advance depot; and four large balloons would remain in readiness at home base: DFW to SBF, 5 May 1906.

man-lifting kites within the unit. These devices allowed for aerial observations in winds prohibitive to spherical balloons and required no hydrogen transport to accompany their movements. It is difficult to determine precisely when serious investigations into the viability of this form of aerostat began. The veteran balloon pioneer, Henry Coxwell, wrote to B. F. S. Baden-Powell in October 1887 indicating that he had proposed the adoption of such a device as early as the 1850s, and his was not the only such claim.[112] But Baden-Powell himself was the first person to initiate practical trials. Well-known as a prominent member of the Aeronautical Society, Captain Baden Baden-Powell was the brother of Robert Baden-Powell and a serving Scots Guards officer. After exhausting the many preliminaries, he finally began a series of man-carrying kite ascents at Pirbright Camp, near Woking in Surrey, in June 1894 – utilising for the purpose his own design of hexagonal monoplane kite. It took more than one season to develop a workable system. Essentially, an observer would sit in a basket suspended from a train of kites. In an effort to combat the more erratic movements characteristic of monoplane kite-flight the apparatus would then be held at its lower apex by two independently secured cables which were anchored to the ground in opposing directions, in the manner of an inverted V.[113] Despite Baden-Powell's assurances to the contrary, however, the problem of erratic movement was never entirely eradicated. He would in fact have achieved more satisfactory results had he imitated Lawrence Hargrave's recent development of the boxkite, the aerodynamic significance of which stretched far beyond the bounds of merely kite technology. He was certainly familiar with this innovation.[114]

Even so, an informal kite detachment, rather misleadingly described in some reports as a section, based on Baden-Powell's methods, was soon established at Aldershot.[115] It made little impact. In April 1896 Templer reported to the Royal Engineers' Committee that he considered that the 'practical difficulties' of employing such devices effectively prohibited their use in war.[116] Consequently no kite detachment accompanied the balloon sections to South Africa. Baden-Powell sought to counter this decision by privately forwarding a set of kites for service with his own regiment, the 1st Battalion Scots Guards, with whom he served throughout the war, but this scheme was thwarted by the inevitable transport restrictions. He managed to construct a few odd kites whilst on active service, but only for the largely experimental purpose of raising Marconi wireless aerials and automatic cameras. As a Scots Guards officer he

[112] H. Coxwell to B. F. S. Baden-Powell, 4 Oct. 1887, RAeS archive; 'Man-lifting war kites', *RAeS Journal*, i (1897), 5–8; Penrose, *Pioneer years*, 44.

[113] Capt. B. F. S. Baden-Powell, 'War kites', *RAeS Journal* iii (1899), 1–6.

[114] In fact he professed to find 'very little advantage' in the boxkite form: a judgement which was queried even at the time. See Capt. B. F. S. Baden-Powell, 'Kites: their theory and practice', *Journal of the Society of Arts* xlvi (1898), 362, 368.

[115] PRO, AIR 1 686/21/13/2245.

[116] Report of sub-committee on kite experiments, 11 Apr. 1896, Templer papers, 7404/59/2.

also witnessed the work of the 1st Balloon Section at Magersfontein.[117] In the end it hardly mattered; meteorological conditions in South Africa were for much of the time reasonably placid, a fact which undermined the original rationale for a kite detachment. Back in Europe, however, the case for their use, *pace* Colonel Templer, remained persuasive.

It was, in part, the example of Baden-Powell and the Australian pioneer, Lawrence Hargrave, that inspired the expatriate American, Samuel Franklin Cody, to embark upon military kite construction. He patented his first device, essentially a Baden-Powell-cum-Hargrave synthesis, incorporating a train of modified boxkites with dihedral wing projections, in November 1901. By 1903, formal demonstrations were being held before representatives of both the services, and a recommendation for its adoption was included in the final report of the War Office Committee on Ballooning in January 1904.[118] Colonel Capper, while on his subsequent visit to the St Louis World's Fair, was to remark that, within its limitations, the device had no equal, and he was to ascend in a man-lifting kite at the following year's army manoeuvres in Oxfordshire.[119] Eventually Cody was engaged as chief instructor in kiting to the balloon companies. The position was regularised in April 1906, when the Balloon School came into being. Following this, his man-lifting contrivance became an integral part of the field unit's equipment. It permitted aerial observations in much stronger winds than even the much-vaunted kite-balloon was capable of sustaining, in addition to being vastly more manoeuvreable. But furnishing the sections/companies with this commodity did not come cheap. The 1906 Army Estimates allocated to ballooning a total vote of £11,500. Of that, the single most expensive item was Cody's kites, valued at some £2,000 [120] – and this was a conservative estimate. The true cost was to rise considerably as Cody began modifying his contraptions for powered flight.

The Russians were also developing a kite observation capability at this time, although, according to the British naval attaché in St Petersburg, with only 'qualified success'. Admiral Makaroff's use of manned kite reconnaissance at Port Arthur in 1904, during the early stages of the Russo-Japanese War, reportedly constituted the device's first employment on active service.[121] However the main emphasis at Farnborough in the post-Boer War period was

---

[117] B. F. S. Baden-Powell to E. S. Bruce, 6 Apr. 1900, RAeS archive. In an effort to promote the value of a kite detachment Baden-Powell indulged in carping criticism of the work of the Balloon Sections, and made himself very unpopular as a result: see, for example, 'The war balloon in South Africa', RAeS Journal vi (1902), 14–15; Lt. A. H. W. Grubb to Maj. F. C. Trollope, 20 Mar. 1902, PRO, AIR 1 728/176/3/9; and F. C. Trollope to E. S. Bruce, 29 Nov. 1902, RAeS archive.
[118] Walker, *Early aviation at Farnborough*, i. 89ff; PRO, AIR 1 728/176/3; WO 32/6930. The Admiralty went so far as to sanction the purchase of four sets of kites for shipborne use in 1903.
[119] Ibid. AIR 1 1608/204/85/36; *Royal Engineers Journal* lxix (1955), 307.
[120] PRO, AIR 1 728/176/3/27.
[121] Report on visit to St Petersburg by Capt. A. G. Calthorpe RN, 15 Sept. 1904, ibid. AIR 1 728/176/3/18; *RAeS Journal* viii (1904), 64.

on the creation of an army dirigible. This bore fruit in September 1907 with the launching of the 'Nulli Secundus' military airship.

This event coincided with Cody's transition from glider to aeroplane design. The previous year he had sought permission to convert one of his modified kites into a power-driven machine, and to this end had arranged for the purchase of a 50 h.p. Antoinette engine. When it arrived, however, the War Office directed him to incorporate it instead into their military airship, which was then nearing completion. In other words, this airship came to be powered by what was ostensibly the army's first aeroplane engine.[122] Cylindrical in shape and of approximately 55,000 cu. ft capacity, the 'Nulli Secundus' airship first emerged from its hangar at Farnborough on 10 September 1907. Its public debut occurred the following month, when Capper elicited a predictable media response by, with Cody's help, piloting the aircraft over the War Office building in Whitehall and round St Paul's Cathedral. It then alighted at the Crystal Palace where, some days later, its envelope was slashed open in a desperate attempt to prevent strong winds tearing the vessel from its moorings. Sadly, it never flew again, although much of the aircraft's material was subsequently re-incorporated into the incongruously named 'Nulli Secundus II' airship.[123]

In the meantime, in April 1908, Templer's services as an adviser to the Balloon Factory were finally – and permanently – dispensed with. Thus ended more than thirty years at the forefront of military aeronautics. Such was his influence at the Factory, however, that sections of the press were still unaware that he had been officially retired two years before and kept on since then in a purely advisory role.[124] Templer was evidently still quite active in retirement, but inevitably increasingly out of touch. He caused some mild controversy, for example, at the Mansion House meeting of the Aerial League in April 1909, when he suggested that the citizens of London raise the money for an airship to present to the nation. It was not, he was informed, the job of the league to 'relieve [the government] of its responsibilities';[125] but, in fact, both parties were arguing from the same outmoded premise. Within months Blériot had flown the channel, the world's leading aviators had met at Rheims and everything looked different. This slight dispute was no more than a last spasm of the old military aeronautical tradition. Like a moth emerging from its cocoon, army aviation had, under the surface, been developing new wings.

---

[122] See Walker, *Early aviation at Farnborough*, i. 190; ibid., ii. 91.

[123] We merely skim the surface of this aspect of the Balloon Factory's work. The reader is directed to Walker, *Early aviation at Farnborough*, i, where the whole airship saga is discussed in some detail. (The 'Nulli Secundus II' was completed in July 1908, and lasted little longer than its predecessor.)

[124] Note on Templer's second retirement, with reference to *Automotor Journal*, 4 Apr. 1908, and others, Templer papers, 7404/59/2.

[125] *Flight*, 10 Apr. 1909.

5

Haldane's imposition

The story so far has been of the forging of the emerging aeroplane industry into what was primarily an extension of the armaments industry, with the military acting as essentially the only market and private manufacturers becoming increasingly dependent on government orders to offset escalating research and development costs. The military, however, had long experience of the practical application of lighter-than-air aeronautics. Its interest in aeroplane development was therefore initially limited to any support it might give to established aeronautical units. Thus, while the industry viewed the military as market consumers, the military tended to view the early aviation entrepreneurs as merely empiricists, whose work might or might not be exploited as needs be. It could play a waiting game. Within the old aeronautical framework aviation experiments went on, but not as part of any defined policy. They were simply incorporated within the work of the Royal Engineers' Balloon Factory at Farnborough, which was administratively the same in 1909 as it had been in 1890.

On to this blank page was to be stamped the political philosophy of R. B. Haldane, Asquith's Secretary of State for War. Haldane had no faith in the entrepreneurial pioneers then beginning their aviationary careers; he dismissed them without so much as a second glance. What, after all, could a handful of motor mechanics achieve against the might of Göttingen's mathematicians and physicists? Were not Zeppelins traversing hundreds of miles while British aeroplanes struggled to hop a hundred yards? What was needed, he surmised, was a properly constituted state-based and, above all, scientific research centre, by which he meant one composed essentially of university-trained men.

It was the divisions born of this decision and its implementation that dominated the development of aviation up to and into the war. A separate research and design department directly under War Office control was to become an integral part of the emerging military air service. Successive administrative field units would then be devised in its wake.

The Balloon Factory, Farnborough: Dunne and Cody

It was Capper's responsibility, as OC, Balloon Companies, to launch the army's investigations into heavier-than-air powered flight. These, in their earliest stages, took place against the background of the protracted negotiations with the Wright brothers which had resulted from Capper's visit to the United

States in late 1904 to view the aeronautical section of the St Louis World's Fair. On his return to England Capper had initiated glider experiments as part of the Balloon Companies' work, but these were just preliminaries until a more definite approach to the subject could be formulated: the Balloon Companies (or Balloon School as the organisation became) were really in no financial position to undertake such research work and, besides, still had their regular duties to perform.[1] It was only when Capper inherited the dual command of both field sections and Balloon Factory, on Templer's retirement in 1906, that this situation changed.

It was in the role of Superintendent of the Balloon Factory that Capper first recommended to his superiors the work of A. V. Roe, as displayed at the *Daily Mail* model aeroplane flight competition of April 1907. By then the Factory was already actively preparing for its own full-scale flight trials, using an aircraft designed to obviate the need for skilled pilot control: a major drawback to the practical application of the Wrights' Flyer. (As a contemporary Smithsonian report coyly put it, the Wrights' machine required an 'accuracy of manipulation out of the ordinary'.)[2] This Factory aircraft was based on the ambitious ideas of a young invalided soldier who had joined the staff the previous year: John William Dunne.

Dunne's interest in aeroplane flight can be traced back to the turn of the century. During the period 1900–1 he saw active service in South Africa, both as a trooper with the Imperial Yeomanry and as a subaltern in the Wiltshire Regiment. It was, he later told Snowden Gamble (an early historian of *The air weapon*), the difficulty of effecting useful reconnaissance in the face of long-range smokeless enemy fire that brought home to him the need for an aerial scouting device.[3] Thus, when temporarily invalided back to England, he began to investigate the matter.[4] Dunne was aided in these early investigations not by any informed aeronautical acquaintance, but by his friend, the novelist H. G. Wells, who was later to caricature Dunne in several short stories.[5] Together they tried to interest both Sir Hiram Maxim and the distinguished physicist, Lord Rayleigh, in their inchoate theories, with, however, little success.

Dunne was fired by the notion that military aircraft in particular should be inherently stable, for reconnaissance duties. With this in mind he had devised various models with swept-back, V-shaped wings and no tail – not dissimilar to Weiss's gliders. The genre was generally known as the 'zanonia' type as the designs were vaguely based on the seed of the zanonia plant, which drops with a steady glide.[6] By employing this configuration Dunne was convinced that he

1   PRO, AIR 1 728/176/3/26.
2   Pierre-Roger Jourdain, *Aviation in France in 1908* (annual report, Smithsonian Institution), Washington, DC 1909.
3   Gamble, *Air weapon*, 91–2.
4   J. W. Dunne to B. F. S. Baden-Powell, 2 May 1904, RAeS archive.
5   N. MacKenzie and J. MacKenzie, *The time traveller: the life of H. G. Wells*, London 1973, 222.
6   C. G. Grey, notes on early aviation, PRO, AIR 1 727/160/1. For a précis of the evolution

had solved the problem of achieving inherent stability in aeroplane flight. After four years' work, however, he was still an isolated and largely unknown figure, so he turned for help to his father, who, as General Sir John Hart Dunne, proceeded to pull a few of the many strings at his disposal.

The first of these took the form of a letter of introduction to the president of the Aeronautical Society, B. F. S. Baden-Powell, in April 1904. It accompanied a technical memorandum and revealed that Dunne was at that time still on sick leave following the South African war. Even so, he was evidently quite active and Baden-Powell was asked, as a friend of the family, at least to speak to him.[7] J. W. Dunne's own memorandum makes it clear, however, that he was not so much after a hearing as prospecting for a working partner. He had, he stated, 'threshed out' the problem of stability in 'gliding aeroplanes' and now wanted to exploit that knowledge for military purposes. 'I know nothing of materials, stresses, etc.', he confessed in a passage that would have admirably satisfied all Haldane's worst fears, 'little of engines and not much of mathematics. Therefore I want a partner . . . *au fait* with the present development of aeronautics.'[8] Work could then begin on the construction of a full-sized machine, incorporating the means of inherent stability.

By this time, of course, Baden-Powell had already built his water-chute at Crystal Palace, and was conducting his own aeronautical experiments there. On receiving this communication from Dunne there was some discussion as to whether or not to investigate the young pioneer's ideas using this facility,[9] but Baden-Powell apparently wavered on the question of providing anything more – such as financial support. Dunne, in turn, baulked at anything less. He wrote to Baden-Powell on 2 May 1904 saying that if he was unwilling to help, General Dunne knew plenty of other people.[10] Baden-Powell – the old man-lifting kite pioneer of the 1890s – was, however, clearly intrigued by the military implications of Dunne's work and was reluctant, as yet, to remove his finger completely from the pie. Consequently, Dunne continued to pour forth, in a series of long, rambling letters, expositions of his ideas, such as were soon to hypnotise Capper at the Balloon Factory. There was, he wrote on 9 May 1904, no point in searching for a machine which would be unable to fly in wind, or a machine which would take years to master 'with even approximate safety' – as was evidently the case with the Wrights' Flyer. Instead, 'we must have a machine . . . which will go up in rain or wind . . . that will come down safely if anything breaks, and . . . that can be made to balance automatically by an aeronaut who has lost his nerve'.[11] They should be concentrating, in other words, on developing an aircraft of battlefield utility.

of Dunne's V-shaped arrow-head design see J. W. Dunne to J. T. C. Moore-Brabazon, 28 Oct. 1904, Brabazon papers, box 52(5).

[7] Gen. Sir J. H. Dunne to B. F. S. Baden-Powell, 27 Apr. 1904, RAeS archive.

[8] J. W. Dunne, memo to B. F. S. Baden-Powell, 27 Apr. 1904, ibid.

[9] J. W. Dunne to B. F. S. Baden-Powell, 2 May, 29 June 1904, ibid.

[10] 2 May 1904, ibid.

[11] 9 May 1904, ibid.

Dunne reiterated these ideas to Brabazon, whom he had met at Baden-Powell's Crystal Palace workshop. The Wrights, Archdeacon and Chanute dare not turn their devices side-on to a strong wind, he maintained in October 1904; 'but a flying machine which can only travel head to wind is just as bad as a balloon which can only travel down wind'. These pioneers were seeking to add motors to airframes still inadequate for flight. He, by contrast, would perfect his gliding frame and only then set about adding a motor.[12] Baden-Powell, he tells Brabazon, had by then left for America, where he apparently had some vague notion of introducing Chanute to Dunne's ideas. Dunne himself, however, remained committed to furthering aviation within the British army, so the pair effectively parted at this point, both figuratively and literally.

Little was heard of Dunne for the next year, although there is evidence that he again tried to interest Lord Rayleigh in his ideas – with no more success than before.[13] It seems that his health took a turn for the worse, curtailing his aeronautical activities. In April 1906, two years after his original memorandum, he wrote again to Baden-Powell to inform him that he had now retired from all regimental duties and was officially on half-pay.[14] There followed some vague discussion about Baden-Powell starting a business to which Dunne might lend assistance, but evidently General Dunne was formulating other plans for his son. As the latter reported it, his father was anxious that he should maintain some official association with the army, 'and realising that the game is up so far as ordinary soldiering is concerned, is trying to get me attached to the Balloon Corps [sic] to do office work, drawings . . . and that sort of thing'. The question was whether his invalid status would now debar the move. Colonel Capper had, apparently, already been informed of the plan and was looking into the matter.

Capper had first had his attention drawn to Dunne's work the year before by Colonel John Winn, of the Royal Engineers' Committee,[15] but it was only with Dunne's retirement from regimental duties that the idea of having him attached to the Balloon Factory arose. In fact, as Capper himself only became Superintendent of the Balloon Factory – as well as Commandant of the Balloon School – in April 1906, there was little he could have done before then. Nevertheless, he was clearly impressed by what he was told, and it became his aim to bring Dunne into the new command with him, and there initiate the first full-scale heavier-than-air powered flight trials in the British army. Inevitably the proposal took time to work its way through Whitehall's system. So early in June Capper, who was sitting at his new desk waiting for the appropriate order, decided to start anyway. 'We are wasting an awful lot of time waiting for . . . official sanction', he wrote impatiently to Dunne; 'if you do not mind working quietly . . . and saying nothing about it, it would be just as well for you

12 J. W. Dunne to J. T. C. Moore-Brabazon, 7 Oct. 1904, Brabazon papers, box 52(5).
13 Lord Rayleigh to 'Capt. Dunne', 25 Jan. 1905, RAeS archive.
14 J. W. Dunne to B. F. S. Baden-Powell, 2 Apr. 1906, ibid.
15 PRO, AIR 1 686/21/13/2245; Gamble, Air weapon, 95.

to come down [immediately].'[16] Thus Dunne joined the Factory at Farnborough on 7 June 1906, a week before his official appointment.[17]

The ambiguity both of Dunne's status as an invalid and his subsequent relationship with the Balloon Factory, administered as it was by the Royal Engineers, was to cause Capper concern throughout the remainder of his term in command at Farnborough, but he never wavered in support of his protégé.[18] Under Capper's patronage Dunne was to gain nearly three years of uninterrupted, state-aided flight research. On the results of that research Capper had effectively pinned his aeronautical career. The mandarins of Whitehall stood their distance and waited.

After some preliminary experiments, work was begun on a V-shaped, arrow-head, tail-less biplane glider, with the wings swept back at a slightly negative angle of incidence. Named simply the D1, the frame was built with a view to an engine being installed at some future date. For initial take-offs, the plan was that the craft should simply be propelled down a slight incline off the back of a detachable wheeled trolley.[19] The question was where should such trials occur? By the time the frame was ready, in June 1907, a veil of secrecy had been drawn over the whole project, and the search for a suitably remote test-site had begun. The formula of inherent stability was felt to be of such military importance that trials should be conducted away from the public gaze.[20] In the event, ideal terrain was found on the duke of Atholl's estate, at Blair Atholl, in the Perthshire-Grampian highlands. During the Boer War Capper had become friendly with the duke's heir, the marquess of Tullibardine, and he was only too happy to place the area at the Factory's disposal.[21] For the marquess it was the start of a close association with Dunne.

In Whitehall, meanwhile, negotiations with the Wrights dragged on. In September 1906, just as the work with Dunne was getting under way, Capper was asked for his views on a renewed offer from the brothers. As with previous interruptions to Dunne's work, he took little trouble to hide his impatience. The prices asked by the Wrights were out of all proportion to the benefits likely to be gained from the use of their machine, he told the Director of Fortifications and Works, his War Office superior. Moreover, the situation had changed. The Factory would soon be able to produce its own flying machine, superior to the

---

[16] Col. J. E. Capper to J. W. Dunne, 6 June 1906, ibid. AIR 1 1613/204/88/17.

[17] DFW to SBF (Capper), 15 June 1906, ibid. Dunne was retrospectively officially attached to the Balloon Factory as from 7 June 1906.

[18] See, for example, Col. Capper to DFW, 3 Sept. 1906, ibid.

[19] Walker, *Early aviation at Farnborough*, ii. 181ff., carries descriptions of all Dunne's early Farnborough designs.

[20] PRO, AIR 1 686/21/13/2245; Gamble, *Air weapon*, 96. See also Col. Capper to DFW, 6 Sept. 1906, AIR 1 728/176/3/33.

[21] Katharine, duchess of Atholl, *Working partnership*, London 1958, 51–2; Broke-Smith, 'Early British military aeronautics', 32. There is no evidence that Haldane took the initiative in this, despite Snowden Gamble's statement to that effect (*Air weapon*, 97), later reiterated by Gollin in order to suggest that Haldane had taken up Dunne as his particular protégé in opposition to the Wrights: *No longer an island*, 233.

Wrights' in several essential details, and at a fraction of the cost. There was no point in pursuing their offer.[22]

Dunne's position was at this time still uncertain, and Capper may have been overbold in his reply in an effort to strengthen it. Should Dunne now be seen to fail, however, that failure would inevitably be Capper's also. He was, in fact, lengthening the rope with which he was later to be hanged. At the same time as Capper was moving into his office at the Balloon Factory, Haldane was establishing himself at the War Office. At first this 'young and blushing virgin'[23] watched, waited and said nothing; but as the smokescreen of Dunne's rhetoric cleared he was not impressed by what he saw: nothing more than the haphazard flounderings of untrained novices. Over the next year or two he resolved to clear such empiricists out of what he felt should be a temple of science. He would recast the Factory in his own image, and it would be 'rational', 'systematic' and 'academic'; above all, there would be no place for the likes of Dunne or Capper. Thus, had he but known it, Capper was digging his own grave in these early months.

The D1 finally went north in July 1907, followed by a discreet party of plain-clothed Royal Engineers – reportedly from the Kite Section of the Balloon School – led by Lieutenant Francis Westland.[24] A base camp was duly established in the Atholl hills some five miles from Blair Castle. After various preliminaries, the first man-carrying gliding trial was then undertaken in the presence of the Master General of Ordnance, Major-General C. F. Hadden, and his subordinate at the Directorate of Fortifications and Works, Brigadier-General R. M. Ruck. Colonel Capper himself acted as pilot. Haldane, it later transpired, was also a witness to this event, but probably not as part of the official delegation, as no mention is made of his presence in Capper's report.[25] The Atholl estate being situated near his family home at Cloan, in Perthshire, he may have visited from there; and such an aloofness would seem apt. At the end of the previous summer, from 29 August to 4 September 1906, Haldane had visited Germany as a guest of the Kaiser during the annual German army manoeuvres. He returned to Whitehall even more convinced of the need to reorganise the administrative base of the War Office on German departmental lines.[26] In aeronautical terms this would mean divorcing the research and administrative base of the army's balloon establishment from its field companion and restructuring it on a wholly independent scientific basis. The effect of imposing such abstractions on the everyday practicalities of aeronautical

---

[22] Col. Capper to DFW, 6 Sept. 1906, PRO, AIR 1 728/176/3/33.

[23] Haldane's own description of himself on assuming responsibility at the War Office: *Autobiography*, 183.

[24] This Blair Atholl party (Westland aside) was said to have subsequently formed the nucleus of the Aeroplane Coy of the Air Battalion: PRO, AIR 1 686/21/13/2245.

[25] See report and proceedings of Aerial Navigation Sub-Committee of the CID, p. 47, ibid. CAB 16/7 (quoted in Gollin, *No longer an island*, 276); for Capper's report see AIR 1 729/176/4/2/1.

[26] Haldane, *Autobiography*, 206.

development will be considered in due course; suffice to say at this point that by the time Capper and Dunne began their trials at Blair Atholl in the summer of 1907 Haldane was probably already wondering how to wipe the slate clean and start again.

On this occasion the assembled dignitaries saw Capper rise on an up-current of air and then gradually drift into a drystone wall. Neither the pilot nor his glider was seriously damaged, but it looked bad, and Capper knew it. In his report he therefore sought to offset any impression of failure. He again explained precisely what they were endeavouring to achieve – not just flight, but automatically stable flight. The trial was seeking to ascertain whether the results achieved to date with models could be reproduced on full-size man-carrying aircraft. The outcome, 'though to an unskilled eye merely disastrous', in fact proved Dunne's calculations to have been fundamentally correct. The machine had remained poised for a period of eight seconds, during which time it showed 'no tendency to upset' either 'with, across or against' the wind. The trial's unfortunate conclusion, he maintained, was due merely to a misplacement of the operator's weight as he hung suspended from the 'top plane' (i.e. the upper wing).[27]

Content with this initial result, two rather inefficient 12 h.p. Buchet engines and adjacent pusher-propellers were then incorporated within the framework.[28] What Capper describes as a 'track' made of planks was also laid down a hillside to reduce friction and facilitate the machine's launching off the back of the wheeled trolley. However the extra propulsion provided by the propellers caught the Engineers by surprise and, on its first launch, the trolley ran straight off the track and slid the machine onto its nose. Repairs being required, experiments were subsequently abandoned for the season. Nothing daunted, Capper emphasised again that valuable lessons had been learnt regarding methods of lift and control. If these lessons could be assimilated over the winter there was every reason to expect success the following year. It was, he affirmed, now evident that Dunne had indeed 'discovered a law of stability', and his continued retention was imperative.[29] This was duly granted:[30] by the following year, however, the whole basis of aeronautical research in the British army was being reconsidered.

Despite Capper's initial optimism it was September 1908 before experiments were resumed in Scotland. In the intervening months work had begun on two new aircraft: another test-glider, the D3; and a second powered machine, the D4 – although this was essentially just the original D1, re-vamped with a 30 h.p. REP engine and a sprung, wheeled undercarriage. Trials were to

[27] PRO, AIR 1 729/176/4/2/1.

[28] Broke-Smith, 'Early British military aeronautics', 32. The latter recalled that the motors gave a combined strength of just 16 h.p.

[29] PRO, AIR 1 729/176/4/2/1.

[30] Dunne's case came before a medical board again in April 1908. Capper, as would be expected, was fulsome in his support (Capper to DFW, 4 Apr. 1908) and Dunne's retention was duly sanctioned (6 May 1908): ibid. AIR 1 1613/204/88/17.

begin with the glider over the same ground as the year before; a decision would then be taken on the use of the powered craft – this, being a self-launching machine, to be run over the lower grounds around Blair Castle itself. Two potential pilots were detailed to undertake the work; Lieutenant Westland again, and a Lieutenant Lancelot Gibbs, who had been commissioned with the Duke of Connaught's Own Hampshire and Isle of Wight Regiment, a Royal Garrison Artillery (Militia) Regiment, in March 1906 and who, as shall be seen, was later to win some renown as a pioneering instructor and exhibition pilot. The press were officially requested to keep away, and most papers, unlike the previous year, this time complied. 'This shyness', commented the *Daily Telegraph*, with some bemusement, 'is quite a new feature of our time and people.'[31]

There is a welter of source material relating to these trials, for nearly all the principal protagonists kept diaries or made reports. They form a record of frustration and failure. Some useful glides were achieved by Gibbs on the man-launched D3 in the Atholl hills, but the D4 powered machine, when finally constructed, barely scraped off the ground at the very close of the season. The best Capper could show for, by then, two years intensive work was a hop of 40 yards. Once again he knew it looked bad, but argued in his report that this was just a superficial impression. They were, he maintained, on the verge of success, and were a smoother ground, lighter chassis and more efficient engine at the Factory's disposal, they could by now have achieved a great deal.[32] As it was, he was still hopeful of good results when back at Farnborough. By that time (December 1908), however, decisions were being considered at a higher level than the Directorate of Fortifications and Works. In this wider context, reports of these trials served only to confirm innate suspicions of the Factory's whole speculative approach to aeronautical research.

With the ending of the trials there arose the question of what to tell the press. 'During this season the papers have been exceptionally kind to us in refusing to publish any details', Capper wrote to the Director of Fortifications and Works.[33] He recommended they be informed simply that the experiments were now concluded and that no 'sensational flights' had taken place, merely studies of 'balance and construction'. By the following year the press were to be told of the departure of Dunne and Capper from the Balloon Factory.

Meanwhile, aeroplane experiments had been continuing at Farnborough itself, as an extension of the work of the chief instructor in kiting to the Balloon School, Samuel Franklin Cody. These experiments were again under Capper's general supervision; but although a sympathetic patron to Cody, Capper never

---

[31] *Daily Telegraph*, 8 Sept. 1908 (Butler collection, vii. 2).

[32] Report dated 23 Dec. 1908, PRO, AIR 1 729/176/4/2/1. For Lt. Dunne's report (dated 19 Dec. 1908) see AIR 1 1613/204/88/10. The same file also contains Lt. Gibbs's and Capt. Carden's diaries and much of the correspondence that passed between Capper and Dunne.

[33] 23 Dec. 1908, ibid. AIR 1 729/176/4/2/1.

showed him quite the same personal commitment that he had lavished on Dunne.

Cody's background need not be detailed: one of the most colourful figures to emerge from the pioneering years of aviation, he has been the subject of two full biographies and innumerable portraits.[34] Like his namesake, William Frederic Cody, 'Buffalo Bill', with whom he has often been confused, S. F. Cody (who had actually been born with the surname 'Cowdery') was a genuine cowboy from the American west, who turned, after a period of gold prospecting in Alaska, to Wild West showmanship to earn a living. In keeping with this *persona*, he sported shoulder-length hair, a waxed moustache, a goatee beard and in later years was seldom to be pictured about the Hampshire plains without a broad Texan sombrero. He had been born in Birdville, Texas, in 1861 and had spent his youth driving cattle across the southern states. His skill with horse, rifle and lasso never left him. He had seen his home attacked by Indians, his brothers killed, and had lived the life he afterwards cultivated for show. It was when contracted to round up and drive 500 horses to the east coast of America and then ensure their passage to London that his life changed, however, for in London he met and married – bigamously, as it later transpired – his employer's daughter. Gradually she was written into his cowboy act, and it developed into a music-hall melodrama entitled 'The Klondyke Nugget'. It was while on the road with this entertainment that he first appeared in contemporary reports in connection with his man-lifting kite experiments.[35]

Cody's kites were taken up by the Balloon Sections on Capper's recommendation. He had reported in June 1904 that he could not 'speak too strongly' as to their excellence. Used in conjunction with captive balloons, he urged, the equipment would 'render the Balloon Sections of our Service superior to those of any . . . in the world'.[36] This was true, but the fact was somewhat diminished by the news that the Wright brothers had flown an aeroplane six months before. None the less, Cody became instructor in kiting to the Balloon Sections, soon to become Balloon School, with permission to supervise the construction of new kites within the Factory.[37] It was in this context that his interest in powered flight grew.

There followed a complicated series of developments leading to the emergence of a Cody-inspired Factory aeroplane. The first was the testing of a glider-kite in 1905. Here a hybrid man-carrying biplane gliding-craft was first flown tethered like a kite, and then – once in the air – released as a glider. By this expedient Factory personnel presumably aimed to circumvent the need to launch potential service gliders off steep inclines, seldom likely to be available

34 G. A. Broomfield, *Pioneer of the air: the life and times of Colonel S. F. Cody*, Aldershot 1953; A. Gould Lee, *The flying cathedral: the story of Samuel Franklin Cody*, London 1965.
35 S. F. Cody to E. S. Bruce (sec.), 14 July 1902; S. F. Cody to B. F. S. Baden-Powell, 8 Dec. 1902, RAeS archive.
36 Col. Capper to GOC RE, 1st Army Corps, 30 June 1904, PRO, AIR 1 823/204/5/62.
37 Memo, Col. Capper to Clerical Staff, Balloon Factory, 3 Sept. 1906, ibid. AIR 1 1613/204/88/15.

as required on campaign, but little is known of the device as aeroplanes soon made the idea redundant.[38] Then at some point over the period 1906–7 experimentation began with a motor-kite, in which a 12 h.p. Buchet engine was incorporated within a modified pusher-propeller Cody-kite structure and a wheeled-undercarriage added.[39] No pilot was carried, however, and it was flown along a horizontal cable suspended between two masts or posts placed approximately 100 ft apart. Both devices were intermediary stages in the evolution of a Factory aeroplane. By the close of 1907 Cody was at work on the first full structure.

This was essentially a Wright-derived biplane, with a forward-elevator and rear-rudder. An auxiliary rudder was later attached to the upper plane for 'roll control', in preference to the original ailerons. By this means, i.e. by pilot control, it was hoped that the subsequent experiments would form a useful contrast to Dunne's theories of inherent stability. The aircraft was to be powered, when an engine became available, by two pusher-propellers. A 50 h.p. Antoinette engine had been acquired for aviation purposes some time before, but it had subsequently been diverted to the 'Nulli Secundus' airship. When this was wrecked in October 1907 it was then incorporated into Farnborough's 'Nulli Secundus II' airship. Thus, although this latest aeroplane frame was completed by early 1908, it was September before full trials could begin. By that time the Factory had purchased a second Antoinette engine. Given the title 'British Army Aeroplane No.1' on account of its Balloon Factory ancestry, it first rose from the ground on 29 September 1908, in an unsustained flight of about seventy-eight yards.[40] Conflicting duties and further modifications, including the permanent removal of the wing-tip ailerons and the repositioning of the engine-cooling radiators away from the cockpit, then delayed matters until the resumption of full flight trials on 14 October. Two days later, with Cody as pilot, it made what is generally agreed to have been the first sustained aeroplane flight in Great Britain.

Attempts were again made to prevent the press from reporting these events, but never to the obsessive extent that characterised the Blair Atholl inherent stability investigations. Capper, in fact, imagined the Dunne trials to be considerably in advance of Cody's work, which, it appeared, embodied no great innovative qualities. He made this clear in a letter to the editor of the *Daily Express*, who had written to complain that while his paper was following the Factory's reporting injunction others were not. The reporting restrictions, Capper indicated, related primarily 'to some trials being made in Scotland,

---

[38] The device is described in Walker, *Early aviation at Farnborough*, i. 127–31, where the information is largely derived from three old photographs. The craft incorporated an early use of ailerons.

[39] PRO, AIR 1 728/176/3; Walker, *Early aviation at Farnborough*, i. 131–4.

[40] PRO, AIR 1 728/176/3; Walker, *Early aviation at Farnborough*, ii. 101, 112–13. Subsequent claims of a flight with this aircraft in May 1908, promulgated by Cody's first biographer, G. A. Broomfield, and included in various unauthenticated summaries of the Royal Aircraft Factory's work, are completely false: ibid. 145ff.

which we are desirous of keeping as secret as possible. The trials that are being made at Aldershot with Mr Cody's aeroplane are not of a particularly secret nature'. He was concerned, however, lest publicity draw crowds to the area. This would both endanger life and impede progress.[41] In any case, he maintained, sustained flight was not yet the aim; they were simply testing the general characteristics of the machine. This was still held to be the case when work resumed on 14 October.[42] In fact the terrain at Farnborough, being bounded on all sides by trees and covered with impediments, was scarcely conducive to anything more. Consequently, when Cody made his impromptu flight on the 16th, Capper, only too aware that his methods were now under the closest scrutiny from Whitehall, was not amused. A sustained flight in such an environment would be inviting an accident under any circumstances; in the experimental context of the time it was culpable. By the crash that resulted the Factory was again put on the spot. It was bad enough when Capper himself piloted Dunne's glider straight into a stone wall before the assembled War Office dignitaries at Blair Atholl the year before; now Mr Haldane would turn from his musings on a 'rational' and 'systematic' scientific research base to find that an illiterate cowboy had piloted his own untried aircraft virtually headlong into a clump of trees. The Superintendent's report was thus primarily an exercise in damage limitation.

'Mr Cody has been running the machine about on a good many occasions in order to get its balance', he informed the Director of Fortifications and Works on the day of the flight, 'but he was instructed to attempt nothing sensational or any long flights.' As soon as he felt confident of a sustained flight, Capper continued, he was to have informed the Superintendent, who would have arranged a proper trial. On this occasion he was taxi-ing along as usual when, 'to his astonishment', the machine rose to a considerable height. He intended to alight as soon as possible, but came upon a clump of trees – which he cleared only to find another clump beyond them. He consequently attempted an evasive turn to the left and struck the ground with his left wing tip, causing the machine to swing round and crash onto its nose.[43] Notwithstanding this accident, however, Cody had, Capper insisted, constructed an aeroplane of 'considerable promise'. It had 'flown and . . . flown steadily' (it covered, in fact, a distance of some 1,390 ft), and he recommended that the Factory continue the aircraft's trials, at least 'until we know . . . we have got hold of a better one'.[44] To this the proviso was added that a more extensive test-ground be found as a matter of urgency. Ignoring this request, he stressed, would be a false economy, merely resulting in further accidents.[45]

[41] Col. Capper to news editor, *Daily Express*, 24 Sept. 1908, PRO, AIR 1 823/204/5/62.
[42] Capt. A. D. Carden, for SBF, to Press Association, 15 Oct. 1908, ibid.
[43] Col. Capper to DFW, 16 Oct. 1908, ibid. WO 32/6933. P. B. Walker argues convincingly that, despite what Cody subsequently felt it expedient to tell Capper, this first sustained flight was in fact premeditated: *Early aviation at Farnborough*, ii. 135–7.
[44] Memo, Col. Capper to DFW re. Cody's trials, 16 Oct. 1908, PRO, WO 32/6933.
[45] Memo, Col. Capper to DFW re. further trials of Cody's aeroplane, 16 Oct. 1908, ibid.

Believing the D4 to be at a similar stage of development Capper penned a warning to Dunne in Scotland not to try anything similar. 'It is useless trying to turn on a restricted area, without having learnt to do so on a big open space', he insisted. Such bravado simply cost the Factory money.[46] Despite Capper's fears, however, his immediate superior, the Director of Fortifications and Works (by this time, Brigadier-General Rainsford-Hannay), in forwarding the latest report to the War Office, supported all its carefully-framed recommendations without qualification. Thus the dossier went forward with the related requests that trials be continued, Cody be retained, and a larger flying-ground be found.[47]

Approval of these recommendations would normally have been a formality, but on this occasion other forces were at work and the requests were effectively dissolved in a new War Office policy, the ingredients of which were at that moment being stirred by the Secretary of State. It was consequently to be some months before the Master General of Ordnance finally responded to Capper's report. In the meantime the latter continued his work with Cody in the apparent belief that his recommendations had been accepted, at least in principle. He released a circular on 8 January 1909 stating that 'from now onwards we hope to be making constant trials'.[48] On the 20th, however, he was again having to inform the Director of Fortifications and Works that Cody had disregarded instructions and flown his machine across Farnborough Common. The result was another crash. This time a note of exasperation can be detected in Capper's communiqué on the event. Cody was, he maintained, distinctly told not to leave the ground except for short hops. Extended trials were to be conducted elsewhere (i.e. on Laffan's Plain, further to the south). If 'Mr Wright' could be induced to test pilot this Factory machine he might work it safely, Capper wrote gloomily, 'but . . . we shall have a good many smashes before Mr Cody has learned to manipulate the controls'.[49]

Whitehall was scarcely likely to look kindly on this last suggestion; Capper himself had downgraded the military value of Wright-derived unstable aircraft in his eagerness to promote Dunne's ideas. Now, far from an inherently stable reconnaissance machine, he was defending a conventionally-derived biplane, the 'manipulation' of which was seemingly beyond its own designer, and which he admitted 'could not hope to be advance[d]' without a series of breakages.[50] The decision to sweep the Factory clean had already been taken; if there remained any doubts, this quashed them. There also followed a good deal of unfavourable comment in the newspapers.[51] Capper angrily sprang to Cody's defence, informing the gentlemen of the press that they too had had to crawl

46 Col. Capper to Lt. Dunne, 19 Oct. 1908, ibid. AIR 1 1613/204/88/10.
47 DFW to MGO, 21 Oct. 1908, ibid. WO 32/6933.
48 Ibid. AIR 1 823/204/5/62.
49 Col. Capper to DFW, 20 Jan. 1909, ibid.
50 Ibid.
51 See Butler collection, vii. 128.

before they could walk;[52] but by now Farnborough had a credibility problem. Even the normally ebullient Cody began to display signs of doubt. On being asked if Lord Montagu of Beaulieu could attend a trial of his aircraft, he wavered. It was, he thought, 'somewhat premature' to attempt anything too ambitious; he did not feel confident of his balance. Perhaps in response to certain opinions emanating from the War Office, however, he was at pains to maintain that his aircraft had been constructed on 'the right and scientific system possibly more so than any other machine known to man [sic]'.[53] 'I can certainly ... prove', he insisted, 'that there are scientific points in the structure of the Cody Army Aeroplane that none other ... can boast.' These points were recorded in 'the secret locker of the War Office', and one day such information would vindicate his name. As it was, he was still hoping the nation would 'display good judgement' and continue to entrust its aeroplane experiments to him. Presumably he hoped Lord Montagu might convey these sentiments to Mr Haldane; if so, he was to be disappointed.

Science to Haldane was not the tinkerings of empirical mechanics; to him the very word was an incantation, invoking the rational principles of great German thinkers. This student of Hegel had left Göttingen, he tells us, convinced that 'the way to ... truth lay in the direction of idealism'. The 'essence' of 'idealism' had led him, he suggests, 'to the belief in the possibility of finding rational principles underlying all forms of experience, and to a strong sense of the endeavour to find such principles as a first duty in every department of public life'.[54] This, he claims, was the 'faith' that prevailed throughout his army reforms. Now he saw that philosophical 'idealism' in the establishment of a state-controlled, academically-based, aeronautical research department. Clearly, the 'rational principles' underlying aeronautics were to be found in the appliance of academic science; that was the 'essence' of the 'ideal'. In this abstract vision, the likes of the Wright brothers, or Farman, or Roe, or more to the point Cody and Dunne, were nowhere. Bicycle manufacturers, a motor mechanic, a locomotive apprentice, a kite-flying cowboy, a quackish invalid soldier: collectively they did not seem to add up to much. That, at least, was the impression. In fact, Dunne and Cody apart, they added up to the birth of heavier-than-air powered flight. It was precisely through such people that aviation emerged: empiricists with a basis in mechanical engineering developing with and through the general technological advances of the period. Haldane, at this time, not only lacked faith in such people, he also lacked faith in the aeroplane as against the airship. But without knowing it, in rejecting the aviation pioneers and all their works in favour of a state-based research department, he was sowing the seeds of future aviationary strife. In time, the need for aeroplanes would become imperative. The state research department,

[52] Col. Capper to Associated Press, 17 Feb. 1909, PRO, AIR 1 823/204/5/62.
[53] S. F. Cody, draft reply to letter from R. P. Hearne (*The Car Illustrated*), 19 Feb. 1909, ibid. AIR 1 1613/204/88/15.
[54] Haldane, *Autobiography*, 19, 352.

as founded by Haldane in the guise of the reconstituted Balloon Factory, would then recruit its own mechanical pioneer, in the person of Geoffrey de Havilland, and turn out military designs in competition with the private sector (the private sector being essentially the entrepreneurial manifestation of those early pioneers Haldane had rejected).

It must be remembered that the military market was the aviation market, that a military–industrial complex had developed. In promoting its own designs against those of the private sector the government began to undermine that relationship. Competition soon gave way to something closely approaching a state design monopoly, culminating in near design-paralysis within the industry and the Fokker scourge controversy of 1915–16. Arguing from Haldane's premise, the results of his policy of initiating state-based research and design can perhaps be seen in terms of the old Platonic distinction between the 'ideal' and its actual manifestation in reality. Divorced from its philosophical presentation, however, the policy never appeared to be much more than an abstract dogmatism imposed on a dynamic environment.

Haldane was not unique in his admiration for German philosophy and German administrative methods; indeed, H. A. L. Fisher recorded that, in the last two decades of the nineteenth century, experience of the German universities was considered a prerequisite to the highest academic distinctions.[55] Where Haldane differed was in the depth of his admiration, the manner in which he wore it on his sleeve, and the way in which he sought to apply it to practical political problems, like education and army reform. By the time he came to focus his mind specifically on the aeronautical problem[56] his bent in this direction was already well-established – and, to an extent, distrusted. By April 1908 Lloyd George was having to defend him against Northcliffe's xenophobic attacks.[57] So rooted was he in this world-view, moreover, that when obliged to explain his thinking following his removal from office in May 1915, Haldane retracted nothing. Numerous 'inventors' had come to him as Minister for War, he said, and he had examined many 'specifications', but he could see that such people were merely 'clever empiricists', and that 'we were at a profound disadvantage compared with the Germans', who, he claimed, 'were building up the structure of [an] Air Service on a foundation of science'.[58]

This last idea became an article of faith with Haldane, but it is not immediately clear where he got it from for, as we shall see, the German approach was characterised by its pragmatism if by anything. However, the same line was fed to the official historian of the air war, Sir Walter Raleigh, who, whatever his qualities as Merton Professor of English at Oxford, was certainly not chosen for his aeronautical knowledge. The official history consequently regurgitates arguments from this same premise. There we find

[55] H. A. L. Fisher, An unfinished autobiography, London 1940, 79.
[56] Haldane's own term for it: Autobiography, 232.
[57] Lloyd George to Northcliffe, 9 Apr. 1908, Northcliffe papers, Add. 62157.
[58] Autobiography, 232–3.

that the incipient British air service owed its 'power' and 'efficiency' to those who had 'endowed it with the means' of confronting 'the organised science of all the German universities'.[59] We are deep in the realms of Haldane idealism here. It was with this notion of an infallible academically-based German science being systematically applied to military aeronautics that he justified his actions; but it has little basis in reality. Yet with its appearance in the official history the idea hardened into orthodoxy. Thus Professor Alfred Gollin, in an article culled from his study of the Wright brothers' negotiations with the British authorities, can repeat Haldane's theories and then triumphantly justify them by citing the conclusions of the official history.[60] Most of this can be cleared away root and branch. The way in which Haldane presented his ideas must be analysed in terms of a particular political context. In 1915 he had been excluded from cabinet office and subjected to much vilification in the press as a result of his allegedly pro-German views. His political reputation lay in ruins. Far from accepting this philosophically, as is often maintained, Edward Spiers has conclusively demonstrated that Haldane almost immediately undertook to vindicate his record in office.[61] This initially involved simply eliciting expressions of support from friends and colleagues, but as the issue grew more urgent to him it developed into the composition of a full-fledged apologia. Thus there emerged in April 1916, for private circulation, a 'Memorandum of events between 1906–1915'. In this Haldane claimed that, far from being blindly enamoured of the Germans, he alone perceived the true nature of their pre-war administration. He had identified a war faction within the German government and endeavoured to isolate it by reducing Anglo-German tension and encouraging an anti-war faction. As a corollary of this policy – lest, as was to happen, the war faction gain the upper hand – he sought to re-establish Britain's defences on the same 'modern and scientific' principles as Germany's. His critics, therefore, were quite wrong: he was not spellbound by German methods; he was seeking to counter them.[62]

Haldane's contemporary public utterances on aeronautics were littered with vague references to science, and, indeed, to German science, but it was only with this memorandum that he systematically set out to justify his decisions in terms of countering a direct threat – i.e. in this instance Germany's creation of a scientifically superior air service. This was consistent with the didactic tenor of the piece, which sought to explain all his decisions with reference to a German precedent – with little regard for historical objectivity. Thus instead of just admitting that he had no faith in 'uneducated' empiricists and wanted to re-establish military aeronautics on an academic basis, as was essentially the case, Haldane intimates that he was compelled to do this in order to counter

---

[59] Raleigh, *War in the air*, i. 160.
[60] A. Gollin, 'The mystery of Lord Haldane and early British aviation', *Albion* ii (1979), 51, 65.
[61] E. M. Spiers, *Haldane: an army reformer*, Edinburgh 1980, 12ff.
[62] Ibid. 14.

the Germans. But while the idea of a systematically or scientifically structured air service was certainly the ideal he sought to impose on the balloon establishment in England, there is little evidence that anything like it existed in Germany in 1908–9. This interpretation is, nevertheless, to be found in all Haldane's later writings, most notably the section on aeronautics in his 1928 autobiography – which on this subject at least is in fact little more than a verbatim reproduction of his apologia of 1916.[63]

Haldane's strictures generally received a mixed reception from contemporaries. He had in fact already earned a reputation for being 'economical with the truth' when it suited him, as for example when he had been caught doctoring the Hansard record of a statement he had given the Lords on the Ulster situation shortly before the war.[64] Political colleagues consequently recognised his memorandum for what it was; the retort of a disappointed man. Subsequent historians have seldom proved so discerning.

Given his Olympian views it is little wonder that those actually concerned with aeronautics first hand found Haldane an aloof and inscrutable figure. Even Frederick Sykes, who was to command the Royal Flying Corps and who professed an admiration for Haldane, had to admit that his 'doctrinaire attitude' and 'slightly pompous manner' readily alienated people.[65] He seems to have taken little trouble to explain his ideas at ground level. Interviewed after the war, Major Alexander Bannerman, who had commanded the short-lived Air Battalion, suggested merely that Haldane had returned from Germany considerably impressed by the Zeppelin and anxious to see such airships developed in the British army.[66] This was a superficial view of Haldane's aims, but it was probably what came across, certainly to ordinary soldiers. Haldane equally cared little about the structure of any actual field units. His verbose apologia gives no indication of how he intended to relate research to field work. In fact the problem was left to his Parliamentary Under-Secretary and successor, J. E. B. Seely.

Haldane's intention, then, was essentially to reconstitute the Balloon Factory as an academically-based research department. Before following his reforms through, however, a clearer image of the continental background should be developed.

The continental background

It is difficult not to feel that Haldane's profound enthusiasm for Teutonic metaphysics cast rather an artificial glow over his view of German arms. Far from being modern, innovative or scientific, as Haldane would say, Fritz Fischer

---

[63] 'Memorandum of events between 1906–1915', 143, Haldane papers, MS 5919; *Autobiography*, 232–3.

[64] See H. H. Asquith to V. Stanley, 18 Apr. 1914, in *H. H. Asquith: letters to Venetia Stanley*, ed. Michael Brock and Eleanor Brock, Oxford 1982, 67.

[65] Sir F. Sykes, *From many angles: an autobiography*, London 1942, 86.

[66] PRO, AIR 1 725/100/1.

has shown us an inherently conservative military organisation, racked by 'glaring anachronisms' and slow to adjust to technological change.[67] One wonders how far Haldane really examined German military aeronautics at this time; or whether his search for essential principles precluded such particulars. Looking at it now one is struck by how similar the German organisation was to that of the British. Whereas in Britain a Balloon Factory and Balloon School (consisting of the field sections) were placed under the control of the Royal Engineers, in Germany a Research Unit and Airship Battalion were placed under the control of the Inspectorate of Transport Troops.[68] Within the War Ministry's financial limits it was then the inspectorate's duty to direct the army's aeronautical operations.

The work of the Research Unit initially centred on balloon and airship developments, but policy was dictated by events rather than by preconceived principles or any form of systematic analysis. The German War Ministry, like its French and British counterparts, had entered into serious negotiations with the Wright brothers, and much attention was being given to the idea of constructing an indigenous military aeroplane when in August 1908 the Zeppelin airship finally established its pre-eminence in German thinking. That it did so was due to more than simply aeronautics. It is often forgotten that before 1908 the German authorities had viewed Count Zeppelin's work with some ambivalence. However, on 4 August of that year considerable publicity surrounded the 24-hour acceptance trial of the massive 400 ft twin-engine 110 h.p. Daimler-Mercedes Zeppelin IV airship – prior to its acquisition by the German government. When the demonstration flight ended in disaster at Echterdingen, where the airship broke free and exploded, a spontaneous wave of sympathy swept the nation and donations totalling millions of marks poured into Friedrichshafen to enable the count to continue his work. The War Ministry could prevaricate no longer; the Zeppelin had become a symbol of the Reich's strength and will. Two more were ordered almost immediately, previous reservations being brushed aside.[69]

This is all curiously at odds with Haldane's interpretation of German developments. What this incident presaged was anything but a rational policy structured on a foundation of science; it was more like the political utilisation of that 'blind natural force' which Clausewitz considered an essential component of war. To that extent Northcliffe was nearer the truth when he asserted that the episode revealed some elemental feature of the German psyche not yet wholly understood in England.[70] The embryonic aeroplane industry could not as yet compete with such a spectacle: public attention in Germany was concentrated elsewhere. But the military never altogether abandoned interest.

---

[67] F. Fischer, *Germany's aims in the First World War*, London 1967, 19–20.
[68] Morrow, *Building German air power*, 14–15.
[69] D. H. Robinson, *Giants in the sky: a history of the rigid airship*, Henley-on-Thames 1973, 33–4.
[70] Northcliffe to Arthur Mee, 22 Aug. 1908, Northcliffe papers, Add. 62183.

They were, in fact, reminded of the alternative just days after the disaster at Echterdingen, for on 8 August 1908 Wilbur Wright made his first public flight at Le Mans. Within the week the Research Unit's aeroplane authority, Captain Wolfram de la Roi, submitted a report on 'The attitude of the military authorities . . . toward the flying machine question'. Immediate action would have to be taken, he suggested, if Germany were not to fall behind other nations in this matter. The question was whether the army should support domestic pioneers or initiate its own experiments. Given the Research Unit's superior resources, he recommended the latter course. The Airship Battalion, however, took the opposite view, arguing that the army should seek to avoid duplicating the work of private manufacturers. It should instead simply announce a set of requirements and let the private sector work towards them. The Inspector of Transport Troops agreed, with the result that nothing happened.[71] The War Ministry soon grew restive. Not only the French, but the British army had begun experimenting with its own aeroplanes, and there they were waiting for some domestic initiative which showed no sign of materialising. Consequently, in January 1909, the ruling was reversed. It is clear from even this brief resumé that Haldane's presentation of developments in Germany was at best selective.

Designed by 'master builder' W. S. Hoffmann, the German military aeroplane, when it did emerge, was as unsuccessful as any product of the Balloon Factory. Serious delays ensued, moreover, while a suitable French engine was procured. The problem of building an indigenous model was apparently beyond the 'organised science of all the German universities'. It was to no avail. The crude triplane was written off after only its second trial.[72] If, as Haldane maintained, Britain was at a profound disadvantage as compared with the Germans, one is left wondering how? Where, in practical terms, was this disadvantage? It was, in fact, a myth. With their fingers burnt, the German authorities went back to relying on private enterprise, which meant, of course, those mechanical engineering empiricists so disdained by Haldane. The intervening year had in fact seen notable advances in this area, although the general consensus was that the Germans remained some way behind.[73] What all this indicated, however, was a policy of pragmatism and adaptability – the very opposite of Haldane's approach, but one which was to work to good effect in the circumstances of war.

By contrast to German developments, Haldane scarcely ever referred to what was happening in France. There would seem to be two explanations for this: first that he viewed the early French pioneers as simply so many more

---

[71] Morrow, *Building German air power*, 17–19.
[72] Ibid. 21–3.
[73] War Office report, dated 15 Nov. 1909, PRO, AIR 1 824/204/5/69. The year had seen the first significant indigenous German aeroplane flights, accomplished by Hans Grade in his Dumont 'Demoiselle'-derived monoplane. The Wrights had also given lessons to German pupils, among them, it is revealed here, the former Balloon Corps Instructor, Capt. A. L. H. Hildebrandt.

empiricists and mechanics – which they were, and which is probably why they succeeded; and second that he had little faith in the military value of aeroplanes anyway, as against airships. Ultimately, however, no matter how reliant Europe was becoming on French aero-engines, or what achievements were forthcoming in either airship or aeroplane design, one is left with the impression that that nation's 'science' – to use the word in Haldane's all-encompassing sense – was simply outside his *Weltanschauung*.

In reality, the French military had an admirable record of broad aeronautical investigation, stretching back many years. As with the British army, captive balloon units had been employed on colonial campaigns, such as those in Tonkin in 1884 and Madagascar in 1895, while experimental work continued at home – in this case at the laboratory of Chalais-Meudon. There, as early as 1884, two officers of the Engineering Corps, Charles Renard and Arthur Krebs, had constructed a near-practical airship, styled 'La France'. Then, in 1892, the French Ministry of War began subsidising Ader's heavier-than-air investigations, a subsidy which continued for five years.[74] The French were thus prepared to investigate aeronautics, in all its diverse elements, long before either the British or Germans.

Aviation research was continued at Chalais-Meudon in the early years of the new century when Colonel Renard got the gliding-flight pioneer, Captain Ferdinand Ferber, of the French artillery, officially seconded to the laboratory. No less a person than General Joseph Joffre presided over the committee which, in 1905, tentatively sanctioned the maintenance of research in this area.[75] There were also the inevitable negotiations with the Wright brothers (similar in all essentials to the British negotiations), but these were soon overshadowed by the work of France's own domestic pioneers, such as Voisin, Farman and Delagrange. It was these empiricists who finally brought practical aviation into being.

This, then, was the objective continental background to Haldane's reorganisation of the British army's aeronautical base. It is not a picture that figures in much of the subject's historiography. Roskill typifies the established view when he remarks that it was the advances made by France and Germany – in particular by Germany – in airships that prompted the British reassessment of 1908–9.[76] So superficial an assumption will no longer suffice. It is true insofar as perceived developments in Germany influenced Haldane (as opposed to service aeronauts) and that he was then in a position to initiate a major policy review; but Haldane himself was not just reacting to external events. This was certainly the impression he gave in his memoirs, and, after the outbreak of war, was the idea he deliberately sought to foster, but it is clear that the reality was considerably more complicated. Haldane was inspired not so much by practical

74 P. Facon, 'L'Armée Française et l'aviation 1891–1914', *Revue Historique Des Armées* clxiv (1986), 78.
75 Ibid. 79.
76 *Documents relating to the Naval Air Service*, ed. S. Roskill, London 1969, i. 3.

events as by his own abstract ideals. That these ideals were filtered through a maze of German metaphysics simply confuses matters. It would be wrong to suppose that the Germans were anything other than pragmatists.

Haldane's aim, ultimately, was to re-establish aeronautics on an academic basis. Such an aim was consistent with his whole philosophy of life. We need look no further for its origins.

The Committee of Imperial Defence Esher Sub-Committee and the Advisory Committee for Aeronautics

Haldane liked to portray his policies as being the rational outcome of reasoned and sober reflection, so it can be assumed that such considerations had long formed an undercurrent of his thought. The first evidence of resultant practical decisions, however, comes with the appointment of the Committee of Imperial Defence Sub-Committee on Aerial Navigation on 23 October 1908, just a week after Cody's first sustained flight.

The sub-committee was placed under the chairmanship of Lord Esher, but Haldane remained the motive force behind it. There was nothing unusual in this procedure; siphoning off more intricate defence questions to specially convened sub-committees was by then an established practice. 'Mr Balfour had founded [the] Committee [of Imperial Defence] with excellent ideas', Haldane later explained, in his 1916 'Memorandum on events', '. . . but we developed it in new directions, setting up a series of sub-committees to deal with special problems'. Nor is it a surprise to find Esher in the chair. He and Haldane had co-operated closely on recent army reforms and, while being close allies, Haldane could shelter behind Esher's political neutrality. It was Esher, in fact, who had originally persuaded Haldane to support the CID against radical Liberal calls for its abolition. Haldane was quick to realise that the organisation could be used as a front for some of his more controversial policies, sheltering them from less tractable elements within the Liberal government.[77] In choosing Esher as chairman, then, the main CID was again following an accustomed practice. Other sub-committee members included the Chancellor, Lloyd George (although he only attended half the meetings); the First Lord (Reginald McKenna); the Chief of the Imperial General Staff (General Sir William Nicholson); and both the Master General of Ordnance (Major-General Hadden) and the Director of Naval Ordnance (Captain Reginald Bacon). In addition, a number of expert witnesses were called to give evidence. These included the Superintendent of the Balloon Factory, Colonel Capper; Major B. F. S. Baden-Powell; C. S. Rolls; and, slightly incongruously, Sir Hiram Maxim.

---

[77] P. Fraser, *Lord Esher: a political biography*, London 1973, 22–3. See Haldane papers, MS 5907 (correspondence 1906–7), for evidence of the close working relationship between Haldane and Esher.

The sub-committee's lengthy deliberations and conclusions need be considered only insofar as they carried Haldane a step nearer his ultimate purpose.[78] The committee's brief was threefold: to consider the dangers to which Britain might be exposed by any developments in aerial navigation; the military advantages which Britain itself might derive from such developments; and the funding which should consequently be allotted for official aeronautical investigations. From the first, however, it was evident that the whole inquiry was being directed down a previously prepared furrow. It was given to Haldane to make an introductory statement and he used the occasion to extol the idea of a permanent Advisory Committee for Aeronautics, to consist of those perceived to be the most distinguished academics in the country. His logic was simple; the military needed scientific guidance: this was the essential prerequisite to any real progress. Needless to say this notion was wrapped in reams of opaque jargon, but the underlying message remained curt to the point of brutality: research had to be directed from a higher authority, not left in the hands of soldiers and waifs. The army, he claimed, with characteristic magniloquence, had never addressed the 'big questions . . . at the root of the whole', yet unless somebody apply their mind to 'the large view' the services would simply continue 'pottering on and accomplish very little'.[79]

An advisory body, constituted along these lines, was to come into being soon after the sub-committee's close. However the suggestion that the army had till then just been 'pottering on' reveals a pretty dismissive view of the work of the Balloon Factory. There is no doubt that its personnel had been marked for some time: the sub-committee was now to provide Haldane with a useful cloak behind which to finally draw his dagger.

It is no surprise, then, that the sub-committee recommended the discontinuance of aeroplane trials. By this means Dunne and Cody and all they represented (to Haldane) could be swept away, leaving the latter free to begin his systematic restructuring of the air service. It is doubtful that the sub-committee's deliberations affected this decision to any significant extent; this element, at least, was pre-ordained. Not that this could be deduced from reading transcripts of the sub-committee's meetings, still less by perusing its eventual report, out of context. Throughout the meetings Haldane was notably self-effacing, and much of the initiative was taken by his 'old friend' General Nicholson.[80] This bluff old cove did not mind how many toes he trod on, and Haldane simply let him run on like a stoat down a rat hole until his carping negativism had cleared the field.

Historians who have examined this report in the past have thus been misled

---

[78] Extracts from the sub-committee's report are reproduced as the first item in *Documents, Naval Air Service*, i. 7–13. The committee's proceedings are summarised in Walker, *Early aviation at Farnborough*, ii. 284ff., and then reconsidered in Gollin, *No longer an island*, 397ff.

[79] PRO, CAB 16/7. This introductory statement is reprinted in Walker, *Early aviation at Farnborough*, ii. 290–1.

[80] For this Nicholson reference see Haldane, *Autobiography*, 236.

into shifting responsibility for its conclusions on to General Nicholson and his subordinates.[81] In reality, Nicholson was Haldane's unwitting tool. The pair had collaborated closely on Haldane's army reforms and Nicholson had already gained a reputation as an acerbic, self-opinionated and uncongenial colleague to those unfortunate enough to differ from him.[82] As at this stage he personally could not give a rap for aeronautics, he was assiduous in attacking those who could. The fact that he was himself a Royal Engineer merely gave added credence to his tirades against Factory personnel. Nicholson came to be seen, indeed, as a myopic figure, occupying a similar position in the folklore of the pre-war advocates of military aviation as General Sir Archibald Montgomery-Massingberd was subsequently to hold in relation to the inter-war pioneers of mechanisation.[83] But only in the context of Haldane's plans can the significance of the sub-committee's conclusions be understood.

With regard to aeroplanes, the sub-committee took the view that, for all the progress that had been made over the previous year, 'they can scarcely yet be considered to have emerged from the experimental stage'. It had yet to be established whether these machines could be employed at high elevations, and the most successful models to date (i.e. those of the Wright brothers) required a mechanical apparatus to launch them; a clear drawback from a military point of view. Not only that but their reliability in unfavourable meteorological conditions was open to question. Lieutenant Dunne's experiments in this area were noted, but 'at present', it was felt, his machine lacked 'practical value'. Indeed, in military terms no machine, or even likely machine, was seen to promise much of any practical value. Given their various limitations it would be dangerous to employ aeroplanes for scouting purposes at sea, whilst cross-country it was thought that the speeds necessary to maintain such devices in flight would prohibit the gleaning of any useful reconnaissance information.

On the subject of offensive operations the sub-committee were equally blunt; aeroplanes had a marginal weapon-carrying capacity, and as it was a matter of conjecture anyway as to whether high elevations could be achieved it was pure speculation to suggest that they might out-manoeuvre airships in combat. The sub-committee had reportedly not received 'any trustworthy evidence' to indicate whether immediate improvements in altitude attainment were to be expected, or whether (as they evidently suspected) the limits of practical usefulness had already been effectively reached.[84] One might have

[81] Walker, *Early aviation at Farnborough*, ii, passim, reiterated all down the line by Gollin, *No longer an island*, passim.
[82] Spiers, *Haldane*, 190, 196; Haldane, *Autobiography*, 189, 198. They were to remain close associates right up to Haldane's departure from the War Office: see W. G. Nicholson to Haldane, 15 Mar. 1912, Haldane papers, MS 5909. Esher, on the other hand, fearing Nicholson's reputation, had in May 1907 secretly lobbied the king to decline Haldane's request that Nicholson be appointed CIGS: J. Lees-Milne, *The enigmatic Edwardian: the life of Reginald, 2nd Viscount Esher*, London 1986, 170.
[83] See, for example, Sykes, *From many angles*, 91.
[84] Report of Aerial Navigation Sub-Committee of the CID, 28 Jan. 1909, PRO, CAB

thought this a good reason for continuing supervised trials, but the sub-committee concluded, on the contrary, that there was no need for the government to continue experiments in this area. Instead, advantage could be taken of what was termed 'private enterprise'. No subsidy was offered to private manufacturers, however; no orders, or even promise of orders, should they develop improved machines. What this statement in fact meant was that, firstly, the Balloon Factory pioneers should be disengaged; and secondly, if by some chance aeroplanes did develop through private initiative this work should then be exploited by the services at no research cost to the War Office. In other words, aeroplane work was to be discontinued, this last option remaining open as a sort of safety-valve, should pressure again build up within the system.

The decision reflected a lack of faith in aviation on the part of both Nicholson and Haldane. When the question of leaving aeroplanes to private initiative arose Haldane had pointedly remarked that he had all along felt they should be left to 'other people'. Airships, however, were another matter. Unlike aeroplanes, he had been 'much impressed' with the possibility of 'the dirigible' becoming 'a very potent factor in war'.[85] In retrospect it is clear that he had already concluded that the state's aeronautical research should be concerned primarily with airships. The sub-committee's report was effectively clearing the way for this. Aeroplanes, it stated, could look after themselves; they had little military value and could survive in the market through the demands of 'sport and recreation'. Airships had no such market: their future development was wholly dependent upon the extent of their military employment. The government was therefore obliged to build them.

This view of the incipient aeroplane industry was a delusion, as the Committee of Imperial Defence was soon to discover; for present purposes, however, it was enough to suggest that some form of private enterprise would be there, to be taken advantage of, should the need arise. The aim was simply to obviate the need to continue the aviation experiments at Farnborough.

In sum, then, the sub-committee concluded that Britain could only ascertain the dangers to which she might be exposed from developments in aerial navigation by possessing her own airships; that the evidence suggested that these held a powerful reconnaissance and offensive potential; that consequently £35,000 should be included in the Naval Estimates for the purpose of building an RN rigid airship; that £10,000 should be included in the Army Estimates for further military airship experiments; and that government-sponsored aeroplane trials should be discontinued at the first opportunity.

16/7. Haldane would have to concede that aeroplanes were capable of 'high elevations' when the CID reconsidered the matter some year-and-a-half later: minutes of 107th meeting, 14 July 1910, PRO, CAB 2/2.

[85] Report of Aerial Navigation Sub-Committee, 62. Gollin quotes this passage (*No longer an island*, 420) and then (p. 428) affirms that the discontinuance of aeroplane trials was a 'bad rejection' of Haldane's advice. This is a direct contradiction of the evidence.

In fact the question of a naval airship had been under investigation since before the sub-committee met. Efforts were being co-ordinated by the Director of Naval Ordnance, Captain R. H. Bacon,[86] and, indeed, Vickers had been invited to tender for the Admiralty contract as early as August 1908.[87] Consequently, Bacon's principal aim throughout the deliberations of the sub-committee had been to win additional support for this project and thereby gain a recommendation for suitable funding. (The Treasury had, before then, begun to backslide on the scheme.)[88] In this he was entirely successful, and Vickers's tender, involving an estimate of £30,000, was provisionally accepted the following March.[89] Bacon was obliged to resign soon after, however, in the wake of the Beresford-Fisher dispute, and it was left to Murray F. Sueter, the newly-appointed Inspecting Captain of Airships, to see the project through. It ended prematurely in the 'Mayfly' disaster of September 1911. The result, slightly ironically in view of the sub-committee's recommendations, was that the Admiralty temporarily abandoned airship work altogether.

The sub-committee report itself was dated 28 January 1909, and a little under a month later, on 25 February, it came before the full CID for confirmation. Here it was approved without amendment; such debate as there was being overshadowed by an offer from one of the sub-committee's witnesses, C. S. Rolls, to place his services and a Wright aeroplane at the disposal of the government, provided they furnish him with suitable training facilities. Rolls was referring here to the first of the six Wright biplanes that Short Brothers had been contracted to build at their new workshop on the Isle of Sheppey. He made this offer in a letter to Lord Esher, who then brought it to the attention of the CID with a recommendation that it be accepted. Given that the sub-committee had just recommended that army aeroplane trials be discontinued 'provided advantage be taken of private enterprise' it would have been contradictory to refuse – as the War Office had refused A. V. Roe's request for flying facilities the year before. On the other hand, the 'private enterprise' clause was always hopelessly ill-defined and this incident only further confused matters. What the offer represented was not investment in any embryonic industry; it was merely a proposal for taking advantage of the generosity, as it then appeared, of a private individual, which was an entirely different thing. Yet the offer was accepted as fulfilling the original provision, leading Churchill, at that time President of the Board of Trade, to complain of amateurism.[90] To Haldane, however, it was a further useful decoy with which to draw off the fire of angry aviation enthusiasts. It had no effect on his reorganisation of the Balloon Factory.

---

[86] Later Admiral Sir Reginald Hugh Bacon.
[87] Admiralty to Vickers Sons & Maxim Ltd, 14 Aug. 1908, PRO, AIR 1 2306/215/15. Bacon invited Vickers to tender for the naval airship contract because the firm had already successfully developed Admiralty submarines.
[88] Treasury to Admiralty, 9 Sept. 1908, ibid.
[89] Admiralty to Vickers, 9 Mar. 1909, ibid.
[90] CID, minutes of 101st meeting, 25 Feb. 1909, ibid. CAB 2/2.

Rolls may, in fact, have made some similar offer a few months before. In July 1946 the Royal Aircraft Establishment issued a report entitled 'A historical summary of the Royal Aircraft Factory and its antecedents, 1878–1918', by S. Child and C. F. Caunter. This provided a chronological list of developments at Farnborough. Unfortunately it was compiled in a very perfunctory manner and is in many respects unreliable, but the authors may have had access to documentation no longer surviving. The summary suggests that in October 1908 Rolls offered to 'bring to the Factory a Farman-Delagrange biplane', but that Cody's recent accident, attributed to lack of manoeuvring space, had discredited the ground at Farnborough and that acceptance of the offer had been deferred. If true, Rolls had presumably been offering himself as an intermediary between the Balloon Factory and Farman or Delagrange, or, more likely, the Wright brothers themselves, for he had yet to acquire an aeroplane of any description. He had, however, recently been in France with Wilbur Wright and had long been seeking to represent the brothers in the UK. Indeed, there had been talk of the Wrights supplying Rolls with a Wright Flyer with which Rolls would act as the brothers' agent in Britain, in advance of which he was planning to receive some practice on a Voisin biplane.[91] As an expert witness at the CID sub-committee inquiry Rolls had doubtless surmised that the War Office would now be obliged to welcome such a proposal – despite the fact that the Short-Wright biplane embodied all the shortcomings of which the committee had so recently complained. It required, for example, the Wrights' cumbersome catapult-start mechanism and had not even been completed, let alone tested, yet. Indeed, the Short-Wright contract had only just been agreed.

Rolls acknowledged these difficulties in a letter to Lord Esher written in March 1909, after the CID's provisional acceptance of his offer. Here it was understood that the government were to await practical demonstrations with the aircraft before embarking upon the search for a suitable flying ground and other facilities; but, in reality, matters had already gone to Rolls's head. He immediately offered to ask the Wrights, when next he saw them, on what terms more machines could be supplied to the War Office and under what conditions they might be induced to visit England to arrange for the tuition of selected officers.[92] He had no authority to go this far, but having been given an inch he thought it a pity not to try for the mile. Soon he was stumbling over the inch. Orville Wright confirmed to Griffith Brewer that arrangements had been made to have Rolls taught to fly on a Wrights' machine in France, but added that he was only sixth on the list and that it would be 'some months' before he could begin.[93] At this point Rolls lost patience and ordered Shorts to build him his

[91] Ibid. AIR 1 686/21/13/2245; Gollin, No longer an island, 308, 356, 369, 381–2. Raleigh, who evidently had access to the same report as Child and Caunter, had already incorporated the information without comment: War in the air, 156. See also Bruce, Aeroplanes of the Royal Flying Corps, 613.

[92] C. S. Rolls to Lord Esher, 22 Mar. 1909, Esher papers, ESHR 5/29.

[93] O. Wright to G. Brewer, 31 Mar. 1909, RAeS archive.

own training glider. This did little to advance the army's position, but in the end it hardly mattered. Due to delays in acquiring the prescribed Wright-designed Bollée engine it was October 1909 before the much-vaunted Short-Wright biplane was even delivered to Rolls. Only then did the search for a suitable flying ground actually begin.[94] By that time the CID sub-committee's conclusions had begun to look increasingly outmoded anyway.

One event in particular, the Rheims aviation meeting of August 1909, had seemed to make a nonsense of the claim that aeroplanes had no practical value. This meeting, wrote one English military representative just three months later, had 'done more to focus the attention of the world on the possibilities of the aeroplane than any other event in the history of its evolution'.[95] Ironically, plans for the meeting were being reported in the British press on the very day the Prime Minister was announcing the formation of the CID Sub-Committee on Aerial Navigation.[96] By the time the Rheims extravaganza took place, however, the sub-committee had long since reported and Haldane was poised to reconstitute the Balloon Factory as an autonomous research department. There is no evidence that his views were in any way modified by the events at Rheims. Aeroplane trials had been discontinued at Farnborough and no provision for them was incorporated into the reformed Factory organisation when it was launched in October 1909.

The French, in similar circumstances, showed a slightly more flexible attitude. Battalion Leader Renaud, head of the French army's Laboratory of Aviation Research, reported that while it was true that existing aeroplanes did not yet 'unite all the qualities required for military usage' ('les aéroplanes existants ne réunissent pas encore toutes les qualites requises pour les usages militaires'), nevertheless it was evident that even as they were they could give valuable service in time of war.[97] He consequently recommended that examples of existing models be acquired by the French army, with a view to training a group of selected officers in their use. By the end of the year the Minister of War had sanctioned the purchase of three machines; a Farman biplane, a Blériot XI monoplane and a Wright Flyer. They were all seen as performing an essentially reconnaissance role.

While no immediately discernible change occurred in Britain in the wake of the Rheims meeting it is clear in retrospect that cracks were beginning to emerge in the official consensus. There was incessant criticism which, in light of events, was virtually irrefutable. Lord Montagu of Beaulieu spoke for many when he alleged that the military 'badly want[ed] educating' in the possibilities of aerial warfare.[98] For Haldane none of this really mattered; he had resolved

---

94 C. S. Rolls to Lord Esher, 25 Oct. 1909, Esher papers, ESHR 5/32.
95 Col. F. G. Stone, 'The Rheims aviation week and its value from a military point of view', *Journal of the Royal Artillery* xxxvi (1909), 353.
96 *Morning Post*, 23 Oct. 1908 (Butler collection, vii. 49).
97 Facon, 'L'Armée Française', 80–1.
98 'Aerial machines and war', lecture dated 22 Feb. 1910 to Aldershot Military Society, Montagu of Beaulieu papers, MI/J18.

to sweep the Factory clean of its existing personnel, the CID sub-committee report had begun that process admirably, and he was now free to re-establish the organisation on a scientific basis. Other sub-committee members, however, found it increasingly difficult to look back upon their recommendations without a sense of profound misgiving. Chief among them was Lord Esher.

Esher had expressed no strong views in the course of the sub-committee's deliberations. This was no doubt primarily because he was not *au fait* with the subject, but it was probably also because he was mindful of maintaining a detached political neutrality. Either way, he was soon not only embroiled in the whole aeronautical controversy, but embroiled in such a way as to endanger his position as a disinterested servant of state. The trouble arose out of his presidency of the Aerial League. After this ill-judged involvement, in which he unwittingly got his fingers burnt, he began to view the air question in a more critical light.

The Aerial League of the British Empire had been formed in January 1909 as both an analogous movement to the successful Navy League and as a response to Germany's Aerial League, founded some three months before.[99] Esher was first approached on the subject of becoming its president in early July 1909, five-and-a-half months after the CID sub-committee had reported. With his habitual prudence he raised the suggestion with the Secretary of State for War before committing himself. Haldane was effusive in his support.[100] Esher therefore consented, but one wonders with how much personal commitment. He had always maintained rather an aloof stance, and it is doubtful if he ever intended to play an active part in the league's affairs. When Blériot flew the channel later the same month, for example, Esher was invited to a celebratory luncheon, but declined, his journal records, because he hated such 'hysterical' scenes.[101] This hardly denotes an enthusiasm for popular movements.

Following Esher's acceptance of the presidency, the executive committee began canvassing for closer service affiliations. Using Esher's name (albeit indirectly), letters were despatched to the Admiralty, the War Office and the Board of Trade, encouraging them to nominate representatives to be elected members of the league's council.[102] Somewhat surprisingly, the defence services reacted positively, especially the Director of Naval Ordnance, Captain Reginald Bacon, on behalf of the Admiralty. If well run, he commented, the league could be 'of the greatest assistance' to his department; and the best way to ensure that it was well run and kept from developing into simply an 'agitating body' was to have government representatives on its council. The War Office

99 'British Aerial League: new movement', *Naval and Military Record*, 28 Jan. 1909 (Butler collection, vii. 132).
100 Haldane to Esher (draft copy), 8 July 1909, Haldane papers, MS 6109.
101 Entry for 26 July 1909, *Journals and letters of Reginald Viscount Esher*, ii, ed. M. V. Brett, London 1934, 396.
102 Chairman, executive committee, Aerial League (Capt. R. A. Cave-Browne-Cave RN), to Dept of Admiralty, 16 July 1909, PRO, AIR 1 648/17/122/398.

agreed.[103] No objection was raised, either, to officers on the active list joining the league, provided, again, that there were 'efficient safeguards' to prevent it becoming a 'political' body.[104]

Developments were reported at an extraordinary general meeting of the Aerial League on 30 September 1909, where it was revealed that Captain Stuart Nicholson, Assistant Director of Torpedoes, was to be the Admiralty's representative, with Colonel J. P. du Cane, a General Staff officer, acting on behalf of the War Office. The following day stories of both Esher's involvement and the services' presumed sympathy with the aims of the league were all over the papers.[105]

It may have been the extent of the publicity that first caused Esher some discomfort. A memorandum, prepared for the First Lord, Reginald McKenna, by the league's Admiralty representative in December 1909, recorded some evident misgivings. It described how, at a meeting of the league's council in November 1909, Esher had asked what the organisation proposed to do if it felt the government was failing to make sufficient provision for the aerial defence of the empire. So long as the council incorporated representatives of government departments, he reminded them crisply, it would remain constitutionally impossible for the league to take any action which would in any sense pressurise the government. Given that, what useful function would, or could, it perform?[106] Captain Nicholson privately concurred with this assessment of the league's increasingly anomalous position, but argued that service representatives need not be withdrawn from the council unless or until it espoused a programme openly prejudicial to government interests. Esher, however, was not prepared to wait: he had already been the subject of attacks in parliament and the press for exercising what was perceived to be an unconstitutional and irresponsible influence on political affairs, and these criticisms were to recur the following year.[107] The last thing he needed was to provide his critics with a further avenue of attack. By the end of December 1909 he had resigned his presidency.

The following March the Aerial League issued what it probably imagined to be a fairly innocuous leaflet containing, amongst other things, some unfavourable assessments of Britain's aeronautical position when compared with other major European powers. Public opinion, it contended, must be roused at once. This was too much for Nicholson. The league had made what amounted

---

[103] Admiralty minute, Capt. R. H. Bacon (DNO), 21 July 1909; War Office (E. W. D. Ward) to sec., Dept of Admiralty (Graham Greene), 9 Aug. 1909, ibid.

[104] Memo, R. H. Brade (sec., War Office) to sec., Dept of Admiralty, 16 July 1909, ibid. AIR 1 653/17/122/490.

[105] For example, *Standard*, *Morning Post*, *Daily Telegraph*, all for 1 Oct. 1909 (Butler collection, ix. 74–80).

[106] 'Notes on the policy and methods of the Aerial League', prepared by ADT (Asst Director of Torpedoes) for the information of the First Lord, 10 Dec. 1909, PRO, AIR 1 648/17/122/398.

[107] See Lees-Milne, *Enigmatic Edwardian*, 168–9, 220–1.

to 'a direct criticism of the government' and he forwarded a memorandum to the First Sea Lord recommending the immediate withdrawal of all state representatives from its council.[108] This course of action was duly followed, but the whole incident had raised a delicate problem for the services: just how far should they be drawn into the concerns of private, non-governmental bodies? In this particular case the tensions were clear and the outcome straightforward: the service departments, after all, did not need the league. When dealing with private bodies respecting arms procurement, however, the positions might not be so unequivocal. Industrialists in the military market had as much interest as the Aerial League in putting pressure on the government, but here the service departments might find themselves in a position of dependence rather than co-operation, particularly where technological advancement was concerned. Haldane's re-establishment of the Balloon Factory as a state-based research and design department would make it possible for the military to effectively by-pass any potentially ambiguous dependence on the private sector and thus impose its own will on the market. The Aerial League imbroglio may unconsciously have helped pave the way for this far-reaching development.

Meanwhile, having managed to extricate himself unscathed from the Aerial League, Lord Esher had begun to reconsider the whole emphasis of the government's position as recently established by the CID sub-committee over which he himself had presided. In a note prepared for the CID in October 1910, he drew attention to the sub-committee's recommendation that state-sponsored aeroplane trials be discontinued and impassively retracted the suggestion. It was, he maintained, 'hardly necessary' to point out the 'immense advance' in the construction and handling of aeroplanes that had occurred since the sub-committee had reported in January 1909. The purely experimental phase of aviation, in which the efficacy of heavier-than-air flight was itself in question, was now irrefutably over: an army unable to call upon aeroplanes for reconnaissance purposes would be seriously disadvantaged in the field. Such machines should be purchased at once and officers trained to fly them. The issue, as he now saw it, was no longer whether an air corps should be formed, but the form it should take – in other words, what autonomy it should have from the present services. This was the problem the CID should be addressing.[109]

Esher was influenced in this view by reports of the previous month's manoeuvres in France, in which aeroplanes had participated for the first time. Aeroplanes had also taken part in the recent British army manoeuvres, but here they had been admitted largely on sufferance. In accordance with the Esher sub-committee's recommendations the War Office was obliged to recognise the assistance of private enterprise. The initiative for their employment came directly from the British & Colonial Aeroplane Company. It can have

---

[108] Admiralty memo, ADT (Capt. S. Nicholson) to First Sea Lord, 9 Mar. 1910, PRO, AIR 1 648/17/122/398 – enclosing copy of offending leaflet.
[109] 'Aerial navigation, note by Lord Esher', 6 Oct. 1910, ibid. AIR 1 2311/221/32.

been no coincidence, however, that shortly after these sets of manoeuvres, and following Esher's retraction of the CID sub-committee's findings, the reconstituted Balloon-cum-Aircraft Factory at Farnborough finally made provision for aeroplane as well as airship design. Thus, paradoxically, just as it was the sub-committee's findings which enabled Haldane to clear the Factory and refound it on a scientific research basis, so it was the chairman's subsequent repudiation of those same findings which helped induce it to take up aeroplane design afresh. This decision was to profoundly affect military–industrial relations in the field of aircraft production until well into the war.

Another blow to the CID's sanguine view of aeronautics came with the International Aerial Navigation Conference, which opened in Paris in May 1910. This had been convened by the French government, after a number of German balloons had illicitly descended on French soil. In Britain an inter-departmental committee was appointed to consider the subject of international air passage in preparation for the conference, but the Foreign Office declined to be represented because it believed the conference itself was to be confined to minor issues, such as permits, identification marks, customs and so forth. This was indeed the original understanding, but Germany then widened the discussion to incorporate the whole question of the legitimacy of free passage over foreign territories. Even then, however, the Foreign Office seemed strangely unconcerned by the possible implications. By this wider provision, at least as it was presented at the conference, a state would retain the right to impose restrictions on its air space, but only to ensure its own security. In other words, as it gradually dawned on the British delegation, Germany was laying the ground for free passage over neutral Holland, Belgium and Denmark in the event of hostilities with Britain and France. And they nearly succeeded. As it was, the British delegation suddenly found itself confronted by questions of strategic and political importance for which it had no brief. When Germany refused to proceed with the rest of the conference without the inclusion of this clause Britain insisted on an adjournment. The following November the conference was abandoned altogether.[110]

Only at this point did some sense of urgency finally enter into the government's deliberations on aeronautical matters.[111] As an incidental result, confidence in the original report of the CID sub-committee, which had failed to predict any such diplomatic difficulty, was further undermined. The failure of the Admiralty's own rigid airship the following year just about broke the camel's back. According to Maurice Hankey, who had been appointed assistant secretary to the CID in January 1908, many people had at the time doubted whether

---

110 'The strategical aspects of certain proposals before the International Conference on Aerial Navigation', 23 June 1910, PRO, FO 368/405; K. Hamilton, 'The air in entente diplomacy: Great Britain and the International Aerial Navigation Conference of 1910', *International History Review* iii (1981), 169–200; Lord Hankey, *The supreme command 1914–1918*, i, London 1961, 110–13.

111 See *Documents, Naval Air Service*, 14ff., where a number of items relating to this episode are reproduced.

the sub-committee had been right in its recommendations, particularly with regard to aeroplanes.[112] The following year had only too amply confirmed these doubts. Indeed, the sub-committee's conclusions had been questioned even within official circles as early as May 1910, when the Home Ports Defence Committee came to consider the question of the defence of magazines, cordite factories and other vulnerable points against airship attack. Colonel Capper's well-aimed suggestion that aeroplanes might in the near future successfully defend such placements led it to advocate both the resumption of aviation experiments and the reconsideration of the air question generally by the War Office and Admiralty. The CID, meeting on 14 July, concurred, but discussions on the subject were already taking place within the Master General of Ordnance's office.[113] Haldane, however, had got what he wanted from the original report, and that, in the end, was about its only major effect. In other ways, it rapidly became little more than an historical curiosity.

Having organised the clearance of the Balloon Factory's empiricists Haldane next set about establishing the promised scientific Advisory Committee for Aeronautics. The appointment of this body was eventually announced by the Prime Minister on 30 April 1909, some three months after the CID sub-committee had reported. Colonel Templer, former Superintendent of the Balloon Factory, and Colonel (retd) J. D. Fullerton RE, secretary of the Aeronautical Society, had proposed their own civil–military Advisory Committee for Aeronautics back in 1906, and this is sometimes held to have been a precursor of the Haldane scheme,[114] but there is no evidence that the two were related. It was Haldane's cherished belief that no useful practical work could be accomplished until the theoretical essentials of aeronautics had first been established; once this had been done, scientific principles could then be applied to the army's own research and development. This was what he meant by structuring an air service 'on a foundation of science'. To this end he enlisted the support of the distinguished Cambridge physicist, Lord Rayleigh, who became the new advisory body's president, and the director of the National Physical Laboratory, Dr Richard Glazebrook, who became its chairman.

Haldane had known Rayleigh for some years. They had worked together as members of the War Office Committee on explosives, set up in May 1900 to investigate, in the wake of the Boer War, the best smokeless propellant for the army, and had obviously retained a close mutual respect.[115] As a former professor of experimental physics at Cambridge and president of the Royal Society Rayleigh had just the sort of academic eminence that Haldane – who judged everyone by the letters after their name – doted on.

[112] Hankey, *Supreme command*, i. 109.
[113] Memo, 19 May 1910, PRO, CAB 13/1; 'Aerial navigation, note by sec.', 6 July 1910, CAB 4/3, CID 117-B; CID, minutes of 107th meeting, 14 July 1910, CAB 2/2. There is a resumé of these developments in CAB 21/21.
[114] PRO, AIR 1 686/21/13/2245; Raleigh, *War in the air*, i. 158.
[115] Haldane, *Autobiography*, 233; Sir F. Maurice, *Haldane: the life of Viscount Haldane of Cloan*, i, London 1937, 104.

Essentially, then, it was Haldane's plan to bring together those he considered to be the best experts in the country to oversee the aeronautical research continuing at Farnborough. This was, at the time, still being conducted under the supervision of Colonel Capper, but plans for his removal – as the next stage of the restructuring process – were already afoot. The Admiralty and the War Office were to be represented on the advisory body by the same men who had served on the CID sub-committee: Captain R. H. Bacon, the Director of Naval Ordnance, and Major-General Sir Charles Hadden, the Master General of Ordnance, though the latter, certainly, had previously shown no aptitude for the subject.[116] Of the other members, only one was not a Fellow of the Royal Society, and yet he, paradoxically, was probably the one figure who understood the practical problems of flight, for he was F. W. Lanchester. It is a further irony, but perhaps not altogether surprising, given that Lanchester had no academic status, that this last appointment came directly through Rayleigh and not through Haldane.[117] Lanchester's aeronautical work was wholly concerned with heavier-than-air powered flight, a fact which would suggest that Rayleigh, at least, was not as restrictive in his views as the non-scientist 'apostle of science', R. B. Haldane. Given the CID sub-committee's report, however, it was going to be some time before either he or any other member of the advisory committee could effectively prove it.

Lanchester's early years have already been touched on, for he was an automobile pioneer of major importance, as well as an early theorist of aerodynamics. Born in Lewisham in 1868, the son of an architect-cum-surveyor, Lanchester, like his near contemporary, H. G. Wells, had attended the Normal School of Science (later the Royal College) in South Kensington, and supplemented his engineering skills by enrolling on evening classes at the Finsbury Technical College, where he would later be followed by such figures as Handley Page and Richard Fairey. However it was primarily through his own efforts that his aeronautical thinking developed. He later claimed to have begun the serious study of mechanical flight as early as the period 1891–2, during which time he was occupying the relatively humble position of works manager with the Forward Gas Engine Company of Birmingham.[118] From the first he realised that the 'solution of the practical problem' depended on the power-to-weight ratio of any engine, and in conjunction with his aerodynamic investigations – conducted through a series of model aeroplane trials in the backyard of his semi-detached suburban villa – he began designing small high-speed motors. He perceived, however, that it would be some time before an effective aero-engine could be created. 'Its evolution', he wrote later, 'depended upon finding some kindred commercial purpose as a nursery for its

[116] According to Sir Henry Tizard, who was attached to the RAeF in the First World War, Hadden simply 'got infected with the [prevailing] scientific spirit' engendered by Haldane: *Nature*, Mar. 1958 (obituary article on Mervyn O'Gorman).
[117] Claim for recognition by Air Ministry (1936), Lanchester papers, 1/12/5.
[118] Ibid. and 'Concise history of my career' (1926) 1/2/1.

development.'[119] It was a crucial point that Haldane never entirely grasped. Yet before having his attention diverted to automobile work (i.e. to that kindred commercial purpose), Lanchester formulated what was to become the fundamental circulation theory of lift on a fixed wing aircraft in motion: the vortex theory.

Lanchester first presented his hypothesis in a paper to the Birmingham Natural History and Philosophical Society in June 1894 – that is some nine years before the Wright brothers flew; but it made little immediate impact. He revised and expanded upon it over the next few years and in 1897 submitted a paper to representatives of both the Royal Society and the Physical Society. Perhaps on account of Lanchester's self-confessed 'inelegant English' (which suggests more than a hint of class diffidence), both these institutions turned him down. After that business commitments absorbed Lanchester's attention, and it was 1907, a full decade later, before the theory was finally published, in the first volume of his book *Aerial flight*. Much had happened in aviation in the meantime. Even in 1909, however, it was evident that Glazebrook and other members of the National Physical Laboratory still viewed Lanchester with considerable misgivings.[120]

Initial reaction to his work came in fact not from England at all, but from Germany. This, one instinctively feels, rather than the inherent merits of his work, was the factor which most influenced Haldane in sanctioning the inclusion of such a lowly figure on the advisory committee. The year before, the eminent Göttingen physicist, Carl Runge, had written to Lanchester seeking permission to translate the two volumes of *Aerial flight* into German. In September 1908 Lanchester had then visited Göttingen to assist in the work.[121] Under these circumstances how could Haldane refuse him a place on the committee? The fact that he would not otherwise have given Lanchester a second glance, that Lanchester's aeronautical work was wholly concerned with heavier-than-air powered flight, that if this engineering-based empiricist could think through the problems of aerodynamics other contemporaries might be in a position to do the same, were matters that caused Haldane not a moment's concern.

Lanchester's vortex theory was later independently rediscovered by Professor Ludwig Prandtl of Göttingen. It is sometimes suggested that Lanchester's work was too difficult to understand, and that it was Prandtl's exposition that made the theory comprehensible and that consequently Lanchester had no

[119] 'History of the Lanchester venture' (1938), ibid. 7/25/2. Lanchester had already demonstrated his skill in engine design with the Forward Gas Engine Co., patenting the world's first automatic gas-engine starter and various engines in the early 1890s: Clark, *Lanchester legacy*, 3–4.
[120] F. W. Lanchester to Prof. A. V. Hill (sec., Royal Society), 2 June 1939, Lanchester papers, 2/15/2 – recounting history of vortex theory.
[121] Kingsford, *Lanchester*, 100–1; 'Notes concerning the position of aeronautics in 1908', Lanchester papers, 1/16/2.

influence.[122] However this neglects the fact that Lanchester's ideas were perfectly well understood by Professor Runge and were widely disseminated throughout Germany – via Göttingen. Prandtl was thus able to incorporate Lanchester's ideas into his own independently-formulated research. Indeed, the formula is now generally designated the Lanchester-Prandtl circulation theory of lift. Notwithstanding this, Glazebrook and the National Physical Laboratory – and therefore Haldane – consistently chose to attribute the formula exclusively to Prandtl, who embellished it with what Lanchester called 'suitable mathematical embroidery'.[123]

Lanchester, being still largely unrecognised, was, for all the difficulties, immensely flattered by being asked to join such a seemingly august body. Even so, he remained something of an outsider. He had that same year become technical adviser to the new aviation firm of White & Thompson, for whom he was planning to design a biplane with a Daimler engine. (Lanchester was consulting engineer at Daimler's Coventry works.)[124] Thus his theoretical thinking intersected reality at points a world away from Haldane's nebulous abstractions. For a time, however, service on the advisory committee would simply channel his creative energy into oblivion. Only when the reconstituted Balloon Factory was obliged to recruit its own aeroplane designer, Geoffrey de Havilland, was Lanchester again to have a creative outlet for his theoretical work. By then his attention was being drawn increasingly to the problems of aviation as an instrument of war, a subject examined in his book *Aircraft in warfare: the dawn of the fourth arm*, first published in serial form in late 1914.[125] His thinking ranged far in advance of the rest of the advisory committee. Nevertheless he remained loyal to the government, supporting it consistently in years to come against the monopoly accusations of manufacturers. As a government servant he had by then come to feel himself a part of what C. G. Grey and others were attacking.

Whatever the constituent elements of the advisory committee, Haldane, at least, was delighted with it. 'We have at last elaborated our plans for the foundation of a system of Aerial Navigation for the Army and Navy', he wrote triumphantly to Northcliffe, one of the War Office's chief critics in the matter, on 4 May 1909. He had constructed, he claimed, a 'real scientific Department of State for the study of aerial navigation'. This, Haldane confessed, had

---

[122] See, for example, Gibbs-Smith, *Aviation*, 125.

[123] Lanchester papers, 2/15/2.

[124] Kingsford, *Lanchester*, 118ff.; Lanchester papers, 1/2/5. Lanchester was adviser to White & Thompson from 1909 to 1911, designing a single-seat – amended to two-seat two-engine – aluminium clad tractor-biplane with tubular steel struts. Constructed at Daimler's Coventry works, it crashed during trials.

[125] The title 'fourth arm' relates aviation to the other three branches of the army – infantry, cavalry and artillery – although in the book Lanchester was to become an early advocate of strategic bombing. He has also been seen, through his use of 'quantitative methods', as a pioneer of what became known as 'operational research', and even as a precursor of the 'formal strategists' of later years: L. Freedman, *The evolution of nuclear strategy*, London 1981, 177.

'particularly engrossed' him, for 'no other foreign Government' had yet got such a department. Next day the Prime Minister would be answering questions in the House on the subject, and Haldane himself would be available for press comment afterwards. He thought, however, that Northcliffe would like to know in advance that 'a real step forward in a subject in which you are much interested is now being taken'. The new department, which was to be con-nected with the National Physical Laboratory, would be under Treasury, rather than War Office or Admiralty control, but by this means it could act as a connecting link between these last two, both of which were to have their own 'construction establishments'.[126]

Haldane's very eagerness to convey all this to Northcliffe suggests that he was expecting his support. If so, he was to be sadly disillusioned. In his reply on 9 May 1909 Northcliffe described the composition of the committee as 'one of the most lamentable things' he had ever considered in connection with 'our national organisation'. Not only could he see no 'practical aviator' in the list of members, but the committee's agenda was such 'as would have been excellently discussed fifteen years ago'. The committee, if it was to have any value, should certainly include some 'practical exponents'. Lanchester, whom he knew as a motor expert, he admitted was something of an exception, but he dismissed him, in this context, as 'one of those unfortunate people who gave themselves away about the Wright aeroplane before they had seen it fly'.[127] This was less than fair, but having seen the brothers' machine in action Northcliffe was still particularly keen to off-load Wright aircraft on to the War Office. He considered it a total waste of precious resources to now start investigating that which was already known: if an advisory committee was to be formed, it should at least, he suggested, be a practical committee concerned with present events.

Northcliffe bandied around the word 'practical' almost as freely as Haldane did the term 'scientific', but he was not alone in his criticism. Frank Hedges Butler, the co-founder of the Aero Club, was quick to make essentially the same point. In a statement to the press he derided the committee's members as being 'by no means experts on the subject' – few of them having so much as seen an airship or aeroplane. The idea of a collection of scientific figures was all very well, but they should at least be joined by a selection of practical men.[128] The Aero Club and Aeronautical Society would have gladly helped here had they been consulted.

These views were widely disseminated in contemporary newspaper re-ports,[129] but Haldane saw no need of amendment: what, after all, did such

126 Haldane to Northcliffe, 4 May 1909, Northcliffe papers, Add. 62155.
127 Northcliffe to Haldane, 9 May 1909, ibid.
128 Quoted in *Evening Standard*, 6 May 1909 (Butler collection, viii. 56).
129 'The new committee: need for practical men' ran a *Morning Post* headline on the subject, 7 May 1909; 'Government committee criticised: too many theorists' confirmed the *Ob-server*, 9 May 1909 (Butler collection, viii. 65, 67).

people understand of science? He defended his position in a letter to North-cliffe on 18 May 1909. The advice which he had received, Haldane was adamant, had convinced him

> of what I was very ready to be convinced, that here as in other things we English are far behind in scientific knowledge. The men you mention are not scientific men nor are they competent to work out great principles: they are very able constructors and men of business. But in this big affair much more than that is needed.[130]

Naval and military experts, he added, had 'demonstrated to the Defence Committee' that dirigibles, 'and still more aeroplanes', were a 'very long way indeed off being the slightest practical use in war'. At Göttingen, 'my old university', he went on (he had, in fact, spent a summer semester there in 1874), there is an aeronautics department to which no student is admitted who is not 'expert with the differential calculus'. The Germans, he concluded from this, knew how much had got to be learned before practical results could be obtained. This, he asserted, was his principle too. Surely, it was inferred, his critics were not questioning the value of Göttingen?

What was poor Northcliffe to make of all this? Haldane here merely smothers all argument with a blanket of academic pomposity. In reality the Germans were singularly confused in their aeronautical proceedings. And while the students of Göttingen were earnestly using differential calculus, motor-engineers from the Finsbury and Crystal Palace technical colleges were designing aeroplanes. It never seems to have occurred to Haldane that he might usefully help these institutions, where some understanding of both aerodynamics and the internal-combustion engine already existed. These colleges were in fact as 'expert' with calculus as any German institute of learning: Sylvanus Thompson, tutor to Handley Page and Richard Fairey at Finsbury, was to publish the standard text-book on the subject, *Calculus made easy*, in 1910. Moreover, if Haldane was tempted to reflect during the Fokker scourge of 1915 on the superiority of German science, he ought to have been reminded that the Dutchman, Anthony Fokker, first went to Germany in 1910 to enrol in the continental equivalents of Finsbury and Crystal Palace. The truth is that the great aviation firms of the period – Avro, Farman, Shorts, Handley Page, British & Colonial, Sikorsky, Fokker, Curtiss, Sopwith – all developed as oblivious of Göttingen as they were of Oxford. With his attitude Haldane was simply widening the gap between state research and industry, and eventually state and industry were to need each other. Of course, Haldane made it clear that he had no faith in the military potential of aeroplanes anyway, but even when referring essentially to airships it was evident that he was establishing a divisive pattern for the future.

On the other hand, Haldane was concerned lest people confuse the role of the advisory committee with the practical work he was envisaging for the

---

[130] Haldane to Northcliffe, 18 May 1909, Northcliffe papers, Add. 62155.

Balloon Factory. The reorganisation of that establishment was to be the next, and final, stage of his aeronautical reforms. To forestall any further criticism he discussed his aims with Northcliffe's erstwhile ally, Arthur Balfour, and he (as Haldane told Northcliffe in that same letter of 18 May) was to address a question to the Prime Minister on the subject so that the situation might be clarified. Asquith's original statement, on 5 May, had simply said that an advisory committee was to be formed for the superintendence of investigations at the National Physical Laboratory and to give general advice on scientific problems arising in connection with military aeronautics. On 20 May, in reply to Balfour, it was further added that it was no part of the committee's general duty to construct or invent but merely to consider problems referred to it by the executive officers of the military construction departments. There was thus, it was implied, no need for practical men on the advisory committee. These lofty scientists would merely consider essential principles, and advise on such theoretical problems as arose in the course of events. They would not initiate design, but simply furnish guidance on work in progress.[131]

Northcliffe possibly realised that Haldane was drawing on wells deep within his own psyche, for he abandoned the correspondence at this point. In referring to the matter some years later he regretted not cornering Kitchener on the subject instead: 'he', Northcliffe assured Lieutenant-Colonel Charles à Court Repington, *Times* military correspondent, 'would have understood it'. One may doubt this, but one may assume that he would not, at least, have embroidered it with his own brand of Hegelian didacticism. Haldane, Northcliffe concluded, was merely 'suave and obviously bored'.[132]

If Haldane could appear so patronising to such as Northcliffe, one may imagine how remote he and his ideas must have seemed to ordinary pioneers, most of whom were without any influence. The great value of the advisory committee, the official history tells us, was that it brought together 'the various bodies concerned with aeronautics' and combined their efforts.[133] Obviously it did no such thing. In fact the committee soon came to be seen as (and indeed, acted as) little more than a branch of the government Aircraft Factory, and as such simply fuelled manufacturers' fears of the emergence of a government monopoly.[134] The man who was to become the industry's principal spokesman was already sounding warnings. This was the journalist C. G. Grey.

Charles Grey Grey, contrary to what might be expected from the fierce

---

[131] 'The formation of the Advisory Committee for Aeronautics', PRO, AIR 1 725/102/1. These arguments were rehearsed again in a sterile Commons debate of 2 Aug. 1909, shortly after Blériot's cross-channel flight: Gollin, *The impact of air power*, 76ff.

[132] Northcliffe to Col. C. à Court Repington, 25 Jan. 1915, Northcliffe papers, Add. 62253.

[133] Raleigh, *War in the air*, i. 159.

[134] See, for example, PRO, AIR 1 1613/204/88/11, for papers dealing with early advisory committee–Balloon Factory co-operation. Capt. Bacon was soon to be replaced as naval representative on the committee by Capt. Murray F. Sueter, and Sueter was to be joined in October 1909 by Haldane's new choice as SBF, Mervyn O'Gorman. In July 1912 Maj.-Gen. Hadden was replaced as army representative by Brig.-Gen. David Henderson, at that time DMT: AIR 1 763/204/4/195.

anti-establishment reputation he acquired during the First World War, came from a comfortable, even distinguished north-country family, imbued with a strong sense of social responsibility. Born in 1875, the third son of Charles Grey Grey of Dilston Hall, Northumberland, manager of the Greenwich Hospital northern estates, he was also a grandson of the noted agriculturalist John Grey, and a nephew of the social reformer Josephine Butler.[135] Educated at the Erasmus Smith School in Dublin on account of his father's work with the Irish Land Commission, he characteristically chose to eschew the alternative careers that must have lain open to him and enrolled at the Crystal Palace Engineering School in south London, where he preceded Geoffrey de Havilland. Following the usual progression he then became a draughtsman with the Swift Cycle Company of Coventry. He first began reporting on motoring matters in about 1904, as a contributor to the *Cycle & Motor Trades Review*. The following year he joined *Autocar*, whose Coventry-based proprietors, Iliffe & Son, had given Northcliffe his first break in newspapers with the editorship of *Bicycling News* some nineteen years before.

While writing for *Autocar* Grey began to specialise in aviation. He reported on the first Paris Aero Show in 1908, and such was the public's evident enthusiasm that, shortly afterwards, he was made joint editor of a new Iliffe penny weekly *The Aero*, first published in May 1909. Grey used this position to question publicly government attitudes towards the development of aviation. He went on to found his own better-known journal, *The Aeroplane*, in 1911 as a further vehicle for his views,[136] but by then he had already become what he was to remain until well into the war, a thorn firmly embedded in the side of the government Aircraft Factory. Grey was an assertive man, who felt none of the intellectual or class diffidence of his contemporaries. Haldane's orotund phraseology left him cold. When, in parliament early in 1910, Haldane reaffirmed his faith in the advisory committee, describing it as 'an organisation containing architects of science' who would put Britain 'in point of science at all events' abreast of all other nations, Grey was there to question the claim. By then Haldane's mannerisms had become well-known. He answered all criticisms in the same way – by invoking the term 'science', and occasionally sprinkling a few references to Göttingen. It was usually enough to brow-beat his critics. Grey, however, merely advised his readers in an editorial comment not to 'take Mr Haldane too seriously'. The government's advisory committee, conscious of its own eminence, had 'like . . . W. S. Gilbert's House of Lords "done nothing in particular, and done it very well" '; and while it did nothing, Grey continued, the NPL occupied itself in 'finding . . . new ways of carrying

[135] *DNB, s.v.* Grey, C. G. and Grey, John.
[136] C. G. Grey to Lord Brabazon, 3 Oct. 1947, Brabazon papers, box 26(2); Hurren, *Fellowship*, 66–7; C. G. Grey, 'The story of "The Aeroplane" ', in A. C. Armstrong, *Bouverie Street to Bowling Green Lane: fifty-five years of specialised publishing*, London 1946.

out twenty-year-old experiments'.[137] Thus, far from being overawed by Haldane's verbosity, Grey simply took a pin to it and burst the bubble.

From then on the Aircraft Factory was to find Grey shadowing all its activities, publicly questioning its postulations and scrutinising its decisions. The enmity with which he came to be viewed by Factory officials testified to his efficiency as their critic, although this was somewhat obscured at the time by the oft-quoted acerbity of his pen. As the military sought to tighten its grip on the industry by issuing sub-contracted orders for the manufacture of aircraft designed by its own research organisation, Grey steadfastly championed the cause of free design initiative in an open competitive market. Grey's stance has invited controversy ever since. The Aircraft Factory attempted to defend itself against his charges by suggesting that they were made primarily 'at the instigation of [The Aeroplane's] advertisers'.[138] This was not a very damaging allegation. Even if it had been true, and there is no evidence that it was, it would not have affected the validity of his criticism. Unfortunately, however, his pre-1914 advocacy of open competition has also tended to be retrospectively dismissed in the light of his later drift towards fascism. This reproof is anachronistic and is usually used as a means of avoiding the real issues. None the less, fascism blighted his subsequent influence. In the inter-war period Grey initially developed a close friendship with Lord Trenchard and vigorously defended the independent Royal Air Force against rival service intrigues for its partition: but, rather like the tank pioneer, J. F. C. Fuller, his political opinions ultimately put him beyond the pale. His reputation never really recovered, so that even today, when his early influence is discussed, Grey is often dismissed as a maverick, or worse,[139] and the opportunity for a more considered assessment of the controversies surrounding his name is lost.

Reform of the Balloon Factory

With the advisory committee in hand, Haldane could turn to the reorganisation of the Balloon Factory itself. This really began with the implementation of the CID sub-committee's report. It will be recalled that in October 1908 the Director of Fortifications and Works had forwarded Colonel Capper's report on Cody's aeroplane experiments with the recommendation that aeroplane trials at Farnborough be continued.[140] This submission was shelved for three months, awaiting the report of the CID sub-committee. In January 1909, the decision to discontinue aeroplane work having been taken, the Master General of Ordnance, Major-General Hadden, finally took up the Cody dossier and,

---

137 *The Aero*, 29 Mar. 1910, 244; 10 May 1910, 364.
138 PRO, AIR 1 686/21/13/2245 (summary for May 1912).
139 See, for example, Bruce, *Aeroplanes of the Royal Flying Corps*, pp. xv–xvi, and Edgerton, *England and the aeroplane*, 57–8.
140 See p. 199.

disregarding Capper's advice, recommended that the employment of both Cody and Dunne at the Balloon Factory be terminated at the earliest opportunity. This was accordingly approved by Haldane – although the full CID had not yet ratified the sub-committee's recommendations.[141]

All aeroplane work at the Factory ceased on 31 March, the end of the financial year, and Dunne was released from service. Cody was kept on as kite instructor until the following September. No claims were made on any designs drawn up by the two men and both were permitted to keep their aircraft, provided they return the engines.[142] Colonel Capper requested, however, that Cody at least be allowed the continued loan of an engine so that while still at Farnborough he might continue his experiments in a private capacity. He was, Capper stressed, 'within very measurable distance of considerable success', and his work would still be of value to the Factory. This was sanctioned.[143] The decision did not affect Haldane's plans. The CID sub-committee's recommendations had been carried through, the 'empiricists' had been removed from government service and the way was clear to re-establish the Balloon Factory itself.

It was, of course, an especial blow to Capper that Dunne's work should have been terminated in so abrupt a manner. He had staked much of his credibility on the latter's success, in which he still believed. From this point on, however, Dunne was just another manufacturer. Within months he had formed the Blair Atholl Syndicate, with an impressive list of sponsors, including the marquess of Tullibardine, Lord Rothschild and the duke of Westminster. Workshops were acquired at the Short Brothers-cum-Aero Club base at Eastchurch – the young Richard Fairey being appointed works manager in 1911.[144] However Colonel Capper's own request to join the syndicate as a director was at this stage refused by the War Office.[145] Even so, he retained a close association and after he had been removed from the Balloon Factory in October 1909 he began reconstructing the original Dunne monoplane – precursor of the Blair Atholl models; and by November 1909 both he and Lieutenant Gibbs were certainly working in some measure with, if not for, the syndicate.[146] Within two years the marquess of Tullibardine was trying to sell both this monoplane and other machines back to the War Office.[147]

---

[141] MGO to Secretary of State (Haldane), minute dated 29 Jan. 1909; Haldane's sanction, 1 Feb. 1909, PRO, WO 32/6933.
[142] Treasury to War Office, 11 Mar. 1909, ibid. WO 32/6934.
[143] Col. J. E. Capper (SBF) to Chief Engineer, Aldershot Command, 4 Mar. 1909; Col. Capper to DFW, 26 Mar. 1909; DFW to SBF, 7 Apr. 1909, ibid.
[144] DBB, s.v. Fairey, Sir Charles Richard.
[145] Col. Capper to Chief Engineer, Aldershot Command, 4 May 1909; War Office to GOC-in-C, Aldershot Command, 28 May 1909, PRO, AIR 1 1611/204/87/22. Capt. Carden also applied to join the syndicate at this time.
[146] Col. Capper to Lt. L. Gibbs, 17 Nov. 1909, ibid. AIR 1 1613/204/88/10.
[147] Marquess of Tullibardine to Maj. A. Bannerman (CAB), 2, 10 Aug. 1911, ibid. AIR 1 1609/204/85/49.

There is no question that Dunne was by then in a position to offer fully practical, inherently-stable aircraft to the government. Even Captain Carden, formerly Assistant Superintendent at the Balloon Factory, a man with only one arm, was able to gain his flying certificate on a Dunne biplane.[148] But what these machines gained in stability they lost in speed, lifting-power and manoeuvrability. Indeed, the very feature of inherent stability, originally conceived as a military necessity, was increasingly to be seen as more of a practical liability. The Dunne aircraft, as C. G. Grey put it, simply lacked 'performance'.[149] Manufacturing rights were acquired by Armstrong, Whitworth & Co. in Britain and Astra in France, and by Burgess Co. & Curtis in America, and a slightly modified Dunne biplane, a D8, was eventually acquired by the War Office in early 1914 – nearly a year after it was ordered; but the type was not carried into the war.[150] Dunne subsequently withdrew from aeronautics altogether. He went on to achieve some notoriety as a writer of speculative psychology and philosophy, notably with his book An experiment with time, published in 1927. Haldane's opinion of the work is not recorded.

Cody, meanwhile, had rehoused his aeroplane on Laffan's Plain, near the Balloon Factory at Farnborough, and having refitted it with an 80 h.p. ENV engine, was soon embarrassing the authorities by making regular flights.[151] In September 1909 he duly left government service and began concentrating solely on his own design and construction. With that flair for publicity that was his trademark, he registered as a British citizen in the middle of this country's first international aviation meeting, at Doncaster in October 1909 – some two months after Rheims,[152] and, by the close of the following year, had won the annual Michelin Cup for the longest close-circuit endurance flight: this after he had installed a 60 h.p. Green engine and circled Laffan's Plain for over four-and-three-quarter hours, covering a distance of 185½ miles. The man who had by that time replaced Capper as Commandant of the Balloon School, and who was soon to command the Royal Engineers' Air Battalion, Major Alexander Bannerman, implored the War Office to purchase this prize-winning machine, and continued to do so for the best part of a year. Indeed, by December 1911 he was urging the procurement of 'four or five of the class' for what he defined as 'training purposes'. Considering what Cody had achieved unaided, Bannerman commented to the War Office on 15 December 1911,

---

148 He was awarded RAeC certificate No. 239, 18 June 1912: A. D. Carden to G. Brewer, 29 June 1928, RAeS archive; Broke-Smith, 'Early British military aeronautics', 34.
149 PRO, AIR 1 727/160/1.
150 Broke-Smith, 'Early British military aeronautics', 34. For the convoluted details of this War Office acquisition see Bruce, Aeroplanes of the Royal Flying Corps, 221–2.
151 A. E. Tagg, Power for the pioneers: the Green and ENV aero engines, Newport 1990, 50.
152 Press reports, for example Daily Sketch and Daily Mirror, 22 Oct. 1909 (Butler collection, x. 86–7).

just think what he might do 'with assistance'. Might they not pay him a royalty to have his designs built at the Aircraft Factory?[153]

There was probably no conscious irony in this suggestion, but the War Office chose not to respond. They hardly needed reminding that barely two years before they had waived any claim they might have on what was then an army machine. The imprudence of this decision was now obvious. Whatever Cody's defects as a designer his efforts would have provided the air establishment with its own training aircraft, whilst the Factory pursued more advanced research. However, the decision had been taken to dispense with Cody's services, and the War Office were not now prepared to reintroduce his biplane. To forestall further entreaties Bannerman was told that the whole question of aircraft procurement was under review.[154]

In fact, the decision had been taken to suspend orders until a military aeroplane competition had been held on Salisbury Plain. This occurred in August 1912 and Cody won it. The War Office therefore had to procure two Cody machines anyway. It had been a freak result, due more to Cody's skill as a pilot and his employment of a 120 h.p. Austro-Daimler engine than to the intrinsic qualities of his aeroplane. As C. G. Grey later commented, if the trials proved anything it was that the human element should be ruled out in such competitions.[155] Further blushes were spared, however, when Cody fell to his death from a newly-devised biplane-cum-seaplane in August 1913. He was buried at Aldershot, with full military honours, by a special dispensation of the district Commander-in-Chief, Sir Douglas Haig.[156] His work died with him.

For Haldane it was enough that these curious individuals were out of the way: there remained the problem of Capper himself. By the time the advisory committee began its work it was obvious that a major restructuring of the Balloon Factory would follow. The new advisory body could only serve its purpose if the Factory was to be subjected to the same scientific rationale. It was hardly likely to be under Capper's supervision. Moreover, Capper was essentially a field soldier; to have such a man in charge of an administrative research unit broke every code in Haldane's book of bureaucratic propriety. 'It was always a matter of principle with me to endeavour to keep the General Staff from meddling with administration', he wrote in his 'Memorandum of events'. 'The insistence of Moltke on this point had made a deep impression on me.' In Germany, he explained, as if expounding some immutable law of

153 Maj. A. Bannerman (CBS) to DFW, 23 Jan., 9 May (by which time he was Cmdt, Air Battalion) and 15 Dec. 1911, PRO, AIR 1 1608/204/85/32.

154 DFW to CAB, 6 Jan. 1912, ibid.

155 PRO, AIR 1 727/160/5. The winning machine, the original Cody V pusher-biplane, was subsequently fitted with a 100 h.p. Green engine and used by Cody to win that year's Michelin Cup, but refitted with the original engine it broke up in the air the following April (1913), killing its pilot, Lt. L. C. Rogers-Harrison RFC: Bruce, *Aeroplanes of the Royal Flying Corps*, 200.

156 PRO, AIR 1 823/204/5/63 (accident report); Leon Cody to Lt.-Gen. Sir D. Haig, C-in-C, Aldershot Command, 11 Aug. 1913, ibid.

nature, all administrative matters came under the *Intendantur*, which was the province of the War Ministry. By this means administration, defined in a rather restricted sense, was dealt with by trained specialists, while strategy and tactics were left to the General Staff.[157] Compartmentalism became essential dogma to Haldane, and was seen at its starkest, and certainly at its most disastrous, in his attitude to military aeronautics. When a suitable replacement was found, Capper was thus to be shunted off to an ineffectual siding, and all association between the field companies and the Factory was to cease as a matter of policy.

A suitable replacement was presently found in the person of the civilian consulting engineer, Mervyn O'Gorman. Something of an aesthete, O'Gorman was a mercurial Irishman – his *Times* obituarist described him as a man of 'Hibernian eloquence' who 'rejoiced in paradoxes' liable to bemuse – who had been educated at Downside and University College, Dublin. He shared with Haldane a disdain for the unlettered and was characteristically described by the latter as 'the best expert in . . . mechanical problems that I could find'.[158] Maintaining the division between administrative and military work, he was placed under the Master General of Ordnance and controlled directly from the War Office.[159] O'Gorman's task as Superintendent of the Factory, Haldane related, was to 'produce' new types of dirigibles, employing for this purpose 'the results of the Teddington Special Committee' (the Advisory Committee for Aeronautics, based at the National Physical Laboratory, Teddington), of which he became a member.[160] But what was meant by 'produce'?; presumably design and construction, that is the production of government-sponsored, scientifically researched, aircraft for the army.[161] If so, it was hardly likely that the producers of these aircraft would then have to tender in open market for army orders. Thus Haldane was undermining the developing military–industrial complex in this area. It might not be wholly accurate to call it an attempted monopoly yet, as there was still to be some private sector procurement, but this was – effectively – the eventual result. From the first, or more particularly from when the Factory was forced back into aeroplane design, private industry was placed in unfair competition with state products. In the years up to, and in the early years of, the war independent firms were to be relentlessly squeezed out of the military market, except as the

[157] 'Memorandum of events between 1906–1915', 120. Note also *Autobiography*, 206. This was not, in fact, a new controversy. Gen. Sir Garnet Wolseley had decried such a system as Haldane now advocated at the time of his celebrated Ashanti expedition of 1873–4: see J. Keegan, 'The Ashanti campaign', in B. Bond (ed.), *Victorian military campaigns*, London 1967, 185.

[158] 'Memorandum of events between 1906–1915', 143; *Times*, 17 Mar. 1958.

[159] PRO, AIR 1 731/176/5/102.

[160] Haldane, *Autobiography*, 233. Haldane was presumably unaware that the Factory had, in fact, already been working in co-operation with the NPL for some time: P. R. Hare, *The Royal Aircraft Factory*, London 1990, 11–12.

[161] O'Gorman actually recorded Haldane as saying that he wanted to separate 'construction' from 'instruction' in military aeronautics: 'Notes on the Balloon Factory', PRO, AIR 1 728/176/1.

constructors of sub-contracted Farnborough designs. Haldane's vision of a government-based research and development department may have appealed as an abstract ideal, but in reality it was to stifle competition and suffocate design initiative.

Capper was informed of these changes rather than consulted about them. A letter dated 15 October 1909 revealed that a decision to separate the Balloon Factory from the Balloon School had been taken, and that O'Gorman was to replace him as Superintendent of the Factory in four days' time.[162] Capper was to remain Commandant of the Balloon School for the time being, but this was scant consolation. The one concession allowed him was that separation of the two commands would be gradual. Given how closely entwined they had become it could scarcely have been otherwise. A similar note was despatched to the Commander-in-Chief, Aldershot Command, informing him that the Factory would in future be under civilian control, responsible directly to the War Office.[163]

Capper took the manner in which all this was done as a personal slight, and made little effort to hide the fact.[164] He requested permission to at least finish the projects he was working on – what he called the 'dirigible No.2' (the 75,000 cu. ft 'Gamma' non-rigid airship) and the 'dirigible No.3' (the 33,000 cu. ft 'Baby/Beta' non-rigid airship, which he had, in fact, designed) – and these two aircraft were designated instructional balloons for this purpose. But he was still required to seek permission from the new Superintendent to use Factory facilities.[165] There was also some dispute as to whether the Assistant Superintendent, Captain A. D. Carden, was to remain at the Factory; however, O'Gorman resolved that problem by installing his own candidate in the position without reference to the Royal Engineers.[166] This action had unforeseen consequences.

In fact the whole protracted division between field and Factory was conducted on a far from amicable basis, with O'Gorman relentlessly coveting all stores, personnel and influence.[167] Only with great reluctance would he countenance any pooling of resources. In this he was doing no more than pursuing Haldane's philosophy of complete administrative separation, but in terms of the continued maintenance of the school and its field units his behaviour was causing chaos. 'There are points coming up daily', Capper complained to the War Office in December 1909, 'where the authority of the

---

[162] War Office to Col. Capper (SBF), 15 Oct. 1909, ibid. AIR 1 1611/204/87/22.

[163] War Office to GOC-in-C, Aldershot Command, 15 Oct. 1909, ibid.

[164] Col. Capper to Chief Engineer, Aldershot Command, 16 Oct. 1909; Col. Capper to M. O'Gorman, 16 Oct. 1909, ibid.

[165] Col. Capper to DFW, 18 Oct. 1909; DFW to M. O'Gorman (SBF), 25 Oct. 1909, ibid.

[166] Capt. A. D. Carden to SO, RE, 21 Oct. 1909, ibid. Carden did, however, remain at the Factory until the end of the financial year (Mar. 1910): War Office to GOC-in-C, Aldershot Command, 10 Dec. 1909, ibid. AIR 1 1612/204/87/26.

[167] SBF memo to War Office, 29 Nov. 1909; caustic comment on it from Col. Capper, memo to Chief Engineer, 16 Dec. 1909; Capper to War Office, 5 Feb. 1910, ibid.

Superintendent of the Factory and the Commandant of the School . . . over-lap.'[168] Separating staff, as was now being done, along military and non-military lines, without hampering the work of both, was simply impracticable. In the past, he insisted, both commands had dove-tailed into one another in both an efficient and economical manner. Now much equipment was having to be duplicated. He recommended that a committee be appointed to examine the problem. Whitehall, not unexpectedly, considered this unnecessary.[169] Disputes were thus destined to continue.

The trouble was that the Balloon School effectively remained dependent on the Factory for the repair and overhaul of its aircraft, while the Factory itself concentrated increasingly on research and design. The Factory no longer had, in the words of the Royal Engineers' official history, a 'co-ordinatory incentive to treat the upkeep [of] aircraft [for] the military side . . . as one of its first responsibilities'.[170] Arguments between the two branches were to run on for years. When service aeroplanes were perforce introduced, facilities for workshop practice in their maintenance and overhaul, which the Balloon School and its successor the Air Battalion did not, beyond a limited degree, possess, could not easily be provided by the Aircraft Factory under the existing policy of administrative separation.[171] In the end one is just left wondering whether Haldane ever seriously considered the practical outcome of his abstract notions. In this instance there is no evidence that he did. Instead Seely was left to fashion a viable air service out of these divisions. 'I had myself . . . little to do with subsequent developments', Haldane admitted in his autobiography.[172] He had done enough.

O'Gorman quickly set about restructuring the Factory on civil lines; following Lanchester's recommendation, for example, F. M. Green was recruited from Daimler as chief engineer.[173] Initially the emphasis continued to be on airships, but whereas Capper had been restricted to these by the recommendations of the CID sub-committee, O'Gorman acted out of conviction. He inherited a workforce of about one hundred, then mainly engaged on the production of Capper's 'Baby/Beta' and 'Gamma' non-rigid airships, and work continued in this direction.[174] The following year construction of the 173,000 cu. ft 'Delta' airship began. The Factory, Haldane reported to his mother in November 1910,

---

[168] Col. Capper (CBS) to War Office, 1 Dec. 1909, ibid.

[169] DFW to Col. Capper (CBS), 9 Dec. 1909, ibid.

[170] Brown, *History of Royal Engineers*, iv. 78. William Baker Brown (1864–1947) had himself served in the Directorate of Fortifications and Works between 1904 and 1908. His volume of the multi-volumed official history of the Royal Engineers was published posthumously in 1952.

[171] PRO, AIR 1 1612/204/87/26; Broke-Smith, 'Early British military aeronautics', 47. Arguments along these lines continued right up to the founding of the RFC: PRO, AIR 1 1608/204/85/38.

[172] Haldane, *Autobiography*, 234.

[173] PRO, AIR 1 686/21/13/2245 (summary for Jan. 1910); Hare, *Royal Aircraft Factory*, 26.

[174] M. O'Gorman, 'Notes on the Balloon Factory'.

was developing splendidly; this would be their fifth dirigible.[175] Yet within three years the army, under Seely's direction, was to abandon airship work altogether, transferring responsibility for this branch of the service to the Admiralty. This decision was taken under the auspices of the Air Committee of the CID, formed in 1912. By then aeronautical work was being further divided between the naval and military wings of the Royal Flying Corps.

Haldane was characteristically disingenuous in admitting this development in his *apologia*. He noted that the War Office subsequently handed all airship work over to the Admiralty, but chided the Royal Navy for then constructing aircraft without reference to the advisory committee. They neglected, he claimed, the 'immense amount of preliminary research . . . indispensible to success', with the result that their first airship project ended in disaster and they were discouraged from going on.[176] This is a tendentious piece of misinterpretation. The Naval Airship No.1 he refers to – the 'Mayfly' – was planned as far back as 1908, before the advisory committee had even come into being. It came to grief in September 1911, having been constructed and tested in close association with the National Physical Laboratory.[177] It was 1913 before the War Office finally abandoned airships, and the Admiralty then continued the work without a hitch. Admiralty airships proved immensely useful in the war, on coastal and anti-submarine patrols, and there was simply no question of any technical inferiority. What these events did demonstrate was how irrelevant many of Haldane's ideas for airships had become within a very few years; and this must have concerned him. Yet, writing as late as 1916, he still claims to have 'wished [that] the Farnborough establishment had been strengthened, and entrusted with the construction of Zeppelins'.[178] As it was, aeroplanes, not airships, were to be produced within the administrative struc-ture he created. This structure, and the inherent divisions it generated, were to be Haldane's real legacy.

The change of emphasis came shortly after Esher, in October 1910, repudi-ated the CID sub-committee's findings. It required only a small drop of pragmatism to get things moving in the new direction, but it was slow in coming. It was O'Gorman's general policy, in his own words, to exclude 'inventors, amateurs, and enthusiasts of all sorts' from the 'calculating and design work' of the Factory – even though they probably 'constituted the bulk of the aeronautical world'.[179] Holding such views one can quite see why Haldane adopted him: but for all his disdain of 'enthusiasts', they had, as Esher had gently pointed out, achieved practical aeroplane flight while the Factory was not only doing nothing, but doing nothing as a matter of policy.

---

[175] Maurice, *Haldane*, i. 276.

[176] 'Memorandum of events between 1906–1915', 144. Much of this was later reworked into Haldane's *Autobiography* (p. 233), with suitably vague embellishments regarding the necessity of correct 'mathematical and physical conditions'.

[177] Sueter, *Airmen or Noahs*, 115–16.

[178] 'Memorandum of events between 1906–1915', 144.

[179] 'Notes on the Balloon Factory'.

By the close of 1910 it was evident that aeroplane design could no longer be ignored. It was equally evident that the Factory lacked the personnel to embark upon it. At this point O'Gorman decided that there might, after all, be room for one or two inventors – perhaps for a spark of creativity – in his technocracy. This spark came from a young enthusiast who had spent the previous summer testing his own privately constructed aeroplane on the sloping Downs south of Newbury: Geoffrey de Havilland. When Haldane's administrative structure became charged with de Havilland's creative skill, the Aircraft Factory was conceived as a competitive force in aviation.

Geoffrey de Havilland: the development of Farnborough aeroplane designs

Geoffrey de Havilland had been born at Wooburn, near High Wycombe, in 1882, the son of, by his own account, a slightly neurotic clergyman – the Revd Charles de Havilland – at that time curate of the Buckinghamshire village of Hazlemere. Most of his childhood was spent, not altogether contentedly, with his family at Nuneaton, in Warwickshire.[180] He was educated at St Edward's School, Oxford, and it was the immediate wish of his family that he should follow his father into the Church, but the fascination of 'independently powered . . . transportation' had, he later confessed, seized a hold of his imagination – as it had 'so many of the young of my generation'.[181] Consequently, in 1900 he enrolled at the Crystal Palace Engineering School on a three-year course in mechanical engineering. Work there, particularly in the construction of a motor-cycle engine, provided him with essential practical experience. He continued such work on his own initiative after leaving the school. While an apprentice with Willans & Robinson Ltd of Rugby, a well-known firm of steam-engine manufacturers, he managed to design and build his own 450 c.c. motor-cycle engine and frame. Two fellow employees, Cecil and Alick Burney, were much taken by it and prevailed upon de Havilland to sell the patent to them for just five pounds. Its exploitation subsequently became the basis of the highly-successful Blackburne motor-cycle engine company.[182] Determined to continue in this field, de Havilland next secured a job as draughtsman with the Wolseley Tool & Motor-Car Company of Birmingham – which at this time, 1905, was passing from the control of Herbert Austin to John Siddeley. Finding few creative outlets with that firm, however, he did not stay long, and transferred after about a year to the post of

---

[180] de Havilland, *Sky fever*, 19–20. Charles de Havilland's half-brother was the father of the actresses Olivia de Havilland and Joan Fontaine: *DNB, s.v.* de Havilland, Sir Geoffrey.
[181] de Havilland, *Sky fever*, 31.
[182] Ibid. 40–2; R. Venables, 'The history of the Blackburn Company', *Motor Sport* xxiv (1948), 69–71.

designer with the Motor Omnibus Construction Company of Walthamstow, London. While there he met Frank Hearle, a like-minded contemporary from the nearby Vanguard Omnibus Company. They became firm friends and were to remain business associates for most of the rest of their lives.

In 1908, while at Walthamstow, de Havilland first learnt of Wilbur Wright's aeroplane flights in France. From that moment, he later acknowledged, '[I] knew that . . . this was the machine to which I was prepared to give my life'.[183] He immediately set about sketching his own aeroplane designs, and then, after some hesitations, persuaded his grandfather to back an aircraft construction venture to the tune of £1,000. Hearle was taken on as an employee at 35s. a week, enabling him, like de Havilland, to leave his job and begin aviation work full-time.

Construction of a prototype biplane began in earnest when de Havilland arranged to have a modified 45 h.p. engine built to his own design by the Iris Motor Company of Willesden. One of the founders of this firm, Guy Knowles, had been a friend of de Havilland's talented elder brother, Ivon, who had indeed designed the Iris car of 1904, shortly before his death. Meanwhile, a workshop was rented in Fulham, where the aircraft's frame itself could be assembled. Here de Havilland and Hearle employed whatever materials were commercially available; the undercarriage, for example, was constructed of ordinary bicycle steel tubing. The main guiding principle, according to de Havilland, was simply to build something 'not too novel'.[184] Hence there emerged what was essentially a Farman-derived, twin-propeller, pusher-biplane, with aileron control, a forward elevator and fixed tailplane.

Shortly before completion of the biplane a suitable test-ground was found just a few miles from the de Havilland family's latest home at Crux Easton, near Newbury. This was at Seven Barrows, an area of secluded grassy downlands on Lord Carnarvon's Highclere Park estate in north Hampshire. Carnarvon was subsequently to win fame for his part in the discovery of Tutankhamun's tomb, but at this particular point he was better-known as a founder member of the Automobile Club of Great Britain and Ireland. Through this association Brabazon had received permission to erect a pair of hangars on the Seven Barrows site, with a view to housing his Voisin biplane there on his return from France. In the event he joined the Short brothers on the Isle of Sheppey, and was only too happy to sell his sheds to de Havilland.

The first full trials finally began in December 1909, but met with little immediate success. At length, a breeze blowing up a slight incline did lift the aeroplane, but de Havilland had so little experience of the lever-operated controls that he soon crashed it, leaving little to salvage beyond the engine. This, none the less, was the essential component, and work began on a successor almost immediately. This time a lighter, more streamlined structure

---

[183] de Havilland, *Sky fever*, 47.
[184] Ibid. 53.

was built around a single propeller unit.[185] The summer of 1910 saw de Havilland and Hearle back at Seven Barrows. By September regular ascents were being achieved under full operational control.

Constructing a practical aeroplane was one thing; what to do with it afterwards was another. No sooner had de Havilland completed his first full flight trials than it registered with him that he was now both penniless and, to all intents, unemployed. He lacked any means of marketing his aircraft, and it looked for a time as if the entire project would have to be abandoned. This was the situation in November 1910 when de Havilland met Farnborough's chief engineer, Fred Green, at the Olympia Motor Show. The pair had known each other while working in the automobile industry in the Midlands, Green with Daimler and de Havilland with Wolseley. Knowing that certain changes were imminent at the Balloon Factory, Green suggested that de Havilland offer both his aeroplane and services to that establishment, via the Superintendent. This he did, and O'Gorman, presumably briefed by Green, took the matter up immediately.[186] After a preliminary trial, an offer to purchase de Havilland's aircraft – the de Havilland biplane No.2 – was sanctioned just before Christmas 1910, and de Havilland was asked to join the Factory as aeroplane designer and test pilot. Hearle was invited to act as his mechanic. As a Factory aircraft the de Havilland No.2 was redesignated the FE1, or Farman Experimental No.1.[187]

Barely two years before, it had been decided to rid Farnborough of its empiricists. Now the reconstituted Factory was compelled to readopt a perfect example of the type. In aeroplane design the scientists of Haldane's imagination were thus to follow in the wake of the engineer, not *vice versa*, and this would generally remain true even after the war. Men like Sydney Camm, designer of the Hawker Hurricane, and Reginald Mitchell, designer of the Supermarine Spitfire, were again essentially intuitive mechanical engineers, who had learnt their craft on the shopfloor and through the technical college.

Even after de Havilland started at the Factory, however, the immediate emphasis remained on lighter-than-air research, and it was some time before he was freely accepted as an integral part of the organisation.[188] This soon proved somewhat ironic. Within a year or two the entire establishment – formally restyled the Aircraft Factory in April 1911 – was being carried along on the back of de Havilland's aviation trials. Farnborough itself was, at this time, little more suited to test flying than it had been in Cody's day, and O'Gorman's diary, from January 1911 on, records repeated attempts to get the

---

[185] Ibid. 56ff.

[186] O'Gorman later recorded that it was actually as a proposed engineering assistant to F. M. Green that de Havilland was first introduced to the Factory: PRO, AIR 1 731/176/5/102.

[187] de Havilland, Sky fever, 66ff.; PRO, AIR 1 686/21/13/2245. Preliminary testing of de Havilland's aircraft occurred at Farnborough on 15 Dec. 1910 (AIR 1 732/176/6/24) and de Havilland flew the official hour-long acceptance trial on 14 Jan. 1911 (AIR 1 2404/303/2/1). The machine was purchased for £400.

[188] de Havilland, Sky fever, 71.

War Office to clear the ground properly,[189] but the workshops, at least, provided de Havilland with undreamt of facilities. With the help of the Factory's technical staff the FE1 was continually tested and modified over the summer of 1911, and a second model, the FE2, was built.

This new model incorporated a broadly similar mainplane-section and undercarriage to the FE1, but was prefixed by a pilot-encasing nacelle. Powered by a 50 h.p. Gnôme engine, it was first flown on 18 August 1911. Early the following year it was temporarily equipped with floats and test flown from the Fleet Pond, near Farnborough,[190] following which it was again modified to carry a Maxim gun mounted in the nacelle. In this form it approximated to Vickers's own contemporary EFB1, which had been specifically designed to carry a Vickers-Maxim gun within the body of its nacelle. Somewhat confusingly, a second FE2 variant with a 70 h.p. Renault engine was constructed in 1913, with no suffix to distinguish it from its predecessor,[191] and the Farman Experimental configuration was to be continually developed at Farnborough over the coming years. Indeed, the Lewis-gun-armed FE2b was later to provide the Royal Flying Corps with a Factory-designed product of some combat utility. More generally, however, the process of extemporisation on a given design was soon augmented by a plan to explore a variety of aeroplane forms, beginning with the 'canard' or tail-first type, in which a fuselage and elevator were built forward of the mainplane section.[192]

At this time the Factory was still being restricted by the CID sub-committee's recommendation that no money be channelled into aeroplane research. (O'Gorman was, of course, directly responsible to Major-General Hadden, the Master General of Ordnance, who had been a principal member of the sub-committee.) The only way research and development costs for this new work could consequently be sanctioned was in the guise of repairs to existing machines. So when it came to the assembly of a 'canard' biplane the Factory ostensibly reconstructed and modified a decrepit Blériot Type XII monoplane, which it had recently taken over from the Royal Engineers' Balloon School. Known colloquially as the 'man-killer', this aircraft had been presented to the War Office in June 1910 by Colonel Joseph Laycock – chairman of the ENV engine company – and the duke of Westminster.[193] In reality, the only part of the old Blériot to be reused was its 60 h.p. ENV engine. In other words, under this cover a totally new machine was constructed to de Havilland's own design.

This new aircraft was designated the SE1, or Santos Experimental No.1,

---

[189] PRO, AIR 1 732/176/6/24.

[190] Ibid. AIR 1 2317/223/24/3; Hare, *Royal Aircraft Factory*, 200. In its float-form in April 1912 the aircraft was initially piloted by Copland Perry, and then, following the installation of a 70 h.p. engine, by de Havilland himself.

[191] Ibid. 201–2.

[192] O'Gorman later suggested (somewhat curiously) that this was initially in imitation of the new-type Wrights' biplane, but, as seen, de Havilland was already designing pusher-biplanes anyway, so it was really only a continuation in that line: PRO, AIR 1 731/176/5/102.

[193] Ibid. AIR 1 727/152/6.

after the canard biplanes of Santos-Dumont.[194] It was essentially a twin-rudder pusher-biplane, with wings and propeller fixed at the back of what developed into a long narrowing nacelle-cum-fuselage, itself prefixed by a forward-elevator. It was reported as being near completion in April 1911,[195] and a series of tentative flights occurred the following June. By August, however, major modifications were still being effected. On the 18th, before these could be completed, the inexperienced Assistant Superintendent of the Factory, Theodore Ridge, defied de Havilland's objections by taking the machine up for a test flight. A bad sideslip occurred on a short turn and Ridge died from the injuries sustained in the resulting crash. (He had, in fact, been attempting to evade a clump of bushes. The accident thus tragically fulfilled both Capper's and O'Gorman's warnings about the Farnborough terrain.) The canard configuration was consequently abandoned. Heckstall-Smith presently replaced Ridge at the Factory.

By this time de Havilland had, in fact, already begun exploring the alternative possibilities of the propeller-first tractor-biplane, the configuration that was to become known at Farnborough, rather misleadingly, as the BE, or Blériot Experimental type. (The title was chosen simply to identify the prototype as being – like the Blériot monoplane – of a tractor rather than pusher configuration.) De Havilland was, moreover, now assisted by a skilful young draughtsman named Henry Folland, who was subsequently to become one of the industry's most outstanding designers. The first tractor-biplane, the BE1, was ostensibly another reconstruction, this time of an old Voisin pusher-biplane presented to the War Office in May 1911 by the duke of Westminster, and transferred to Farnborough for repair the following July.[196] Again, however, only the Voisin's 60 h.p. Wolseley engine was incorporated into the new two-seater tractor design, and even this was soon replaced. Much has been claimed for the de Havilland BE1, which was the forerunner of a whole series of Factory-designed reconnaissance machines, but as O'Gorman at least had grudgingly to admit, the type was to some degree derived from the Avro Type D biplane.[197] Thus if anyone was the originator of this modern tractor genre it was the independent, unsubsidised figure of A. V. Roe, not the staff of the government Aircraft Factory. Geoffrey de Havilland was simply correct in

---

[194] Ibid. AIR 1 731/176/5/102; AIR 1 686/21/13/2245.

[195] O'Gorman's diary, 26 Apr. 1911, ibid. AIR 1 732/176/6/24. For an examination of early drafts of the design see P. Jarrett, 'Farnborough's first: the story of de Havilland's SE1', *Air Enthusiast* xlii (1991).

[196] The duke of Westminster, who had also jointly donated the previous Blériot to the army, was himself an avid aviator: see *The Tatler*, 13 Apr. 1910, 'The enthusiastic Duke at the Aero Club nest . . . Sheppey Island' (Butler collection, ii. 89). A sponsor of Dunne's Blair Atholl Syndicate, he was no mere dilettante in mechanical matters. He had been a BARC founder member, and subsequently commanded a pioneering armoured car detachment (Royal Naval Division, 1914–16), eventually becoming personal assistant to the Controller, Mechanical Dept, at the Ministry of Munitions.

[197] Ibid. AIR 1 731/176/5/102.

perceiving that, at the time, this particular biplane tractor-configuration was structurally more reliable than the Blériot monoplane form.[198]

The prototype BE1 was initially test piloted by de Havilland himself, at Farnborough, on 4 December 1911. Among the first passengers then to accompany him were Captain C. J. Burke, of the Royal Engineers' Air Battalion, and Heckstall-Smith on 27 December, and Mervyn O'Gorman on 1 January 1912.[199] Numerous minor modifications followed before the prototype was handed over to the Air Battalion in March 1912, with a certificate of airworthiness verifying a mean speed of 59 m.p.h. and a climbing rate of 155 ft per minute.[200] Confidence in the design was then reinforced by a succession of impressive results with a second Factory model (an alleged reconstruction of an unidentified existing army machine). In June 1912, for example, de Havilland reportedly flew this aircraft, fitted with a 70 h.p. Renault engine (thereby earning the designation BE2),[201] to an altitude of 10,000 ft, and on 12 August, at the start of the military aeroplane competition on Salisbury Plain, he climbed to a height of 10,560 ft, with the addition of Major F. H. Sykes, Commandant of the Royal Flying Corps, in the passenger's seat. This was a new British altitude record;[202] indeed, it remained unequalled for the next three years. Speeds of up to 72 m.p.h. were also being achieved.[203] But most conclusive of all, at least to those who had arranged it, was the BE's overall success in the military trials themselves.

In fact the Royal Aircraft Factory (as it had become in April 1912) did not formally enter this competition, as O'Gorman and his staff were responsible for organising the event. But although not officially entered, the BE2 was put through the prescribed trials *hors concours* and, under normal circumstances, would probably have won the contest.[204] O'Gorman saw this as a vindication of the advisory committee, of which he was by then a member;[205] but it is difficult to see how this body affected the issue. De Havilland himself was to attribute his success as a designer to the experience he gained as a pilot.[206] He made no mention of the advisory committee. In reality, the result was not that startling anyway; the competition had from the first been planned around the performance of the BE, which by January 1912 had been decreed the 'standard

---

[198] de Havilland, *Sky fever*, 82. Roe had defended his original tractor-triplane configuration against advocates of the celebrated Blériot XI tractor-monoplane shortly after the latter's successful channel-crossing: A. V. Roe to B. F. S. Baden-Powell, 4 Aug. 1909, RAeS archive.

[199] Aircraft Factory diary, 4 Dec. 1911, PRO, AIR 1 2404/303/2/2; Aircraft Factory diary, 1 Jan. 1912, AIR 1 2405/303/2/4; Bruce, *Aeroplanes of the Royal Flying Corps*, 343.

[200] BE1 certificate, 14 Mar. 1912, PRO, AIR 1 1608/204/85/38.

[201] PRO, AIR 1 730/176/5/92. It was first flown with a (60 h.p.) Renault engine on 1 Feb. 1912, piloted again by de Havilland himself: AIR 1 2405/303/2/4.

[202] Ibid. AIR 1 686/21/13/2245; RAeC Committee minutes, 1 Oct. 1912, RAeC papers.

[203] de Havilland, *Sky fever*, 77.

[204] PRO, AIR 1 731/176/5/102; AIR 1 686/21/13/2245.

[205] Ibid. AIR 1 731/176/5/102, 10.

[206] Ibid. AIR 1 724/85/1.

of excellence' for all competition machines.[207] As a sop to private manufacturers, Colonel Seely (who, as Under-Secretary of State for War, had ministerial responsibility for the competition) had invited the Aero Club to nominate two 'practical aviators' to assist in the planning of specific tests, and G. B. Cockburn and Alec Ogilvie – as neutrals – had duly been recommended.[208] However, no outside influence materially modified Farnborough's disjointed and overspecialised management of the trials. In the end it scarcely mattered. There was always an implicit dichotomy between the Factory's designs and the rest, irrespective of who won the formal contest. Indeed, the official results were soon seen as little more than an historical curiosity. The event's principal effect was simply to confirm the Factory in its presupposed superiority. As O'Gorman's fallacious apportioning of the credit to Haldane's advisory committee indicates, they had always believed themselves to be in a different class from the general run of constructors. Were they not, after all, an organisation based on a foundation of science? Sub-contracted orders for twelve government Aircraft Factory BEs followed.[209] This was the thin end of what was to be an extremely contentious wedge.

In view of the later controversy surrounding the combative quality of these aircraft it is interesting to note that criticism of their manoeuvrability, or rather, lack of it, was evident from the earliest days. De Havilland subsequently acknowledged this in his autobiography, published in 1961, when he admitted that the wings had a tendency to 'take charge' in volatile conditions and 'warp' the pilot, instead of allowing the pilot to warp the wings. It was the re-emergence, under Factory auspices, of the principle of inherent stability.[210] Commander C. R. Samson, Commandant of the RFC (Naval Wing), complained of this feature in May 1913, when he accused the machine of giving 'the most violent oscillations' in the slightest gusts. The aircraft would effectively 'right' itself against the will of the pilot. A good many flying officers, Samson insisted, were very uneasy with this 'self-warping' control mechanism.[211]

O'Gorman attempted to brush aside this criticism, recording on a docket

---

207 Aircraft Factory report, 'Suggested aeroplanes for construction 1912-13', 26 Jan. 1912, ibid. AIR 1 730/176/5/92. The Judges' Committee had consisted of Brig.-Gen. David Henderson, Capt. Godfrey Paine RN (Cmdt of the recently established CFS), Maj. Frederick Sykes (OC, RFC, MW) and O'Gorman himself. O'Gorman seems to have guided their findings from the beginning. Even Sykes (not a naturally modest man) subsequently acknowledged how heavily the other judges had leaned on his technical knowledge: Sykes, From many angles, 101.
208 RAeC Committee minutes, 21 Nov. 1911, RAeC papers.
209 The evolution of the Type BE2c, PRO, AIR 1 730/176/5/92; Heckstall-Smith's eulogy for M. O'Gorman, 30 Nov. 1917, AIR 1 730/176/5/100; O'Gorman reminiscences, AIR 1 731/176/5/102.
210 de Havilland, Sky fever, 76. O'Gorman had, in fact, previously favoured investigating the canard configuration precisely because of what he perceived to be its inherent stability: Jarrett, 'Farnborough's first'.
211 Cdr C. R. Samson to CO, HMS Actaeon, 2 May 1913, PRO, AIR 1 2500.

attached to Samson's report that it appeared to him that such difficulties were largely a question of what a pilot was used to,[212] but matters soon worsened. The wing-warping arrangement was dispensed with only for the whole design to be revamped into an even more powerfully stable (and, incidentally, slower) aircraft. One can, perhaps, discern Lanchester's influence here, but the main proponent of inherent-stability was one of the Factory's new university-trained technicians, the Cambridge-educated E. T. Busk. After his modifications, involving the addition of a triangular fin attached to the tail and ailerons incorporated within staggered mainplanes set at a slight wing-tip culminating dihedral angle, the design re-emerged as the BE2c and the concept of control was brought right back to where Dunne had left it.[213] No-one in the organisation seems to have seen any irony in this, but under the stress of combat the whole problem was to re-emerge as a major political issue. By then a virtual monopoly had been imposed on the market and there were allegations in parliament that, by being ordered out in these government machines, airmen were effectively being 'murdered'.

All this broke under the 'Fokker scourge' banner. Historians in the past have tended to view this as predominantly a crisis of tactics and armaments. (The Fokker Eindecker, introduced in the late summer of 1915, incorporated the first propeller-synchronised interrupter-device, allowing for a direct forward-firing machine-gun mounting in aeroplanes of a tractor configuration. It thus became the prototype modern fighter.) In reality, however, this episode was more concerned with the crisis in the aviation industry, that is with the development of what amounted to a procurement-based monopoly and the suffocation of free, competitive, design initiative. It was, in short, a crisis arising out of the air service's inherent inflexibility in a challenging technological environment. They were inadvertently maintaining a course which ran entirely against the grain of growth and development. It was private enterprise having to adapt competitively in the Darwinian sense to a technologically dynamic environment that led to the evolution of such inter-related industries as those of the bicycle, automobile and aviation in the first place: and all within the previous two or three decades. By their constitutional inflexibility the aeronautical authorities effectively stifled technological-industrial development and with it their own war effort.

Factory personnel later tried to defend this policy by suggesting that there was actually no alternative at the time; that in August 1914 they were faced

---

212 19 May 1913, ibid. Despite – at least publicly – adopting a dismissive attitude to criticism of the Factory's BE biplane, O'Gorman was compelled to reiterate his defence of the aircraft in a formal letter to the CID Air Committee. It was printed in the committee minutes for 16 June 1913.
213 The evolution of the Type BE2c; PRO, AIR 1 771/204/ 4/289. The new prototype was first flown on 30 May 1914. Essentially it derived from the previous year's RE (Reconnaissance Experimental) derivatives of the BE2: Hare, *Royal Aircraft Factory*, 240ff. For criticism of the revised design's speed note C. G. Grey in H. Penrose, *British aviation: the Great War and Armistice 1915–1919*, London 1969, 28–9.

with an immediate demand for large numbers of aircraft and that Farnborough designs were the only ones for which 'competent drawings' existed permitting rapid expansion through large-scale sub-contracting to the private sector. According to Major Heckstall-Smith (Factory staff adopted military rank in October 1915), the aeroplane industry in Great Britain was 'totally incapable' of turning out more than 1.5 per cent of the total number of machines required 'for the Front'.[214] Such arguments are unlikely to have convinced anyone. The policy of sub-contracting BE orders to the private sector, with its collateral effect of starving those same companies of a market for their own designs, had begun long before the crisis of August 1914. It in no way originated with the war. Indeed, in his autobiography Frederick Sykes admitted that one of the Factory's central functions had been to 'keep down trade prices'. Moreover, even in 1914 private firms could just as easily have expanded production of their own designs as those of government machines; indeed many did so in supply of Admiralty needs. Far from being incapable of turning out such designs they were, in fact, ordered by the army to discontinue this work in order to concentrate on government machines.[215] In total some 2,000 BEs of various modification were eventually manufactured. The official history of the Ministry of Munitions calculated that the type formed between 70 and 80 per cent of the total number of aeroplanes employed by the Royal Flying Corps up to the middle of 1916.[216]

The procurement of adequate engines and even engine components presented a less contrived difficulty. Over the previous decade the German firm, Bosch, had built up a virtual monopoly of magneto spark-plug production, leaving British manufacturers, at the outbreak of hostilities, dependent upon existing stocks or such supplies as could be obtained through neutral countries – at least until reliable indigenous replacements could be manufactured successfully.[217] Fortunately, Frederick Simms had latterly been Robert Bosch's partner, and in 1910 had established (initially as a sales organisation) the Simms Magneto Co. of New Jersey. This firm became an essential source of supply.

Holt Thomas fully appreciated the significance of the country's lack of an adequate indigenous engine industry and he co-ordinated the English

[214] Eulogy for M. O'Gorman.
[215] See, for example, H. V. Roe, 'Pioneers', echoed in Penrose, *Great War and Armistice*, 65, and the evidence of B&C's company minutes, quoted in ch. 3. Furthermore, far from Farnborough designs being the only ones of which 'competent drawings' existed, the evidence actually suggests that such drawings as were distributed to the private sector were often manifestly incompetent. The managing director of British & Colonial, G. Stanley White, had to report in November 1914 that 'serious delays' in the production of BE2s had resulted from the faulty drawings supplied by the Aircraft Factory, a point that was – at the time – openly conceded: minutes, 30 Nov., 28 Dec. 1914, B&C papers. For the Sykes quote see *From many angles*, 101.
[216] *History of the Ministry of Munitions*, London 1920–4, xii/1, 24.
[217] Ibid. vii/1, 103; M. Cooper, *The birth of independent air power: British air policy in the First World War*, London 1986, 16.

production of French Gnôme and Rhône rotary engines.[218] To this end, he acquired Peter Hooker Ltd, a firm of precision engineers based at Walthamstow. Farnborough, however, had devised its own Renault-derived engine. This they eventually permitted Montague Napier, of D. Napier & Son Ltd, one of their principal sub-contractors, to redesign, and the Napier RAeF 200 h.p. engine came into production late in 1916, but supply remained a desperate problem. In the early years of the war innumerable accidents were to result from the employment of unreliable engines. The eventual development of Rolls-Royce aero-engines, on the other hand, stemmed directly from Admiralty pragmatism.

Ultimately, then, Haldane's reforms had led to administrative inflexibility, excessive standardisation and market control. His ideal of a state-based research and development department had in practice dried the essential lubricant of competition. This process simply accelerated with the onset of war. The charge levelled against the authorities in 1916 was not that airmen were being murdered through futile tactics, but that they were deliberately being sent into action with inadequate aircraft. (The term 'murdered' was actually uttered in the Commons on 22 March 1916, by the 'Independent air member' – and former entrepreneur, aviation journalist and all round hustler – Noel Pemberton Billing; although his demagogic intemperance – given popular credence as a result of his recent exploits as a Royal Naval Air Service pilot – actually proved, in the short term, counter-productive to the cause he purported to serve.)[219] This parliamentary altercation was the culmination of a wider dispute, deeply embedded in the whole development of military–industrial relations in this area. The Admiralty were always more flexible in their procurement, and to them was due the survival of those private firms, such as Sopwiths, who post-1916 were to be given the opportunity of producing aircraft for the Western Front as well. By then responsibility for procurement

---

218 The Gnôme radial rotary engine had been designed by the Séguin brothers in 1908. (The 'Société des Moteurs Gnôme' had been formed two years previously.) It successfully powered its first aeroplane in June 1909, when fitted into Paulhan's Voisin biplane. (Gibbs-Smith, *Aviation*, 143.) The engine spun through the air at 1,100 r.p.m., thereby cooling itself, overheating having previously been the chief cause of engine failure. The alternative cooling system was to let an aeroplane's slipstream fan the cylinders, as in the stationary air-cooled engine developed by Anzani. (The 'in-line' configuration enabled all cylinders to receive the slipstream.) The limited speeds yet achieved by aeroplanes tended to negate this effect. Renault emulated the stationary air-cooled engine, while Le Rhône (a Gnôme subsidiary) developed the rotary.

219 The speech containing the charge is reproduced in Pemberton Billing, *P-B, The story of his life*, 121. For a summary of his political campaign see B. D. Powers, *Strategy without slide-rule: British air strategy 1914–1939*, London 1976, 22–7. Penrose (*Great War and Armistice*, 124) asserts that Pemberton Billing was, in fact, reiterating a charge already made in the House by Col. Walter Faber, but the latter's delivery had nothing like the same effect. Pemberton Billing subsequently won even more notoriety as the initiator of the preposterous 'Black Book' spy scandal of 1918: C. Andrew, *Her Majesty's Secret Service: the making of the British intelligence community*, New York 1986, 188–90.

had been taken out of the army's (and, indeed, Admiralty's) hands and transferred to the Ministry of Munitions. This transfer occurred in January 1917 and represented the largest single extension of the ministry's duties since it was established in May 1915.

The decisive break with the military's inherent monopolistic tendency can be said to have come with the appointment of William Weir as Controller of Aeronautical Supplies shortly after this reorganisation. A revamped Air Board was to determine the general policy of the air services and such aircraft as were required were to be procured by Weir. To ensure the correct correlation between the air strategies pursued and aircraft procured Weir was made a member of the Air Board.[220] He thus became the linchpin of the entire British forces' air programme. In December 1917 he became Director General of Aircraft Production.

Ironically, this administrative upheaval occurred little more than a month after the final report of the Bailhache judicial inquiry into the administration and command of the Royal Flying Corps instigated in the wake of Pemberton Billing's outburst had – largely at the DGMA's, David Henderson's, instigation – effectively whitewashed the Directorate of Military Aeronautics' and Air-craft Factory's restrictive practices. Matters had gone beyond the power of any such inquiry to amend, however sympathetically. The army's air service had become a victim of its own bureaucracy. The Royal Flying Corps was originally to have been administered under general army precepts, personnel coming under the Adjutant-General and procurement coming under the Master General of Ordnance. In practice, however, due to its technical nature, procurement had become the immediate responsibility of the army's Director-ate of Military Aeronautics, founded in September 1913 under the command of David Henderson. This body was established with three branches: MA 1, dealing with General Staff concerns, such as policy, training and mobilisation; MA 2, dealing with design, equipment and procurement; and MA 3, dealing with finance. Major Sefton Brancker headed MA 1 and became, *ex officio*, Henderson's deputy. Colonel Walter Macadam RE, and Captain J. T. Dreyer RA, took over MA 2, but through inexperience were dependent on Brancker's advice. A civil servant, A. E. Turner, took over MA 3. The CID Air Committee co-ordinated the work of the directorate with the Admiralty's Air Department, founded the previous year under Captain Murray Sueter. Following Hender-son's departure for the continent in August 1914 to take field command of the Royal Flying Corps Brancker was left in charge of the directorate at the War Office, officially as Deputy-Director (temporary Lieutenant-Colonel), but with full responsibility. Major D. S. MacInnes, who had been responsible for the original Training Directorate branch 3b, concerned with aviation, sub-sequently replaced Dreyer. From here the procurement 'monopoly' had been controlled.[221]

---

[220] *History of the Ministry of Munitions*, ii/1, 57–8.
[221] Ibid. xii/1, 44; Macmillan, *Sir Sefton Brancker*, 34–5, 40–1, 65–6, 123–4. For an objective

An earlier inquiry inspired by the Fokker scourge and its implications, this time into the administrative efficiency of the Royal Aircraft Factory and held under the chairmanship of Sir Richard Burbridge, was rather overtaken by this larger Bailhache inquiry, but its principal recommendation, that Farnborough concentrate upon research rather than design, ultimately carried more weight.[222] Thus neither the flawed proceedings of the Bailhache inquiry, nor its soon to be published and entirely predictable conclusions, were sufficient to prevent O'Gorman from being replaced as Superintendent of the Royal Aircraft Factory as early as September 1916. This was done very quietly: his contract was simply allowed to expire. Doubtless conscious of the irony, O'Gorman subsequently joined Holt Thomas's Aircraft Manufacturing Company. He was followed into the private sector by a number of other Farnborough employees. In January 1917, for example, Major Fred Green, the Factory's chief engineer, joined the aviation division of the Siddeley-Deasy Motor Company (subsequently Armstrong-Siddeley) of Coventry, where he was assisted by his former Farnborough engine designer, S. D. Heron. Meanwhile, Henry Folland, previously de Havilland's junior draughtsman at Farnborough, who had gone on to design the SE5 fighter, joined the Nieuport & General Aircraft Company of Cricklewood, where he worked in conjunction with Major Heckstall-Smith and Henry Leonard Hall, formerly manager of Farnborough's mechanical engineering department. Hall went on to become the chief engineer of Imperial Airways.[223]

The break-up of the Farnborough log-jam coincided with the development of improved fighter pilot skills, best illustrated by the Royal Flying Corps' adoption of Major Robert Smith Barry's celebrated Gosport system of flight training, through which the Avro 504 finally came into its own.[224] But the end of the old procurement system did not mean it was all roses for the pioneers. Private manufacturers were soon enmeshed in a labyrinth of excess profits duty legislation: and beyond general sub-contracting, through which the more innovative firms received low returns for the loss of their patent rights, there were even plans towards the end of the war for the establishment of national aeroplane works, to be controlled by the Department of Aircraft Production at the Ministry of Munitions. In other words, by the end of the war unfulfilled ministry quotas, compounded by increasing labour unrest, were in danger of precipitating a new system of state control. Even then, however, past experience would ensure that the system would stop short of fully-fledged nationalisation. Design would be left to private firms, who would effectually manage

summary of the Bailhache inquiry's proceedings see Cooper, *Birth of independent air power*, 44–5.

[222] Penrose, *Great War and Armistice*, 123, 159, 176; Hare, *Royal Aircraft Factory*, 113.

[223] Penrose, *Great War and Armistice*, 217–19; E. H. S. Folland, 'The life and work of H. P. Folland', *Aerospace* i (1974), 14.

[224] See F. D. Tredrey, *Pioneer pilot: the great Smith Barry who taught the world to fly*, London 1976, passim.

the national works. The official history of the Ministry of Munitions, written shortly after the war, emphasised this point to a degree which must have appeared puzzling to those unfamiliar with the earlier controversy:

> The new emergency state factories owed their inception not to any definite plan or policy of state monopoly but to the immediate stress of practical necessity. They came into being as additional to existing sources of supply, not in substitution for them. Throughout the war they worked side by side with trade factories, and . . . their products were the same as those produced under contract.[225]

The key was expediency. As a result of the increasing demand for aircraft, including, by the final year of the war, the need for heavy bombers, it was felt that the management of a number of large scale factories should be formally co-ordinated in an effort to facilitate urgent mass-production.

The decision to go ahead with state factories was taken in September 1917. Three national aircraft factories were initially planned, at Croydon, Liverpool and Manchester. (The Sopwith workshop at Richmond was also originally chosen to form the basis of a government-sponsored national factory, but this proposal was not followed through.)[226] In the event none of them prospered. Manufacture began at Holland, Hannen & Cubitt's Waddon factory at Croydon in March 1918, and at Crossley Motors' Heaton Chapel factory in Manchester (which undertook the construction of DH9s and DH10s) the following April, but it was June before the new Cunard factory based at Aintree near Liverpool began production, and in every instance output failed to match estimates. The recruitment of skilled labour remained a desperate problem and the state factories were racked from the first by a debilitating degree of labour unrest.[227] Plans to convert several other works into national factories were consequently abandoned.

Certain engine works were also taken under government control. Mitchell, Shaw & Co.'s works at Hayes, in Middlesex, were, for example, transformed into a national aero-engine factory in October 1917, while the works of the Motor Radiator Manufacturing Company, based at Greet, near Birmingham, and Sudbury, Suffolk, were also given national factory status, as heavy bombers employed water-cooled engines which required radiators of a particular construction. But once again results were disappointing and serious difficulties were experienced in recruiting sufficient skilled labour. Given the government's maintenance of a broad industrial base, however, no crippling crisis resulted.

In the final analysis the later plan for state factories was simply guided by pragmatism. Once the gap between theory and reality proved unbridgeable

---

[225] *History of the Ministry of Munitions*, viii/1, 34.
[226] Ibid. viii/2, 197.
[227] Ibid. 202ff.; xii/1, 85.

plans were amended to meet the facts. There was never any question of private firms being nationalised in the socialistic sense.

De Havilland designed just one more aeroplane for the Royal Aircraft Factory after the BE2: the BS1 or Blériot (tractor) Scout biplane of early 1913. This was an advanced, single-seat, single-bay, wing-warp controlled staggered-wing biplane, incorporating a monocoque (single-shell) circular-section fuselage. When employing a 100 h.p. Gnôme engine it reached speeds of up to 92 m.p.h., but it crashed in March 1913, injuring de Havilland himself, and was refitted with an 80 h.p. Gnôme engine. In this form it was formally redesignated the SE2 (Scout Experimental No.2). Even then, however, its continued lack of speed differential and poor field of vision made it unpopular with the Royal Flying Corps; so, despite further modifications undertaken with great skill by Henry Folland, it never went into production.[228]

In June 1914 de Havilland left Farnborough to join Holt Thomas's Hendon-based Aircraft Manufacturing Company, Airco. Six months previously he had been coerced into joining the Aeronautical Inspection Department, and this diversion from design work seems to have prompted his resignation.[229] (An additional attraction may have lain in the fact that Airco's general manager was by this time Hugh Burroughes, a former Farnborough colleague.) Former Avro and Bristol pilot, R. C. Kemp, became the Aircraft Factory's new chief test pilot, but, in the event, the war was to see de Havilland again working under general government direction. As a reservist, he was, of course, subject to RFC control.[230]

By early 1916 the war in the air had so far advanced that it was a firm rule that reconnaissance machines such as the BE2 should be escorted by fighting-biplanes developed for the purpose under the auspices of the Directorate of Military Aeronautics and the Royal Aircraft Factory: notably the two-seat FE2b and the single-seat Airco DH2 Gunbus-type pushers. Both of these aircraft derived from de Havilland designs. French Nieuport tractor-scout biplanes, with a top-wing mounted machine-gun, were also in use by the opening of the battle of the Somme. Indeed, somewhat ironically, the Admiralty were asked to provide the Royal Flying Corps with a variety of additional aircraft for this offensive, and a number of Sopwith 1½ Strutters were among those transferred. For their part, naval pilots attached to the Royal Flying Corps refused to fly the BE2c.[231] Meanwhile, the staff of the Royal Aircraft Factory

---

228 Report by Lt.-Col. Sykes, OC, RFC, 7 Mar. 1914, PRO, AIR 1 121/15/40/106; Hare, *Royal Aircraft Factory*, 190–2, 272–4; Bruce, *Aeroplanes of the Royal Flying Corps*, 464–71.

229 de Havilland, *Sky fever*, 93–4; PRO, AIR 1 686/21/13/2245. He was already regularly performing aircraft inspection duties for the Factory (AIR 1 729/176/5/69), and had been obliged to become a 2nd Lt. in the RFC Special Reserve the previous year (official correspondence, Jan.–Feb. 1913, AIR 1 789/204/4/646).

230 de Havilland, *Sky fever*, 96–7.

231 J. H. Morrow, *The Great War in the air: military aviation from 1909 to 1921*, Shrewsbury 1993, 167.

went on to design the BE-derived RE8 two-seater reconnaissance biplane, and eventually, in 1917, the Folland SE5a fighter: this last being, by then, one of a number of this new genre, including the Sopwith Camel and the two-seater Bristol F2b – better known as the Bristol Fighter.

In the post-war slump Airco were to be taken over and liquidated by the Birmingham Small Arms Company. Out of the remnants of Holt Thomas's team the de Havilland Aircraft Company was formed in 1920. (The Gloucestershire Aircraft Company had already branched out as an Airco subsidiary in June 1917.) De Havilland were to remain one of the world's foremost aviation firms until well into the jet age.[232]

---

[232] de Havilland, *Sky fever*, 111 passim.

6

The formation of a flying corps

It was Haldane's view, as has been shown, that aeronautics could not be effectively utilised for military purposes until the essential principles underlying the new technology had been properly understood. Only then could an air service be built on what he called 'a foundation of science'. To this end, much energy had been expended in drawing worthy academics into the field of aeronautical research and the army's own Balloon Factory had been reconstituted as a civilian state research department under direct War Office control. The corollary of this position, however, was that until a scientific base had been properly laid it would be premature to speculate on the form any actual field units should take. Haldane consequently declined to give much thought to the matter. He had little faith in the practicability of aeroplanes anyway, believing large airships, which he would characteristically refer to as Zeppelins irrespective of their origins, to be immeasurably more significant. But he did not labour the point. Once research had been established on a correct basis he felt that the question of organisation and personnel could safely be left to his deputy and successor, J. E. B. Seely.[1] The upshot was that while a government aeronautical research institution was established in 1909 with great deliberation, a corresponding army aviation corps evolved some way behind it – separately, slowly and fitfully.

As this would suggest, the first army aviators (that is serving officers on the active lists) were therefore essentially only enthusiastic individuals within the service, rather than officially sanctioned precursors of any potential air corps. These figures took it upon themselves to forge links with aviation's practical innovators. They had little option if they wished to pursue this activity. The CID Aerial Navigation Sub-Committee had, as we have seen, recommended the discontinuance of all government-sponsored aeroplane trials, with the proviso that, in their place, advantage be taken of what was ambiguously described as private enterprise. The first indication of how this was actually being interpreted came with the full CID's endorsement of the sub-committee's findings in February 1909, when Rolls successfully offered to place a Short-Wright biplane at the government's disposal, on condition that they provide him with adequate facilities for trials. Concurrently, Rolls's Aero Club associate, Frank Hedges Butler, was also lobbying the War Office to allow ordinary Aero Club members access to War Office land, and this too was granted.[2] But

---

1 'Memorandum of events between 1906–1915', 144.
2 Col. E. W. D. Ward, Permanent Under-Secretary for War, to Frank Hedges Butler, 1 Feb. 1909, Butler collection, vii. 144.

it was October 1909 before Rolls was actually in a position to deliver a fully operative aircraft to the authorities. Only then did the search for a suitable army flying ground actually begin.

Ideally, Rolls informed Esher (his contact in the CID), any site should consist of a smooth, flat surface area of about half a square mile, with 'fairly open country' around it, making longer flights feasible.[3] He contrasted this with the Short Brothers-Aero Club ground on the Isle of Sheppey, which he depicted as being crossed with dykes. His request that the War Office draw up a list of alternative locations, however, met with little immediate response. This was mainly because the War Office initially took the view that their existing ground at Farnborough was itself adequate for preliminary trials. Rolls's former chauffeur-cum-manservant, Tom Smith, recalled towing the Short-Wright biplane to the Farnborough site shortly after the Olympia Aero Show of March 1910.[4] The aim was for Rolls to eventually instruct selected officers in the art of aeroplane flight: but when actually confronted with the terrain at Farnborough (Cody's former stamping ground), Rolls apparently refused to go up at all.[5] The War Office was therefore forced to find an alternative location. (This procedure may simply have been a formality that it was necessary to go through.)

There was a good deal of press speculation during this time, much of it ill-informed. In early June 1910 *The Aero* reported that the proffered biplane was in storage at the army's Hounslow barracks west of London – with apparently little prospect of its ever seeing service. In fact, however, at that time it was at the Balloon Factory, where Rolls joined it on the 20th and 21st. The previous February the army had transported an aeroplane-shed from the Isle of Sheppey to the War Office manoeuvre ground on Hounslow Heath, and an adjacent launch-pylon and mono-rail had been provided by Short Brothers, but nothing followed until Rolls's new French-Wright biplane was towed to the site after Rolls's double channel-crossing of 2 June. This was presumably what *The Aero* had learnt of. There is no evidence of army involvement in Rolls's brief June flights at Hounslow, and, indeed, it remains unclear what the official level of authorisation was for this activity. (The Hounslow shed was only purchased by the War Office, for £150, after Rolls's death the following month.)[6] Soon after the unofficial Hounslow flights and Rolls's visit to Farnborough, the decision was taken to build Rolls a shed at Larkhill, on Salisbury Plain. Here a small group of servicemen had already begun to collect for the

[3] C. S. Rolls to Lord Esher, 25 Oct. 1909, Esher papers, ESHR 5/32.
[4] T. O. Smith memoir, Rolls papers. In fact, on 16 March 1910 the War Office agreed to buy the aircraft for £1,000, this price including the cost of its delivery to Farnborough. Precisely when and, more importantly, why it was decided to purchase a machine Rolls was reported to have been placing at the War Office's disposal anyway is unclear. Perhaps Rolls's near-legendary parsimony had got the better of him: Bruce, *Charlie Rolls*, 31–2.
[5] PRO, AIR 1 731/176/5/102.
[6] 'Aviation and the army', *The Aero*, 7 June 1910; Bruce, *Charlie Rolls*, 31, 35; Montagu of Beaulieu, *Rolls*, 212, 226.

purpose of carrying out experiments in military aviation. Foremost among these was Captain John Fulton (usually referred to as J. D. B. Fulton), of the Royal Field Artillery, 65th (Howr) Battery.

Fulton's imagination had been fired by Blériot's cross-channel flight the previous year. After the idea of building his own aeroplane had proved impractical, he had ordered from Grahame-White a monoplane of the conventional Blériot type, fitted with a 28 h.p. Anzani engine. With this he began experimenting on Salisbury Plain. (The Royal Field Artillery were based at nearby Bulford Camp.) This work was indirectly financed by the War Office, inasmuch as Fulton's aptitude as a mechanical engineer had led him to patent several field gun improvements, for which he was officially rewarded. Fulton then used these proceeds to maintain his aeroplane. In November 1910 he was to become the first regular soldier to gain his aviator's certificate.[7] Both in terms of location and terrain, then, it was soon apparent that a suitable air station had emerged.

No sooner had a shed been erected for Rolls's use, however, than he was killed at the Bournemouth aviation meeting of July 1910. The shed was allocated to Fulton instead.[8] The Short-Wright Flyer presented to the army by Rolls was never used. It soon became warped and obsolete, and was eventually written off.[9] Rolls's place as (unofficial) instructor to any assigned army officers was taken by G. B. Cockburn – whom one of those originally selected, P. W. L. Broke-Smith, was later to describe blandly as a 'philanthropic private aviator'.[10] Cockburn had learned to fly at the Farman School at Mourmelon the previous year, and had subsequently won renown as the only British competitor at the 1909 Rheims meeting. On this basis, and with the active backing of Northcliffe, he had been permitted to erect a shed at Larkhill and to continue his work there.[11] This was again accepted as being in accordance with the 1909 CID sub-committee's recommendations. Once established, Cockburn began working in close co-operation with Fulton. It was on the former's Farman, indeed – the first Henry Farman III, known by contemporaries as 'the father of all Farmans' – that Fulton took his aviator's certificate. (The pair would remain close associates; Cockburn subsequently served under Fulton as a wartime AID Aeroplane Inspector.)

These two aviation enthusiasts were then joined at Larkhill by another artillery officer, Captain Bertram Dickson. The Royal Artillery had long been closely associated with the Balloon Sections of the Royal Engineers and, since

---

7 RAeC certificate No.27, 15 Nov. 1910: PRO, AIR 1 727/160/5; AIR 1 725/100/1; reminiscences of G. B. Cockburn. Lt. Lancelot Gibbs, who had qualified for certificate No.10 on 7 June 1910, was a militia officer. Capt. Bertram Dickson, who had qualified for Ae.C. de France certificate No.71 on 19 Apr. 1910, was not on the active list.

8 Reminiscences of G. B. Cockburn.

9 Broke-Smith, 'Early British military aeronautics', 44.

10 Ibid. 51.

11 PRO, AIR 1 727/160/5; reminiscences of G. B. Cockburn; A. E. Widdows (Haldane's private secretary) to Northcliffe, 15 Sept. 1909, Northcliffe papers, Add. 62155.

before the Boer War, its officers had regularly trained in methods of artillery observation from captive balloons. Unfettered by any restraints resulting from the abolition of aeroplane experiments under the auspices of the Royal Engineers' Balloon Factory or Balloon School, it was almost solely artillery officers who now maintained the momentum behind introducing practical military aviation into the British army. Indeed, this branch of the service would remain a fruitful source of airmen, as demonstrated in the emergence of so influential a figure as Sefton Brancker shortly afterwards. Brancker had attended the Lydd siege artillery camp, which incorporated balloon observation, as early as 1897.[12]

Bertram Dickson had, of course, established his reputation on the European exhibition circuit. In July 1910 he made the acquaintance of Sir George White, founder of the Bristol-based British & Colonial Aeroplane Company. British & Colonial had negotiated flying rights over Salisbury Plain, and consolidated this advantage, in June 1910, by establishing a flying school at Larkhill.[13] This was the month before Rolls's death. The arrival of the Bristol school confirmed Larkhill as the army's principal pre-war aviation base. Dickson became an advisor to British & Colonial and, through this association, did much to publicise aviation within the services.

Another gunner associated with the practical development of British military aviation was Lieutenant Lancelot Gibbs. Gibbs had been closely involved in Dunne's Blair Atholl trials some two years before. Like Cockburn and Dickson, he had, however, learned to fly to certificate standard at the Farman school at Mourmelon, from where he had procured his own Farman biplane.[14] Using this, and a Sommer biplane purchased from C. S. Rolls, he had rapidly established a reputation as an adept exhibition pilot, forming, in the summer of 1910, his own charter company, L. D. L. Gibbs & Co., which provided instruction or mounted flying exhibitions according to the needs of its customers.[15] (Gibbs had received a good deal of pre-launch publicity as a result of becoming among the first British aviators to fly for more than an hour without alighting.)[16]

That the army should now have to rely on such figures as these – Fulton, Cockburn, Dickson, Gibbs and Rolls, until his death – was a direct result of the CID sub-committee's recommendation that official army aviation trials be discontinued. Thus although Fulton, Dickson and Gibbs were all ostensibly artillery officers, they had each to move outside the service to some degree before they could introduce aviation into it. They had to function in a private capacity, on their own, not the army's, initiative. Haldane had little time for such figures, but if they wanted to work off their own bat he had no objection.

---

[12] Macmillan, *Sir Sefton Brancker*, 6.
[13] See ch. 3, pp. 107–10.
[14] Reminiscences of G. B. Cockburn.
[15] *The Aero*, 21 June 1910.
[16] PRO, AIR 1 727/160/5; *The Aero*, 12 Apr. 1910.

Indeed, such work helped assuage precipitate calls for action, and left him free to, as he saw it, re-establish military aeronautics on a scientific basis. The result was that while Fulton, Cockburn and the others were establishing themselves at Larkhill, the revised Balloon Factory at Farnborough remained wholly preoccupied with airships.

The first positive evidence of the inclusion of aviation in the army's order of battle came with the summer manoeuvres of 1910. This was true of the French as well as of the British army. In fact the French War Ministry had been undertaking its own bureaucratic investigation into the development of military aviation and a dispute had arisen between their Engineering Corps and the artillery over who should have administrative control of the new arm. As in Britain, captive balloons had long been used in conjunction with field guns for ranging on hidden targets. Moreover the artillery already had at its disposal workshops experienced in mechanical engineering. Control of the automobile service of the French army had already been allocated to the command for this reason. With experience in aeronautics, mechanical engineering and engine maintenance, the artillery argued strongly that the administration of aviation should also fall to them.[17] The arming of aeroplanes would raise ballistic and pyrotechnical problems which they alone were capable of solving. This was reason enough for the French Minister of War, General Brun, to organise an aviation section at the close of 1909 and entrust its command to the artillery. However this decision simply exacerbated divisions. So bitter was the Engineering Corps' opposition, indeed, that it was eventually decided to create two separate aviation departments, general research falling essentially to the Engineering Corps and practical aviation falling essentially to the artillery. This did nothing to quell arguments over where the true responsibility for aviation lay.[18] Consequently in February 1910 General Brun conferred with his service chiefs. They recommended unification. As a result it was decided in June 1910 that the whole of the aeronautical service should be placed under what was called the 4th Command, that is the Engineering Corps. Improved performances over the previous six months were deemed to have made the practical application of long-range autonomous aeroplane reconnaissances feasible. It was increasingly evident that aeroplanes need not be tied to immediate ground forces. Once he had conceded this General Brun had little option but to sanction a complete reversal of his previous policy. An aviation service would self-evidently have more autonomy under the Engineering Corps.

The Picardy manoeuvres of September 1910, the first in which aviation played a part, confirmed the increasing range and practicability of aeroplanes when skilfully piloted and properly co-ordinated. The opposing forces employed both Blériot and Antoinette monoplanes and Farman and Sommer biplanes, and their performance completely overshadowed that of the dirigibles. The army aviators, watched by such famous pioneers as Paulhan and

17 Facon, 'L'Armée Française', 81.
18 Ibid.

Latham, flew with a surprising degree of assurance and success. As a result a permanent Inspectorate of Military Aeronautics was formed in October under the command of General Roques.[19]

The British army manoeuvres took place a couple of weeks after those of the French, but this time the initiative in the use of aeroplanes came not so much from the authorities as from the semi-autonomous Larkhill pioneers. (The manoeuvres took place on and around Salisbury Plain.) In particular, Captain Dickson, on behalf of the British & Colonial Aeroplane Company, placed his services at the army's disposal. He piloted a Bristol Boxkite (No.9) and was joined by the actor and aviator, Robert Loraine, on another Bristol Boxkite (No.8) and Lieutenant Gibbs (in his role as an officer in the reserve) on a Farman biplane of the clipped-wing Paulhan variety. Colonel Capper was also present, piloting the 'Beta' non-rigid airship. The results, in the words of one observer (G. B. Cockburn), 'could not be said to be striking'.[20] The various aircraft were modified for passenger flight, with reconnaissance, liaison and communications as their envisaged tasks, but a combination of engine failure, bad weather and inexperience among both the aviators and the army's direct- ing staff reduced any advantage to be gained from their employment to a minimum. Compared with French efforts, in fact, the whole affair was dispirit- ingly amateurish. Dickson was compelled to improvise a sketch-map of the area for his thirty-mile dawn reconnaissance along the Wylye valley on 21 Septem- ber – the first significant flight of the week.[21] Yet, despite the fact that his projected landing site had not been secured and that he was forced to alight in disputed territory, he managed to file successfully a report of the position of the opposing Blue army's advance cavalry units. Later that morning he took off again, and on alighting telephoned his report from a nearby house. Friendly Red army officers proceeded to the spot but were deemed 'killed' by local enemy forces, so any directly communicated information was disallowed. Sir John French and Winston Churchill – the latter in the uniform of a Major of the Oxfordshire Yeomanry – also arrived to consult with Dickson, who had achieved at least a local celebrity. Two further sorties occurred near dusk, Lieutenant Gibbs supplementing Dickson's efforts.

The Blue army had had the use of the wireless-linked 'Beta' airship during this time. Much popular attention, however, focused on Robert Loraine's somewhat contrived transmission of wireless messages from an aeroplane. Also assigned to the Blue army, he sent the first of these whilst passing over Stonehenge, thereby linking what were taken to be Britain's most ancient and most novel technological features.

George Holt Thomas had witnessed the French manoeuvres and done much

---

[19] Ibid. 83.

[20] Reminiscences of G. B. Cockburn.

[21] The map is now displayed in the Museum of Army Flying, Middle Wallop. The best account of the manoeuvres from an aviationary perspective is in Munro, 'Flying shadow', 16–21.

to publicise their success in Britain. For this reason he was asked to report for the *Daily Mail* on the British manoeuvres.[22] As an official correspondent he became, in his own words, 'a severe critic'[23] (although by this time he was already acting on Farman's behalf, and would soon establish the Aircraft Manufacturing Company for the production of Farman aeroplanes:[24] so he was not an altogether disinterested party). In his articles, and through personal influence, he did all he could to combat the notion, gained in some quarters, that aeroplanes had shown themselves to be impractical under battlefield conditions. He was able to place the difficulties faced by Dickson and Gibbs – whose work was undertaken largely on their own initiative, with little support from the War Office – in perspective.[25] Unlike in France, no practical developments resulted immediately from these events; but one can see in retrospect that beneath the surface attitudes had begun to change.

Shortly after these manoeuvres Esher issued his retraction of the recommendations of the 1909 CID sub-committee, which had, of course, urged that War Office trials with aeroplanes be discontinued and that advantage be taken, instead, of private enterprise – which, contrary to what Esher was now implying, had meant simply private or individual initiative, rather than the utilisation of any formal industry, of which there was then virtually none. The way this policy was being interpreted, wrote Esher in October 1910, was clearly no longer sufficient. He had been most forcibly struck by Colonel Charles à Court Repington's report of the French manoeuvres, published in the *Times* on 3 October 1910. From this it was evident that the experimental stage of aviation was over. It was now imperative that the British services acquire their own heavier-than-air aircraft and set about forming an air corps. In other words, they must take the initiative and work towards some defined objective, not simply utilise philanthropy.[26]

This was plainly the logical direction in which to proceed if the army now seriously intended to deploy aeroplanes in battle; and despite previous reservations it was almost inconceivable that such devices would not be employed in some form. Consequently Esher's advice does seem to have been widely accepted. By this time Haldane had restructured the Farnborough Balloon Factory on his own lines and cared little after that for the form of any practical field unit. As a result he had paid little attention to the old Balloon Sections, the remnants of which survived in some disarray in the form of the Balloon

22 Reminiscences of G. Holt Thomas. This was in addition to Northcliffe's regular – rather inane – air correspondent, Harry Harper.
23 Ibid. 12.
24 Ibid. 18.
25 Ibid. 14; *The Aero*, 28 Sept. 1910.
26 'Aerial Navigation, note by Lord Esher'.

School, still operating under the command of Colonel Capper. Following the implementation of the CID sub-committee report this establishment had been left to train with captive balloons and small dirigibles. This work was of marginal consequence, but there was little else the school could now do. It could only watch, with mounting frustration, as a few individuals, in an essentially private capacity, attempted to infuse an aviationary element into the 1910 manoeuvres. Once it was accepted that this was insufficient and that some sort of service air corps would be necessary attention focused again on the anomalous appendage known as the Balloon School, Royal Engineers. This ill-defined formation was to provide the basis of the new command.

In fact discussions as to the organisation of a corps of airmen had been taking place within the Master General of Ordnance's office for some months. These had been preceded by successive complaints from Colonel Capper regarding the inadequate maintenance of the Balloon School. By November 1909 he was even questioning whether drawing personnel solely from the Corps of Royal Engineers was sufficient. 'It is practically a new branch of the Army', he insisted in a memo to the Chief Engineer, Aldershot Command (Brigadier-General Scott-Moncrieff), when asked for his views on a projected structural reorganisation.[27] The Chief Engineer agreed. 'The time has now arrived when aeronautics should be opened to volunteers from the whole Army', particularly the Royal Artillery, he suggested.[28] A decision to form some kind of corps from the Balloon School was taken in principle at an MGO conference on 8 March 1910. On 12 July the matter was again debated at an MGO conference. Present, as well as Major-General Hadden, were the Director of Military Training, Brigadier-General A. J. Murray, the Director of Fortifications and Works, Brigadier-General Rainsford-Hannay, and the Director of Staff Duties, Brigadier-General L. E. Kiggell. They recommended the adoption of the title Air Corps to replace the Balloon School, this body to be commanded by the former Commandant. A tentative establishment estimate was to be drawn up presently, on approval of the proposal by the Army Council.[29] As events transpired a new organisation was to come into being, but not yet as a corps.

Even as these discussions were proceeding, however, the Balloon School, which included men who had previously worked on the Dunne and Cody projects, was tentatively beginning the resumption of aeroplane trials (or training, as such work would have had to have been labelled, to distinguish it from research), under the guise of exploiting private enterprise. This resulted from the presentation to the War Office of a Blériot Type XII two-seater monoplane by Colonel Joseph Laycock and the duke of Westminster. The Balloon Factory having abandoned aeroplane trials, it fell to Lieutenant R. A. Cammell, of the Balloon School, to collect the machine directly from the

[27] Memo, 4 Nov. 1909, PRO, WO 32/6936.
[28] Memo, 10 Nov. 1909, ibid.
[29] Memo, 13 July 1910, ibid.

Blériot aerodrome at Étampes.[30] What the War Office does not seem to have been told was that this Type XII was, in fact, none other than the 'White Eagle', previously relinquished by Grahame-White following confirmation of its structural imbalance.

Lieutenant Reginald Cammell had been commissioned as an officer of the Royal Engineers as recently as 1906, and had already gained a considerable amount of ballooning and kiting experience with the Balloon School. Indeed, Capper had identified him two years previously as a possible pilot for the Dunne Blair Atholl trials.[31] He would certainly have already gained some degree of aviation experience, and this was doubtless why he was selected to collect the proffered aeroplane from France. The Blériot XII, however, was inherently unstable, with what C. G. Grey described as a dangerously low centre of gravity.[32] It was, moreover, powered, as before, by an ill-tuned inefficient 60 h.p. ENV engine. (The promotion of this marque of engine appears to have been one of the motives behind the presentation: Laycock was the publicity-conscious chairman of the ENV motor syndicate.) Presumably over the previous months Blériot had modified the chassis in some manner, but precisely how is difficult to determine. Cammell, only learning of the machine's history on his arrival in France, registered no significant structural revisions since Grahame-White had, as he understood it, rejected the model on account of its instability.[33] Indeed, quite the reverse: his report simply confirmed all previous misgivings, and he frankly doubted the existing aeroplane would survive his own debut. It was, he noted on 29 July 1910, 'too heavy', and in the glide gave a dangerously steep angle of descent.[34] The least one could expect from this was a damaged undercarriage. But in the end the aircraft was accepted on condition that the engine be thoroughly overhauled.[35] It became available for service in October 1910, and was stationed at Larkhill. There it acquired the sobriquet 'man-killer'.

It did not remain in service for long. Cammell's fears were realised when engine failure necessitated a forced landing whilst he was flying the machine to Farnborough for yet another overhaul. Too steep a descent resulted in a disabling smash and the aircraft was transferred to the Balloon/Aircraft Factory to re-emerge some time later as de Havilland's SE1. In that form it fully justified its predecessor's 'man-killer' reputation. As far as the Royal Engineers' Balloon School was concerned, however, this simply meant that they had lost their only official aeroplane to date. So Lieutenant Cammell privately purchased a

---

30 PRO, AIR 1 727/152/6; AIR 1 1612/204/87/31.
31 Col. J. E. Capper to J. W. Dunne, 5 Oct. 1908, ibid. AIR 1 1613/204/88/10.
32 The Aero, 4 Jan. 1911, quoted in Bruce, Aeroplanes of the Royal Flying Corps, 134.
33 Lt. Cammell to CBS (Col. Capper), 11 July 1910, PRO, AIR 1 1612/204/87/31.
34 Lt. Cammell to CBS, 29 July 1910, ibid.
35 Daily report, 3 Aug. 1910, ibid.

more conventional Blériot from France, a side-by-side two-seater Type XXI, and piloted this through most of the following year.[36]

The Blériot XII fiasco highlighted the dangers of relying on presentations from private enterprise as a *de facto* method of procurement. After this episode the War Office began to countenance purchases from the private sector in the normal way. (This was before the Aircraft Factory began producing government designs in competition with the private sector.) In late 1910 Whitehall sanctioned the purchase of a Farman 'Type Militaire' and what was described as an experimental Paulhan biplane directly from the manufacturers. By this time it had been decided that an Air Battalion would be formed out of the remnants of the Balloon School, and Captain C. J. Burke, of the Royal Irish Regiment, who had learnt to fly Farmans in France, and the slightly more experienced Captain Fulton, were attached to the school in advance of its formation. In view of their respective degrees of experience Captain Burke was then assigned the new Farman, which became available in November 1910, while Fulton, who, of course, already owned his own Grahame-White 'Blériot' monoplane, was selected to take delivery of the experimental Paulhan.[37] This became available in December 1910.

Cockburn accompanied Fulton to France to collect the Paulhan, and was to describe its composition as 'weird'. It was essentially a jagged-winged, two-seat, pusher-biplane, with the pilot encased in a small central nacelle. Particularly striking were the W-girder front-spars, to which the mainplane's curved ribs were attached, the mainplane wings being joined by four single interplane struts. The tailplane and forward elevator employed the same W-girder construction. The aircraft derived its more unorthodox features from Paulhan's recent collaboration with Henri Fabre, the Marseilles-based boat-builder and seaplane pioneer. It was evidently constructed in a very speculative spirit, its attraction from a military point of view being its perceived portability, meaning its relatively convenient dismantling and transporting in a crate measuring 15 ft 6 in. by 3 ft 3 in. by 3 ft 3 in. Despite passing the requisite duration tests, however, it was never thereafter to fly satisfactorily.[38] It was thus a fruitless acquisition. Having finally decided that it must reinvestigate aviation, Whitehall seems to have been sold this dud (Harry Busteed's description) through the simple expedient of language. The manufacturers reported the model as embodying many 'improvements'. And so it must have appeared; it

---

[36] Brown, *History of Royal Engineers*, iv. 290; PRO, AIR 1 1611/204/87/9. See also Bruce, *Aeroplanes of the Royal Flying Corps*, 135–7.

[37] Broke-Smith, 'Early British military aeronautics', 45–6; PRO, AIR 1 1612/203/87/28.

[38] The Paulhan 'experimental' arrived at Farnborough, after its reception tests in France, on 16 February 1911. Less than a month later, on 13 March, O'Gorman was recording in his private diary that it had not 'prove[d] . . . much good' in preliminary trials: PRO, AIR 1 732/176/6/24. It was subsequently modified, under de Havilland's supervision, but ensuing test flights merely resulted in a succession of accidents. The aircraft was delivered to the Air Battalion in October, but by the close of 1911 the model had been dismantled and was never flown again: Bruce, *Aeroplanes of the Royal Flying Corps*, 340.

was so bizarre it was not merely experimental, it was (one could almost believe) scientific. This doubtless accounts for the interest shown in it by the Advisory Committee for Aeronautics. Within civilian circles, on the other hand, news of the acquisition was greeted with incredulity. 'Mr Holt Thomas', commented the editor of *The Aero* (C. G. Grey) waspishly, 'who now represents M. Paulhan's interests in this country, has done a good piece of business.'[39]

The choice of the Farman 'Type Militaire' was no better received. The military designation was practically meaningless. The machine incorporated strut-braced extensions to the upper-wings and a third, central, rudder between the tailplanes but was essentially no different from an ordinary Farman III, nor from those Farman-derived biplanes already being constructed in Britain by British & Colonial, Short Brothers and Howard Wright. Why buy abroad?[40] Both purchases seemed to reveal a degree of ignorance about the developing aviation industry.

Only when the War Office sanctioned the purchase of four Bristol Boxkites, early in 1911, was the material basis of an aviation unit actually created. These were to provide the emergent Air Battalion with a set of reliable training aircraft soon after it came into being in April of that year. Additional new models, from a variety of manufacturers, were then to be purchased as they appeared, for trial purposes. The British establishment, however, remained at a particular disadvantage compared with that of the French when it came to recruiting a skilled ground staff. When in France to take delivery of the experimental Paulhan, Captain Fulton (who had qualified as an interpreter 1st class in French some three years before) took the opportunity to report on the French Military School of Aviation at Châlons. Here he noted that a staff of experts was retained for the express purpose of maintaining aero-engines, specifically Gnômes. 'It is not easy to realise', he commented, 'how greatly the French military system facilitates the obtaining of men with expert knowledge.' Unlike Britain, France had a conscript army. The highly-skilled staff of the Châlons establishment consisted of professional mechanics doing their military service. Indeed, one of the ground staff when Fulton visited was the son of Louis Séguin, the coinventor of the Gnôme engine and head of the Gnôme engine company.[41] Short of conscription, there was no way the British military aviation establishment could attract such figures from civil life. This was to remain a major problem.

Shortly before the army's Air Battalion came into being Francis McClean offered to place his Short-Sommer biplanes at the Admiralty's disposal. This,

---

[39] *The Aero*, 2 Nov. 1910. See also reminiscences of G. B. Cockburn, and Penrose, *Pioneer years*, 242–3. For the Advisory Committee's interest see Bruce, *Aeroplanes of the Royal Flying Corps*, 340.

[40] *The Aero*, 2 Nov. 1910. In the event, the 'Type Militaire' was wrecked within a month, and the frame had to be entirely reconstructed under the supervision of Geoffrey de Havilland: Bruce, *Aeroplanes of the Royal Flying Corps*, 223–4.

[41] PRO, AIR 1 119/15/40/70.

as noted in chapter 2, first provided the senior service with an aviation capacity. Four naval officers were selected to undergo a course of instruction at the Aero Club's ground on the Isle of Sheppey, and G. B. Cockburn came over from Larkhill to provide flying tuition. This was in March 1911. When Cockburn returned to Larkhill he found the Aeroplane Company of the Royal Engineers' Air Battalion in the course of formation.[42] He was to become unofficial instructor to this unit also.

The Air Battalion was to replace the Balloon School, which during much of this time had been continuing general training at Aldershot. Standards there had dropped markedly since the split with the Balloon Factory. In a report of the unit's 1910 summer training, Colonel Capper, the school's Commandant, complained that his senior NCOs were often ignorant of the technical work involved in the maintenance of airships and small dirigibles. Indeed, he felt that recruits generally were of a 'distinctly lower standard' than before.[43] His complaints went unheeded. The opinions of Capper and the interests of the Balloon School no longer held any weight. In October 1910 Capper completed the regulation period of five years as a substantive lieutenant-colonel. His promotion to full colonel necessitated his leaving the aeronautical service altogether. (The post of Commandant, Balloon School, was considered of insufficient importance for a full colonel.) The position being a regimental one, he was succeeded by GSO, Major Sir Alexander Bannerman RE.[44] Capper was subsequently appointed Commandant of the School of Military Engineering at Chatham, where he remained until the outbreak of war.[45]

Sir Alexander Bannerman was a competent professional soldier, but compared with Capper he had little aeronautical experience. Born in 1871, he had, like Capper before him, been educated at Wellington College and the RMA, Woolwich, and since 1907 he had also been a member of the Aero Club of the UK – under whose auspices he had participated in a number of balloon ascents. However he had really made his name in the Russo-Japanese War of 1904, when he had acted as British military attaché at the Japanese HQ. There he had been interested to observe the limited use made of a captive balloon by the Japanese forces during the siege of Port Arthur. In this role of military attaché he was, ironically, subordinate to Lieutenant-General (as he then was) Sir William Nicholson, who had been appointed the head of a special military mission to monitor the campaign.[46]

---

[42] Reminiscences of G. B. Cockburn.

[43] PRO, AIR 1 1612/204/87/39.

[44] Brown, *History of Royal Engineers*, iv. 77; Broke-Smith, 'Early British military aeronautics', 41; *Daily Telegraph*, 4 Oct. 1910 (Butler collection, xii. 106).

[45] In August 1917 the Army Council would propose Capper (by then Maj.-Gen. – latterly a divisional commander in France and Cmdt, Machine Gun Corps Training Centre) as David Henderson's replacement (DGMA); but the appointment was not popular with the RFC nor, in particular, with the Deputy-Director of Military Aeronautics, Sefton Brancker, and was quickly rescinded: Cooper, *Birth of independent air power*, 112–13.

[46] Brown, *History of Royal Engineers*, iv. 163. Bannerman was questioned on his observation

With Bannerman's appointment as Commandant, Balloon School, came the announcement that the field establishment would over the coming months be reconstituted as a broad aeronautical army-support unit, incorporating aeroplanes, balloons and airships. There would be created, in a much quoted phrase, 'a body of expert airmen', from which the companies of a proposed Air Battalion would eventually be drawn. But while this battalion would remain an integral part of the Corps of Royal Engineers, its officers would be recruited from the service as a whole. This would allow for the reassimilation of those army aviation enthusiasts previously obliged to pursue the subject in a more or less private capacity.

In effect, then, the old Balloon Sections of the British army were dissolved following Capper's departure. Their demise went unlamented. Even before October 1910 it was apparent that, whatever emerged from the reconstituted Balloon School, it would have to incorporate some measure of aviation. No one knew this better than Capper himself. As a parting shot he placed before the Army Council a detailed paper on the way in which military aeronautics should develop after him. 'It is, and must be', he commented, 'particularly difficult for anyone who has not been in constant practical touch with the work' to understand the technical difficulties involved in the maintenance of an air service. In any future expansion there must be a 'proper proportion of expenditure'. Procurement, in other words, must include the provision of 'all necessary accessories for the proper equipment and upkeep' of aircraft – including, he stressed, an adequately trained and maintained ground staff. This in itself would involve the maintenance of training facilities for such personnel. These were not extras, they were essentials. It was imperative that the subject be 'treated as a whole' and not in a 'piecemeal' manner.[47] Bannerman, however, had enough difficulty simply retaining his aeroplanes. The first two machines procured, the Farman and Paulhan described earlier, were partly under the control of the Balloon Factory, which was endeavouring to cut all ties with its erstwhile field counterpart.[48]

By the close of 1910, none the less, the proposed structure of the Air Battalion had been formulated by Major Bannerman. It was to consist of an HQ and four companies: No.1 (Airship) Company; No.2 (Aeroplane) Company; No.3 (Balloon and Kite) Company; and No.4 (Line of Communication) Company. The Aeroplane Company was to have been subdivided into three sections, each possessing four aeroplanes, a mobile field depot and an HQ.[49] In reality, however, Bannerman was in no position to implement such a programme: nor would he ever be.[50] To exacerbate matters, the men of the

---

of balloon-support in Manchuria by the CID flying corps technical sub-committee on 5 Feb. 1912: PRO, CAB 16/16, 101.

[47] 'Notes on military aeronautics', 27 Oct. 1910, Capper papers, III/2/1.

[48] PRO, AIR 1 725/100/1; AIR 1 1612/204/87/28.

[49] 'Notes on proposed establishment, Air Battalion', 5 Nov. 1910, ibid. AIR 1 1608/204/85/35.

[50] Ibid. AIR 1 725/100/1 (Bannerman interviewed in 1919).

proposed battalion were also, theoretically, still to be subject to general duties. With regard to aeroplane work this was plainly becoming impractical. Moreover it would leave those officers attached from other regiments in an anomalous position. So Major Bannerman recommended dropping the old field work course. This, he suggested on 17 January 1911, would put the proposed Air Battalion on a par with the telegraph units.[51] Such a measure was eventually sanctioned, but it failed to alleviate the problem of conflicting responsibilities. Meanwhile Bannerman was forced to modify his immediate plans for the unit's distribution. In fact he now gave way to complete pragmatism, suggesting, 'as a rough outline', that one 'air company' camp on Salisbury Plain for a provisional period of three months, with aeroplanes, kites and captive balloons, whilst another remain at Aldershot with the dirigibles.

It was given to Lieutenant Cammell to formulate proposals for the establishment of a new aeroplane section for this truncated unit, but he again was handicapped by the need to accommodate the complex responsibilities of most recruits. In any detachment, he conceded in a report dated 19 January 1911, officers and men were to be trained essentially for general duties, and most would consequently pass through the aeroplane section in the course of their technical instruction. But even accepting this, it would still be necessary to undertake practical experimental work, involving the testing, improvement and equipment of aeroplanes for military purposes, so a permanent specialist staff of some description would, perforce, have to be formed.[52] The question of whether such a staff could be satisfactorily administered under Royal Engineers' control was not, however, directly addressed. Cammell turned his attention instead to an estimate of what aeroplanes would be required for the proposed section. In keeping with his division of functions he identified two categories: those aeroplanes required for instructional, and those required for experimental purposes. For instruction he recommended the acquisition of representative types of both the biplane and monoplane, identifying, in particular, the Bristol Boxkite biplane and the Blériot Type XI monoplane. For experiment, on the other hand, he recommended the procurement or utilisation of half-a-dozen aeroplanes of various design: the Farman 'Type Militaire', a Blériot, an Antoinette, a Paulhan, a Dunne biplane, and a Valkyrie monoplane or Cody biplane. Preparations, he suggested, should begin at once to equip a suitable flying ground. If sufficient progress had been made, training could then begin the following March (1911). But, he insisted, efforts must be directed towards the maintenance and repair of aircraft in the field, otherwise the section would lack any practical value.

These views were reinforced by no less a person than the Chief of the Imperial General Staff, General Sir William Nicholson, who had done so much to discredit aviation little more than two years before. It was now most

---

[51] Memo to Chief Engineer, Aldershot Command, ibid. AIR 1 1609/204/85/61.
[52] 'Winter scheme 1910: proposals for Aeroplane Branch of Air Corps', ibid. AIR 1 1607/204/85/6.

important that 'we . . . push on with the practical study of the military use of aircraft in the field', he affirmed in a minute of 17 February 1911: 'other nations have already made considerable progress in this training and in view of the fact that aircraft will undoubtedly be used in the next war, whenever it may come, we cannot afford to delay in the matter'. It was his wish, Nicholson continued, that both airships and aeroplanes be employed in any forthcoming manoeu-vres.[53] From this we can see that by the time the Air Battalion actually came into being even aviation's previously most trenchant critic had become an enthusiastic advocate. Such a *volte-face* was later, but no less complete, than Esher's of some four months before.

The Army Order establishing the Air Battalion was finally issued on 28 February 1911. The unit was to consist of an HQ and two companies: No.1 (Airship) Company and No.2 (Aeroplane) Company. Officers were to be selected from any regular arm or branch of the service, provided applicants were under thirty years of age and unmarried. Warrant Officers, NCOs and men were to be drawn from the Corps of Royal Engineers. The new organisa-tion was to supersede the Balloon School, taking effect from 1 April 1911.[54] As Bannerman had recommended, the HQ and No.1 Company were then to remain at the Balloon School's old south Farnborough-Aldershot base, while No.2 (Aeroplane) Company encamped at Larkhill. The entire establishment eventually consisted of fourteen officers and 176 other ranks. The former Balloon School instructor, Captain P. W. L. Broke-Smith, became the battal-ion's adjutant, and among the first officers to be seconded to the unit were Captain J. D. B. Fulton, who assumed command of No.2 (Aeroplane) Com-pany, Captain E. M. Maitland, of the Essex Regiment, who assumed command of No.1 (Airship) Company, Captain C. J. Burke, who, like Fulton, had already been attached to the old Balloon School for some months in advance of the formation of a new air unit, and Captain A. D. Carden, of the Royal Engineers, formerly the Assistant Superintendent of the Balloon Factory at Farnborough, who became the battalion's experimental officer. Six officers in all were seconded from other regiments. Provision was also made for the unit to be supplemented by a reserve, consisting of officers who had qualified as aeroplane pilots at one or other of the civilian flying schools.[55]

For all the preparation, however, the battalion was to have but a brief existence. A bare eight months after its establishment the formation of an unaffiliated flying corps was decided upon. Thus the Air Battalion proved to be merely the final intermediary stage between the Royal Engineers' old ballooning organisations and the creation of, effectively, an autonomous air arm. Broke-Smith, echoing Bannerman, later described it as a 'transitory

53 Ibid. AIR 1 119/15/40/71.
54 Raleigh, *War in the air*, 142; Gamble, *Air weapon*, 126–7; 'New Air Battalion', *Standard*, 1 Mar. 1911 (Butler collection, xiii. 12).
55 Broke-Smith, 'Early British military aeronautics', 46–7; Brown, *History of Royal Engineers*, iv. 77, 292.

training and experimental unit'.[56] But, if nothing else, it did signal the War Office's official readoption of aviation. Responsibility for this branch of the service was never again to be left to the initiative of individual officers working in an essentially private capacity. As a first step in the creation of, specifically, an Aeroplane Company, four Bristol Boxkites had been ordered from the British & Colonial Aeroplane Company in March 1911. Frustratingly, however, none were ready before May and even after then there remained a persistent shortage of available machines. Shortly after its foundation the Air Battalion theoretically had at its disposal the four Bristol Boxkites (although owing to the War Office's insistence that the second two be installed with 60 h.p. Renault engines these were not delivered until July and August), the Paulhan 'experimental', the Farman 'Type Militaire', the notorious Blériot XII (although this machine was at the Aircraft Factory, from where it reappeared as the SE1), Cammell's Blériot XXI, Fulton's conventional 'Grahame-White Blériot', Cockburn's Farman and Rolls's Short-Wright Flyer. In reality, few of these aircraft were serviceable at any one time. A Howard Wright ENV pusher-biplane was subsequently purchased from Captain Maitland, but this acquisition did little to alleviate what remained an inveterate problem. Indeed, it was withdrawn from service in July 1911 – not a month after its purchase.[57] The number of aeroplanes actively in service with the Air Battalion was seldom to exceed four or five at any given moment.

Complaints also derived from the unit's lack of domestic facilities. For any extended period Larkhill was a decidedly uncomfortable billet. Throughout the summer of 1911 Aeroplane Company personnel were obliged to live in tents alongside the flying ground. Only with the onset of winter, when it was decided to make this camp permanent, was proper accommodation found for the men, in barracks at the Royal Artillery Mess, situated at Bulford – some three miles away.[58] This compromise, no doubt achieved as a result of Fulton's connections, was scarcely satisfactory, but, in the circumstances, it was the best they could expect. Since September 1911 the Commandant of the Air Battalion, Major Bannerman, had been arguing that the company should no longer be regarded – and consequently administered – as merely an experimental detachment. 'Enough is already known about the powers of aeroplanes', he insisted, 'to allow of a serviceable unit being formed ready to take the field on mobilisation.'[59] This point was largely conceded and Bannerman's views were

[56] Broke-Smith, 'Early British military aeronautics', 41, echoing PRO, AIR 1 725/100/1 (Bannerman interviewed in 1919).

[57] Bruce, *Aeroplanes of the Royal Flying Corps*, 270.

[58] Maj. Bannerman, Employment of Air Battalion, winter 1911–12, 24 Aug. 1911, PRO, AIR 1 1608/204/85/30; Capt. J. D. B. Fulton to Maj. Bannerman, 29 Jan. 1912, AIR 1 1609/204/85/61.

[59] Maj. Bannerman to Chief Engineer, Aldershot Command, 1 Sept. 1911, ibid. AIR 1 1609/204/85/51.

duly forwarded by Aldershot Command to the War Office.[60] By then, however, the whole structure of the Air Battalion was being reconsidered.

In the meantime, the recruits of No.2 Company, with Bannerman's backing, were demanding new aeroplanes for reconnaissance training. 'The Bristol biplanes bought this summer', Major Bannerman remarked to the Chief Engineer, Aldershot Command, on 24 August 1911 à propos a second batch of four Bristol Boxkites delivered earlier that same month, as well as the original March order, 'have proved invaluable as machines on which officers can gain experience', but technology was moving on. Air Battalion pilots were now of sufficient competence to handle 'faster and more modern machines'. These should consequently be provided at the earliest opportunity.[61] Three new types were expediently procured for comparative trials soon after: a 60 h.p. (Renault) Breguet tractor-biplane, a 50 h.p. (Gnôme) Nieuport monoplane and a long, slender, 60 h.p. (Anzani) Deperdussin monoplane.[62] The Nieuport was considered a particularly fast aircraft, reaching speeds of up to 70 m.p.h. (Edouard de Nieuport was quick to appreciate the monoplane's speed potential, deriving from its reduced head resistance. By further reducing the camber, i.e. convexity, of the mainplane surface, he significantly improved the genre. Unfortunately, however, he was killed in an alighting accident at the Charny military trials, on 16 September 1911.)

The short-lived Hendon-based Aeronautical Syndicate Ltd also presented the War Office with at least one of their peculiarly spartan Valkyrie canard-monoplanes. However this philanthropic gesture served only to hasten the firm's imminent demise. In fact Horatio Barber, stung by his company's unaccountable disbarment from the Parliamentary Aerial Defence Committee's Hendon display in May 1911, had initially offered four Valkyrie monoplanes to the military and naval authorities, and two used models were duly forwarded to Lieutenant Samson at Eastchurch; but documentary evidence survives for only one such aircraft, a newly-built two-seater model, being accepted by the War Office, in August 1911 – and even then the army was to provide the engine.[63] Indeed Lieutenant Cammell supervised the troublesome installation of a 50 h.p. Gnôme motor at Hendon early the following month. On Sunday 17 September he took the aeroplane out for a trial flight, with the intention of flying it on to Farnborough that same day. A fatal crash resulted.

---

[60] 25 Sept. 1911, ibid.

[61] Employment of Air Battalion, winter 1911–12.

[62] Maj. Bannerman to DFW, 20 Nov. 1911, PRO, AIR 1 762/204/4/174. Two of these types, the Breguet and Nieuport, had been under consideration for some time (AIR 1 1612/204/87/37), and Capt. Fulton had previously recommended the acquisition of all three models – plus a Sommer monoplane – in a memo to the DFW dated 8 Sept. 1911: Bruce, *Aeroplanes of the Royal Flying Corps*, 315. The Deperdussin was originally to have been delivered to the Aircraft Factory, but this directive was amended by the War Office and it was sent directly to the Air Battalion: AIR 1 2404/303/2/2, 15 Dec. 1911.

[63] Bruce, *Aeroplanes of the Royal Flying Corps*, 36–7; Wallace, *Claude Grahame-White*, 137–8.

Eleven days later, on 28 September, Samson reported that both the aircraft in his care were also in a potentially dangerous condition, the woodwork being old and strained, and the fabric worn thin and perished. Moreover the 'wire work' had had to be renewed, and he described the aluminium of which the fittings were made as unsuitable for aeroplanes. The engines – in this case presumably the original Aeronautical Syndicate Green engines, although he does not specify – he acknowledged as being in a fair condition, but inherently unreliable anyway, and thus 'useless' for extended flights. Bearing all this in mind, and being also perhaps affected by Cammell's death, he therefore dismissed the worth of either machine to the Royal Navy, concluding that they held no practical value for either experimental or instructional work.[64]

Only the week before Cammell's accident Captain Fulton had complained of the War Office's continuing acceptance of undeveloped machines as part of its utilisation of what the CID sub-committee had called private enterprise. This practice, he insisted, '[could] not be too strongly condemned'.[65] As if it were necessary, Cammell's death tragically exemplified the dangers involved – although many of Cammell's colleagues privately held him responsible for the accident by not having gained sufficient practice on the machine and by being far too over-confident. (He had evidently misapplied the controls.) Clearly, a more systematic approach to procurement had to be adopted. With this in mind, Major Bannerman forwarded a revised estimate to the Director of Fortifications and Works on 19 December 1911. In order to provide the army with two aeroplane sections each operating four aeroplanes, for what Bannerman termed 'co-operative training' the following summer, the battalion should now aim to secure 'five complete aeroplanes' of 'each of the two types' approved as most suitable after comparative trials (i.e. ten aeroplanes in all).[66] The object was to establish two patterns as standard for the next twelve months.[67] The advantages to be derived from maintaining homogeneous groups of aeroplanes were obvious and Bannerman had long advocated this course after suitable trials.[68] But these proposals were to be complicated by the emergence of the Aircraft Factory's own BE biplane at this time, and by preparations for the imminent formation of an autonomous flying corps.

Procurement apart, problems also recurred in the maintenance of a qualified ground staff. Initially, what was described as a cadre of six mechanics was included in the battalion's ranks, but there were soon calls for this number to be doubled. Unfortunately, however, due to what Broke-Smith impassively

[64] Lt. C. R. Samson to CO, HMS *Actaeon* (Capt. Godfrey Paine), 28 Sept. 1911, Samson papers, 72/113/2 E (39). The two Valkyrie monoplanes at Eastchurch were subsequently stored with Short Brothers: Capt. G. Paine to Lt. Samson, 15 Nov. 1911, ibid.

[65] Memo, 'Proposed scheme for aeroplane work', 8 Sept. 1911, PRO, AIR 1 762/204/4/174.

[66] Maj. Bannerman to DFW, 19 Dec. 1911, ibid., reiterated in 'Notes on training scheme – Air Battalion 1912', AIR 1 1609/204/85/61.

[67] Maj. Bannerman to DFW, 6 Jan. 1912, PRO, AIR 1 1609/204/85/61.

[68] See, for example, 'Aeronautics and the army', *Army Review* i (1911), 336.

reported as 'recruiting difficulties' this was never achieved.[69] Capper could have been forgiven for deriving a melancholy satisfaction from this fact. His valedictory warnings on the subject had been entirely justified.

There is only limited evidence of the battalion's training over this period, as although plans had been drafted for the unit's participation in the army's 1911 summer manoeuvres, the whole exercise was subsequently cancelled.[70] From these plans, however, and from other documents relating to the work of the Air Battalion at this time, it is evident that a reconnaissance role was envisaged for however many sections were to be employed in any future campaign. In this capacity they were to serve in close conjunction with existing branches of the service, working as a rule under the direct command of Army Headquarters.[71] In specific circumstances aircraft could be detached to lower, more localised commands, but generally speaking this was not thought advisable. Military aeroplanes could now comfortably range over a wide area and, unless directed from the centre, there was felt to be a danger of, at best, duplication of work, and at worst, friendly machines being mistaken for enemy aircraft.[72] As to whether such machines could ever be employed in direct offensive action against enemy positions and supplies, there was some doubt. Bombs could be dropped from aeroplanes without the consequent weight loss significantly endangering flight, Bannerman conceded, but, in his view, it was 'an open question whether the small results to be expected even from large charges of explosive, dropped more or less wildly, [could] be taken to justify the risk of losing invaluable aircraft'.[73] Consequently, bombing was never considered one of the Air Battalion's major objectives. 'If ever the time comes when the armaments of the Crown are divided into land forces, sea forces and air forces, the number of aircraft may be so great as to render offensive action feasible', Bannerman commented in October 1911, 'but that time is not yet.'[74]

As these deliberations suggest, the Aeroplane Company quickly became the dominant branch of the Air Battalion, despite the initial airship/aeroplane division of the establishment. In fact the army was soon to abandon airship work altogether. By August 1911 Bannerman was complaining to the War Office that an inadequate number of officers were presenting themselves for lighter-than-air work. He reported that there were currently forty new applicants for aeroplane duties and none for airships.[75] Even as he was making this point, however, Bannerman was himself primarily concerned to equip the Aeroplane Company to participate in a continental war. Uniquely among the

69 Broke-Smith, 'Early British military aeronautics', 47.

70 PRO, AIR 1 1609/204/85/56.

71 'Organisation and employment: Air Battalion', PRO, AIR 1 119/15/40/71; Lt. R. A. Cammell, 'Aeroplanes with cavalry', *Cavalry Journal* vi (1911), 197–9; War Office, Field Service Regulations Pt 1, 1912 edn.

72 'Aeronautics and the army', 336–7.

73 'Aircraft for use in war', *Journal of the Royal Artillery* xxxix (1912), 185.

74 'Aeronautics and the army', 336.

75 Employment of Air Battalion, winter 1911–12.

major European powers, Britain, in such an eventuality, faced the disadvantage of having to establish makeshift operational bases overseas. The establishment of such bases constituted an essential prerequisite to the maintenance of any regular air service. It was toward the diminution of this problem, Bannerman insisted, that War Office thought should now be directed.[76] The situation was confused by the fact that army planners were preparing for a campaign of movement; they were not expecting, if war broke out in Europe, land-deadlock, enabling flying corps aerodromes to be established with some permanence. As the Air Battalion's adjutant, Broke-Smith, stressed to the secretary of the Flying Corps Committee in May 1912, shortly before transferring to it authority for military aviation, in the field 'extreme portability and quickness of erection and dismantling are . . . a *sine qua non*'.[77]

It was tensions arising out of the Royal Engineers' continued control of the developing air service that finally led to the formation of an unaffiliated flying corps. As we have seen, the Corps of Engineers' responsibility for British military aeronautics can be traced back to the establishment of the Balloon Equipment Store in the 1870s – if not further. With the advent of the Air Battalion officer recruitment was widened to encompass the attachment of flying enthusiasts from other regiments, but the organisation remained solely under Royal Engineers' jurisdiction. This was increasingly resented by those powerfully motivated seconded officers, the ambiguity of whose status had never been entirely settled. Such figures were always seen as, or felt themselves to be, in some sense, interlopers.[78] This feeling was hardly assuaged by the attitude of the battalion's Commandant. Major Bannerman was an officer of the Royal Engineers who saw his duty as being first and foremost to the Corps. He tended to look upon the Air Battalion, both for himself and his men, as a temporary posting in the normal course of Royal Engineers' duties. In a memorandum to the Chief Engineer, Aldershot Command, dated 24 August 1911, he suggested that all RE subalterns posted to the Air Battalion spend some time with another unit beforehand. If they came to the battalion 'direct from Chatham', he maintained, 'the work is so specialised that they are apt to get into a groove'.[79] In other words, he did not think they should become specialised airmen. This attitude was guaranteed to alienate non-Engineer officers and to reinforce their demands for a more autonomous unit.

Captain Fulton was particularly outspoken in his efforts to get aviation recognised as a separate branch of the army. In his frustration, indeed, he appealed above the heads of his RE superiors (he was himself, of course, a Royal Artillery officer) and contacted directly the then Under-Secretary of State for

---

[76] 'Some problems of aviation in war', *Army Review* i (1911), 123; 'Aeronautics and the army', 332.

[77] Capt. P. W. L. Broke-Smith to sec., Flying Corps Committee, 1 May 1912, PRO, AIR 1 1607/204/85/20.

[78] See, for example, reminiscences of H. R. M. Brooke-Popham, ibid. AIR 1 1/4/1.

[79] Employment of Air Battalion, winter 1911–12.

War, Colonel J. E. B. Seely. Writing from the Royal Artillery Mess, Bulford Camp, on 5 December 1911, he 'ventured' to forward some notes on what he called 'the grievances of the aeroplane officers'. 'I have tried to keep it moderate in tone', he maintained, 'as I am sure that if I wrote half as strongly as we all feel you would think me guilty of exaggeration.' The notes were framed with reference to the CID's imminent review of the army's air service. He was sending them directly, Fulton declared, while there was still time to effect change, for 'we are all convinced that if the aeroplanes are to become a mere dependency of the RE [sic] they will at once sink into the slough of oblivion and general inefficiency which characterises the majority of such RE Units: for example, the Wireless, Telegraph and Searchlight Companies'.[80]

The notes actually amount to an unofficial and highly critical report of the Air Battalion's composition, as maintained under RE direction. The essential charge was that the unit lacked any sense of integration. This defect had to be amended if an efficient fighting force were to be forged out of the disparate elements making up the contemporary air service. In other words, Fulton urged, the unit's latent *esprit-de-corps* had to be more conscientiously fostered. The present situation he described as 'intolerable'. Officers from various regiments were posted to the Aeroplane Company of the Air Battalion and then regulated as mere 'acting Engineers'.[81] The Aeroplane Company was in itself administered as no more than what he called 'No.2 Company, Royal Engineers'. Worse still, the Commandant of the battalion was by nature of the organisation an RE appointee, but for that reason the incumbent need not, and indeed at that time did not, possess an aviator's certificate. Thus he was inherently alienated from his own flying officers and their men. In fact the extraordinary position existed in which the hierarchy in direct control of aviation consisted, for the most part, of Royal Engineer staff officers who did not fly and who maintained conflicting responsibilities. How, Fulton asked, could any aeroplane unit mature under such conditions? To achieve efficiency any future aeroplane corps, such as was then being considered, would have to be entirely extricated from RE control. To belong to an autonomous aeroplane corps, he contended, would be 'a distinction of which anybody would be proud'. To be merely 'attached' to the Royal Engineers' Air Battalion, or some such unit, is to 'get all the kicks and none of the ha'pence'. He even suspected that Bannerman was aiming, through his appointments, to convert the Aeroplane Company back into an 'all-Sapper' section. Should Whitehall decide to uncouple the army's air service from the jurisdiction of the Royal Engineers, as he fervently hoped it would, 'it would be wise', Fulton suggested, to appoint

---

[80] Capt. J. D. B. Fulton to Col. Seely, 5 Dec. 1911, PRO, ADM 116/1278.
[81] This was made clear by the Battalion's Adjutant when answering enquiries from potential applicants. See Capt. P. W. L. Broke-Smith to Lt. E. B. Hawkins, 2nd West Yorks Regt, 23 Nov. 1911, National Army Museum, 8205/103/1. Here he reiterates that officers wishing to join will be seconded to the Battalion for a four year period.

as commander 'an officer who is not a Royal Engineer'. That way the principle of separation would be established 'without possibility of misconception'.[82]

Fulton's memo was immediately forwarded for comment to the First Lord of the Admiralty, Winston Churchill. The previous month (November 1911) the Prime Minister had requested that the Standing Sub-Committee of the CID consider what further measures might be taken to secure an 'efficient Aerial Service' for the armed forces.[83] To this end, the sub-committee, of which both Churchill and Seely were members, was due to meet shortly. It was this that gave Fulton's memo its sense of urgency. With this in mind, Churchill, in turn, urged in his reply of 9 December 1911 that 'Whatever happens the RE must have nothing to do with HM's Corps of Airmen, which shd [sic] be a new and separate organisation drawing from civilian, as well as military and naval sources.'[84] 'Terms and conditions', he went on, must be so devised as to make military aviation 'the most honourable, as it is the most dangerous' profession a young man can adopt. No regard for military seniority should prevent the 'real . . . capable men' from being placed 'effectually at the head of the new Corps of Airmen'.

This directive proved decisive. The Air Battalion was quickly perceived as being merely an interim organisation, prior to the establishment of some more expansive unit. If there remained any doubts as to whether this need be an autonomous air corps Churchill's response to Fulton's communication quashed them. Following this exchange the necessary steps were quickly taken. On 18 December 1911 the CID Standing Sub-Committee (which included Haldane – the *ex officio* chairman – and Esher) met and, through the influence of Churchill and Seely, delegated to a technical sub-committee the task of devising an unaffiliated military and naval aviation service. Colonel Seely was appointed its chairman and he was joined by the Director of Military Training, Major-General A. J. Murray, the Director of Fortifications and Works, Brigadier-General G. K. Scott-Moncrieff, GSO to the Inspector-General of the Forces, Brigadier-General David Henderson, Commander C. R. Samson RN, Lieutenant Reginald Gregory RN and the Superintendent, Aircraft Factory, Mervyn O'Gorman. The secretaries were Rear-Admiral C. L. Ottley and Captain Maurice Hankey. In fact the detailed work of this second sub-committee was itself prepared by a specially-convened 'think-tank' consisting

---

82 Notes by Capt. J. D. B. Fulton, headed 'Air Battalion', PRO, ADM 116/1278. Similar views were being propagated by C. G. Grey and a degree of suspicion fell on Fulton as the possible instigator of press agitation: C. G. Grey, *The Aeroplane*, 17 Nov. 1915, PRO, AIR 1 727/160/8/3. An immaculately turned out and handsome soldier with an incisive sense of humour, Fulton evidently caused some enmity.

83 This review originated with the CID meeting of 14 July 1910, during which the Home Ports Defence Committee's call for a reconsideration of aeronautical policy was endorsed (see ch. 5, p. 218). The delay was to allow time for the careful collation of technical information: CID, minutes of 107th meeting, PRO, CAB 2/2.

84 Minute by Churchill, First Lord, 9 Dec. 1911, ibid. ADM 116/1278, repr. in *Documents, Naval Air Service*, 26.

of David Henderson again, the head of MO5, Colonel George Macdonogh,[85] the GSO, Signals, Major Duncan MacInnes RE and Captain Frederick Sykes – a General Staff officer who, like Henderson, had recently gained his aviator's certificate with the British & Colonial Brooklands school, and who was soon to be appointed OC, Military Wing, RFC. The technical sub-committee adopted this third delegation's recommendations and reported in February 1912. A Flying Corps was constituted by Royal Warrant on 13 April. Divorced from its previous command the structure of the Air Battalion was officially absorbed into this new body the following month.[86]

---

85 Later Lt.-Gen. Sir George Macdonogh.
86 PRO, AIR 1 653/17/122/489; CAB 16/16.

Conclusion

With the formation of the Royal Flying Corps, consisting of a Military Wing, a Naval Wing, a Central Flying School and a Reserve, the last connection between aeronautics and the Royal Engineers was cut. Aviation had come into its own. Commander C. R. Samson, who had been the senior of the four naval volunteers selected to undergo instruction on Francis McClean's Short-Sommer biplanes at Eastchurch the previous year, was placed in command of the Naval Wing, and his immediate superior, Captain Godfrey Paine, was appointed Commandant of the Central Flying School. This was formally opened at Upavon on Salisbury Plain on 17 August 1912, its first four instructors being listed as Captains Fulton and Broke-Smith (Military Wing) and Captain Gerrard and Lieutenant Longmore (Naval Wing). The school's instructional fleet was initially composed of some seven aircraft: two Avro biplanes, two Short biplanes, two Farman biplanes and a Bristol biplane. A separate experimental Naval Flying School was also kept in being at Eastchurch. Meanwhile, towards the close of April 1912, following Sykes's appointment as Commander of the Flying Corps' Military Wing, Captain H. R. M. Brooke-Popham took over command of the Aeroplane Company of the Air Battalion, and the following month this became No.3 Squadron, RFC (Military Wing). No.1 (Airship) Company was in turn converted to No.1 Squadron, RFC (Military Wing), and the HQ nucleus at south Farnborough became the organisation's No.2 Squadron, under the command of Captain C. J. Burke.[1]

Sykes, Henderson and the other members of the Flying Corps Committee's think-tank were responsible for introducing the squadron formation. They recommended that this be composed of eighteen aeroplanes, subdivided into three flights of four machines, with a reserve of two machines per flight.[2] Clearly, however, such an expansion would take time. Each squadron was also to retain its own depot and transport service. The original Military Wing was eventually to have contained eight squadrons, but only five were in being upon mobilisation in August 1914.[3] An Aircraft Park, i.e. a corps depot, had also been established under the command of Captain A. D. Carden.

[1] Reminiscences of Brooke-Popham.
[2] Sykes, *From many angles*, 95.
[3] Ibid. 122. On 1 Jan. 1914 No.1 (Airship) Sqn under Capt. Maitland was, in Brancker's phrase, transferred 'lock, stock and barrel' to the Naval Wing as part of the transposition of all airship duties to the Admiralty: Macmillan, *Sir Sefton Brancker*, 44–5.

Fulton, whose memorandum had so galvanised Seely and Churchill, was appointed the CFS's Chief Instructor in November 1913, with additional responsibility for the school's workshops; and he was subsequently to be placed in charge of the inspection of all army aeroplanes, helping to form the Aeronautical Inspection Department early the following year. With the increase in sub-contracting brought about by the war the work of the inspection staff increased enormously: so much so that Fulton's death due to a throat infection in November 1915 was said to be the greatest blow the RFC suffered during the war.[4] He had been appointed Assistant Director of Military Aeronautics only weeks before and, had he lived, a rather more benign influence might have been exercised over the directorate's procurement policy – as C. G. Grey himself speculated.[5]

After much dispute, and at the instigation of the Board of Admiralty, the Naval Wing of the RFC formally adopted the title of Royal Naval Air Service in 1914 and remained administratively separate for most of the war, but by 1912 military aviation had evolved into the regimentally unaffiliated form in which it would see active service. In November 1914 the squadrons of the RFC were grouped into wings, to co-ordinate directly with the army's corps commanders.

However it is only as a result of an examination of the whole development of aviation up to this point, as in this study, that it is possible to properly understand the difficulties that were to be encountered in the maintenance of an air service in combat, for at the outbreak of war inherent tensions came to a head. The imposition of a virtual state procurement-based monopoly succeeded in suffocating design initiative and meant that the recently formed Royal Flying Corps had to wage the early stages of the air war with inadequate aircraft. State-based design control had led in practice to inflexible standardisation and market control. By drying the essential lubricant of competition it had resulted in an air service which was slow and at times reluctant to adapt to a rapidly changing environment. This was the culmination of a dispute ingrained in the whole development of military–industrial relations in this area.

The survival of the nascent industry had come to depend on the continued exploitation of the military market. Aeroplane production had become primarily an extension of the armaments industry and there had developed an early military–industrial complex. By undermining this linkage in the name of scientific purity the Directorate of Military Aeronautics and the Royal Aircraft Factory inadvertently checked the commercial-technological dynamic through which the bicycle, automobile and aviation industries had sequentially developed. Growth and development were impelled by private

---

4   PRO, AIR 1 727/160/5.
5   Extract from *The Aeroplane*, 17 Nov. 1915, ibid. AIR 1 727/160/8/3.

enterprise competitively adapting in the Darwinian sense to the vicissitudes of the commercial and technological environment.

For both technological and industrial advancement service procurement had to be opened again to the free market: the military–industrial complex had to be restored.

Appendix 1

The 'first British flight' controversy

The catalyst that ignited this dispute was a proposal for a unified Royal Aero Club, Royal Aeronautical Society, Air League of the British Empire and Society of British Aircraft Constructors banquet in 1928, marking the twentieth anniversary of A. V. Roe's alleged first flight of 8 June 1908. Brabazon and Griffith Brewer – prominent members of the Aero Club and of the Aeronautical Society respectively – disassociated themselves from the event because it implicitly endorsed what they considered a bogus claim.[1] Neither accepted that Roe had in any legitimate sense achieved flight on the day in question – nor, indeed, at any time in 1908. The aim of the banquet was subsequently modified to simply honouring Roe's achievements in aviation generally,[2] but this failed to mollify either Brabazon or Brewer, who continued to conspire against any anniversary celebration.[3]

There were two basic, and essentially contradictory, criticisms of Roe's retrospective claim: one, that any hop he may possibly have made could not reasonably be said to have constituted a flight; and the other, that even a hop was sheer invention. Brewer's objection finally rested on the first of these grounds; Brabazon's on the second.[4] Only Brabazon persistently maintained that Roe flagrantly lied about his first flight.

Colonel Semphill, president of the Royal Aeronautical Society, tried to enlist Orville Wright's support for the banquet, and at first Orville responded positively. However he later backed down, claiming that he 'may have been under a misapprehension as to the nature of the dinner'.[5] Someone had clearly got to him in the meantime and this could only have been his long time confidant, Griffith Brewer. It was probably at this time that Orville Wright forwarded Roe's ambiguously-worded letter to Wilbur Wright of August 1908, for by July 1928 Brewer was himself bringing it to the attention of Lord Thomson, chairman of the Royal Aero Club.[6] Brewer now called for an official investigation into the matter. This was the genesis of the Gorell committee on early flights, which met in December 1928 and January 1929. The actual form

---

1   J. T. C. Moore-Brabazon to Sir C. Wakefield, 17 May 1928; Moore-Brabazon to Col. Semphill, 17 May 1928, Brabazon papers, box 25(3); RAeS Council minutes, 21 May 1928, RAeS archive.
2   Col. Semphill to J. T. C. Moore-Brabazon, 23 May 1928, Brabazon papers, box 25(3).
3   J. T. C. Moore-Brabazon to G. Brewer, 19 May 1928; Brewer to Moore-Brabazon, 18, 21 May 1928, ibid.
4   G. Brewer to J. T. C. Moore-Brabazon, 22, 25 May 1928, ibid.
5   O. Wright to Col. Semphill, 12 June 1928, ibid. box 26(1).
6   G. Brewer to Lord Thomson, dated July 1928, ibid. box 25(3).

the inquiry took was suggested by Handley Page. A small committee would investigate all early flight claims (not simply Roe's) in an effort to establish priority. By this means it was hoped that a confrontation for or against Roe could be avoided.[7]

The brisk proceedings of the committee were held under the auspices of the Royal Aero Club.[8] Roe cited the witness statements he had collected back in 1912 to substantiate his claim. (Subsequent assertions that these witnesses admitted that Roe had used a slope for his hops are unfounded. A later testifier, the managing director of Palmers Tyres, Herbert Morris, writing in December 1928 and January 1929, mentioned the 'old finishing straight' at Brooklands, which, as everybody knew, eventually merged with the members banking gradient, but that is all: and he was adamant that the machine had 'certainly' risen 'under its own power'.[9] Walter Windham also subsequently made a statement substantiating Roe's claim, but this was not available to the Gorell committee.)[10] Meanwhile Oswald Short verified Brabazon's 'Bird of Passage' flights, which occurred over the weekend of 30 April–2 May 1909.[11] In the event, these were the only claims that mattered. Cody and Maxim were excluded from the inquiry on account of their American nationality. Both men were American citizens at the time of their alleged first flights.

Roe was subsequently cross-examined, but the minutes of the investigation reveal that he failed to present his own case very convincingly, lacking either the eloquence or influence of Brabazon. He was understandably confused when confronted with his ambiguously-worded letter to Wilbur Wright, written some twenty years before, and had no prepared defence. Great play was also made on the letter he published in the journal *Flight* on 26 June 1909, the year

[7]   G. Brewer to J. T. C. Moore-Brabazon, 4 July 1928; F. Handley Page to Cdr H. Perrin (sec., RAeC), 4 July 1928, ibid.

[8]   The RAeC Committee on Early Flights eventually consisted of three members. The chairman was the Rt Hon. Lord Gorell, and he was assisted by Capt. Geoffrey de Havilland and Lt.-Col. W. Lockwood Marsh. Cdr Harold Perrin acted as secretary. It met at the RAeC on 6 Dec. 1928, and 2, 29 Jan. 1929, and delivered its report on this last date: ibid. box 26(1).

[9]   A. V. Roe, in an article written in April 1912 – just a month before the first Brooklands' sworn witness testimony was recorded – undefensively remarked that he had planned to use the Members Hill gradient for his original 'assisted glides': *The Aero*, Apr. 1912, 96. When a more appropriate engine had been secured these were to be dispensed with and full trials begun.

[10] Lord Gorell to G. V. Roe, 18 June 1958, Brabazon papers, box 26(1); G. V. Roe, 'Was Roe first?: the grounds for appeal', *Flypast*, Jan. 1989, 19–20. Harry Harper subsequently claimed that Windham telephoned him at the time to say that he had seen Roe's wheels leave the ground. The veracity of this statement is uncertain: G. Wallace, *Flying witness: Harry Harper and the golden age of aviation*, London 1958, 77. In December 1954 it emerged that a railway signalman may also have witnessed these early flights/hops from his signal-box overlooking the Brooklands circuit: Ludovici, *Challenging sky*, 52.

[11] J. T. C. Moore-Brabazon to O. Short, 11, 17 Dec. 1928; O. Short to Moore-Brabazon, 15 Dec. 1928, Brabazon papers, box 26(2); Gorell report, ibid. box 26(1). See also Short papers, file RS 4/1.

after his alleged first flight. Here Roe for the first time claimed publicly to have been making short flights with his new triplane. If these were so notable in 1909, his critics argued, why had he made no similar public claim with regard to his supposed flights or hops of June 1908? Such criticism, however, took the event out of context. If Roe's behaviour in 1909 is to be properly understood one must appreciate the circumstances in which he temporarily found himself.

In fact the letter to the press of June 1909 was a cry for help at a time when Roe's experiments had run into desperate financial straits:[12] nor was it the only one of its kind. Humphrey Verdon-Roe later confirmed this;

> My partnership with A. V. commenced 27 April 1909, with the idea of helping him to find somebody to finance his experiments. On 29 June 1909, *Aero* [sic] published a letter from A. V. in which he gave a very fair account of what he was doing and what he had accomplished. He was asking for a financial partner to help. This brought no response.[13]

Whether substantiated or not, however, the Gorell committee chose to disregard Roe's claim on the grounds that what he was said to have achieved in June 1908 did not constitute a flight. They consequently judged Brabazon's initial 'Bird of Passage' flight the first in the United Kingdom by a British pilot. Awarding the decision on these grounds was, to say the least, dubious; nor did Roe accept it.[14] Even so, the ruling tacitly acknowledged that Roe must have been the first to become airborne – if only for what was called a hop. Brabazon would not admit even this. The dispute was thus destined to continue.

A. V. Roe wrote angrily to Commander Harold Perrin, secretary of the Royal Aero Club, on 21 February 1929, mathematically disproving the Gorell committee's contention that his hops of approximately 150 ft could not be said to constitute self-sustained flights, adding that 'The D. H. [De Havilland] Stress Department can work it out for themselves.'[15] This confutation seems to have been accepted in some quarters. In fact on receiving a draft copy of the report Geoffrey de Havilland had evidently already requested that the matter be reconsidered. On grounds of both evidence and definition he questioned the judgement. After all, the Wright brothers' first officially credited powered flight was (whatever the circumstances) just 120 ft.[16] Brabazon, meanwhile (by this time Parliamentary Private Secretary to Sir Samuel Hoare, the Secretary of State for Air), demanded that the introductory volume of the official history of *The war in the air*, by the late Sir Walter Raleigh, be appropriately amended. Brabazon blamed Raleigh for Roe being given any credence in the first place.[17]

From here the whole affair grew slightly farcical. For instance, Brabazon's

---

[12] See ch. 2, pp. 62–3.
[13] H. V. Roe, 'Pioneers'.
[14] A. V. Roe, *World of wings*, 47.
[15] Brabazon papers, box 26(2).
[16] 'Was Roe first?: the grounds for appeal', 20.
[17] J. T. C. Moore-Brabazon to Sir S. Hoare, 18 Feb. 1929, Brabazon papers, box 26(2).

former ally in his 'anti-Roe' campaign, Griffith Brewer, wrote to him in March 1931, claiming that actually he (Brewer) was the first Englishman to 'fly' on account of his having been the initial member of the Brewer, Rolls, Butler, Baden-Powell group to have accompanied Wilbur Wright as a passenger on 8 October 1908, although, ironically, not even this was entirely correct. Wilbur Wright had in fact taken George P. Dickin, the English-born correspondent of the *New York Herald*, up for a flight on 3 October. Brewer was thus the first resident Englishman to fly in a powered heavier-than-air machine.[18] Brabazon himself became obsessed with the affair, writing to anyone and everyone whom he felt had deprived him of the honour of having been Britain's first indigenous pilot.

Ultimately, however, it is simply sad that an issue of such peripheral importance should have caused perhaps the most distinguished pioneer of British aviation, A. V. Roe, such an inordinate amount of unhappiness in his declining years.

---

[18] G. Brewer to J. T. C. Moore-Brabazon, 18, 20 Mar. 1931, ibid.; Bruce, *Charlie Rolls*, 22.

Biographies

Abel, Sir Frederick Augustus, 1826–1902. 1851–5, professor of chemistry, RMA, Woolwich; 1854–88, chief chemist, War Office; 1870, member of Royal Engineers' Committee permanent Balloon Sub-Committee; 1888–91, president, special committee on explosives.

Ader, Clément, 1841–1925. 1890, completed bat-shaped tractor–monoplane with large canopied wings (the Eole). Power provided by light steam-engine. Design lacked any method of flight control. Oct. 1890, achieved a take-off covering some fifty metres, but no sustained flight; 1892, commissioned by French War Ministry; Oct. 1897, the Ader 'Avion III', which employed twin tractor-propellers, underwent two separate, officially observed, practical trials. Neither was successful.

Alexander, Patrick Young, 1867–1943. Son of Andrew Alexander, one-time manager of Cammell's steel works, Sheffield, and RAeS founder member. Alexander sen. died July 1890, bequeathing P. Y. a legacy of some £60,000. Believing that his father, had he lived, would have solved problem of heavier-than-air powered flight, Alexander developed all-consuming enthusiasm for aviation (Griffith Brewer memoir, 23 Aug. 1943, Alexander papers) and became internationally renowned patron of aeronautics. Under delusion that he was not destined to live beyond age of fifty he ultimately bankrupted himself.

Archdeacon, Ernest, 1863–1957. Parisian lawyer, philanthropist and aviation sponsor: president of the Aero Club de France: see ch. 2.

Ardagh, Sir John, 1840–1907. 1859, commissioned, RE; 1874, joined War Office's Intelligence Department. Rendered valuable service during Congress of Berlin, 1878, and relief of Khartoum, 1885. 1888, appointed ADC, C-in-C, India, then private secretary, Viceroy of India; 1895, Cmdt, SME, Chatham; 1896–1901, DMI, War Office; 1899, British military-technical delegate, Hague Peace Conference.

Austin, Herbert, 1st Baron Austin, 1866–1941. Educated Rotherham grammar school. 1884, emigrated to Australia; 1893, made manager of Wolseley SSMC, English division, where began automobile manufacture; 1905, founded Austin Motor Co. Ltd; 1922, produced Austin Seven; 1918–24, Conservative MP, King's Norton. 1917, KBE; 1936, baron.

Bacon, Admiral Sir Reginald Hugh, 1863–1947. 1900–4, Inspecting Capt. of Submarines; 1904, Naval Asst to First Sea Lord (Fisher); 1906, Capt., HMS

Dreadnought; 1907, DNO; 1909, promoted Rear-Admiral. Retired from service, becoming (1909–14) managing director, Coventry Ordnance Works. 1915–17, OC, Dover Patrol: promoted Vice-Admiral; 1918, Controller, Munitions Inventions Department: promoted Admiral. Retired 1919. Author of controversial account of Jutland (1924) and biographies of Fisher (1929) and Jellicoe (1936).

Baden-Powell, Baden Fletcher Smyth, 1860–1937. Member of distinguished military family. 1882, commissioned, Scots Guards; 1883–4, visited Paris to investigate dirigible developments; 1884–5, active service Nile campaign; 1886, elected to council of Aeronautical Society; 1888–91, ADC to Governor, Queensland, Australia; 1894, initiated man-lifting kite trials; 1896, hon. sec., Aeronautical Society: attached to Balloon Section RE (kite detachment); 1897, founded *Aeronautical Journal*; 1899–1902, active service South Africa; 1900, elected – in his absence – president, Aeronautical Society; 1904, retired from army: delegate, St Louis World's Fair; 1907, resigned presidency of Aeronautical Society: co-founded *Aeronautics* journal; 1919, elected FRAeS.

Balfour, Arthur James, 1st Earl Balfour, 1848–1930. 1902–5, PM; 1915–16, First Lord of Admiralty; 1916–19, Foreign Secretary. Something of a motoring pioneer in his own right. 1899, purchased a 4½ h.p. De Dion, which he would subsequently drive to Downing St; 1902, acquired a 9 h.p. Napier, the carriage built to his own design, and contributed message of support on launching of *The Car Illustrated*. In February 1909 Northcliffe invited him to Pau to witness Wright demonstration flights, and he even then sought to accompany Wilbur.

Bannerman, Lt.-Colonel Sir Alexander, 1871–1934. 1891, commissioned, RE; 1899–1902, active service South Africa; 1904, military attaché, Japanese HQ; 1906–10, GSO, Brevet Maj.; 1910–11, CBS; 1911–12, CAB (Maj., Aug. 1911: RAeC certificate No.213, 30 Apr. 1912); 1912, retired with rank of Lt.-Col., becoming an independent consultant; 1915, served with 37th Division; 1917, OC 2nd/5th North Staffs Regt, France.

Barber, Horatio C., 1875–1964. Educated Bedford School. 1909, formed the Aeronautical Syndicate Ltd, initially located under one of the Battersea railway arches. Barber's early aircraft designs were built by Howard Wright and tested at Durrington Down, on Salisbury Plain – including the prototype Valkyrie canard-monoplane; Oct. 1910, moved to Hendon; 22 Nov. 1910, awarded RAeC certificate No.30 (Valkyrie monoplane); Apr. 1912, Aeronautical Syndicate folded; 1913, donated Britannia Challenge Trophy to RAeC, to be awarded each year for most meritorious performance; 1914–18, served in RFC/RAF; 1919–21, chairman, Lloyds technical committee on aviation; 1921–30, chairman, Aero Underwriters Corp., New York.

Barnes, Ronald Gorell, 3rd Baron Gorell, 1884–1963. Educated Winchester, Harrow and Balliol College, Oxford. 1910–15, on editorial staff of the *Times*; 1915–16, Adjt., 7th Btn, the Rifle Brigade; 1917, Maj., General Staff (M.C.);

1918–20, Deputy-Director of Staff Duties (Education), War Office, founding Royal Army Educational Corps; 1921–2, Under-Secretary of State for Air; 1933–4/1938–9, chairman of Gorell committees on civil aviation; 1933–6/1943–6, chairman, RAeC.

Barnwell, Robert Harold, 1879–1917. 1911, in company with his brother, Frank, constructed Grampian Motor Co. aircraft; Sept. 1912, awarded RAeC certificate No.278, B&C school, Brooklands. Subsequently undertook advance instruction at B&C's Larkhill school and joined Vickers's aviation department. As flying school's chief instructor he trained Sefton Brancker and the young Hugh Dowding. 1914, appointed Vickers's chief test pilot. Also test flew for Martin-Handasyde and piloted Martin-Handasyde monoplane at 2nd Aerial Derby, 1913, and Sopwith Tabloid at 3rd Aerial Derby, 1914. Killed test flying Vickers FB26 pusher-fighter, 25 Aug. 1917.

Barnwell, Frank Sowter, 1880–1938. Chief designer, British & Colonial Aeroplane Co./Bristol Aeroplane Co.: see ch. 3.

Barry, Robert Smith, 1886–1949. Educated Eton. 1909, entered consular service (hon. attaché, Constantinople); 28 Nov. 1911, awarded RAeC certificate No.161, B&C's Larkhill school, becoming an unofficial instructor; Aug. 1912, commissioned 2nd Lt., RFC Special Reserve: attended CFS, graduating Dec. 1912; July 1913, transferred to RFC (MW), No.5 Sqn; Nov. 1915, Flt-Cdr; July 1916, Sqn-Cdr (as OC, 60 Sqn, he was a sympathetic CO to Lt. Albert Ball); Aug. 1917, Wing-Cdr; Jan. 1918, Brig.-Cdr; Apr. 1918, transferred to RAF. Originator of Gosport/Avro 504 system of flight training.

Battenberg, Admiral of the Fleet Prince Louis Alexander of, 1854–1921. 1902–5, DNI; 1908–10, C-in-C, Atlantic Fleet; 1910–11, Vice-Admiral, Home Fleet; 1911–12, Second Sea Lord; 1912–14, First Sea Lord; 1917, assumed surname of Mountbatten: created marquess of Milford Haven.

Beaumont, Lt.-Colonel Frederick Edward Blackett, 1833–99. Educated Harrow and RMA, Woolwich. 1852, commissioned, RE; 1854–6, active service Crimea, and 1857–8, Indian Mutiny; Apr. 1859, Capt.; 1861–2, travelled from Canada to become an observer with Gen. McClellan's Army of the Potomac; 1862–73, active campaigner for adoption of balloon unit within British army; July 1872, Maj.; 1877, retired from service with rank of Lt.-Col.; 1868–80, sat as Liberal MP for South Durham. Considered by contemporaries an inventive genius. His revolver trigger patent *c.* 1854, which automatically turned the breech-block and recocked the weapon, was one of the major arms innovations of the century. Equally notable was his diamond-point drill patent. He later investigated bicycles and motor engines. Ill-advised speculations curtailed promising military-political career.

Bennett, James Gordon, 1841–1918. Born New York, son of James Gordon Bennett (1795–1872, noted newspaper proprietor of Scottish origin). 1866, managing editor and chief executive officer, *New York Herald*. Sponsored H. M. Stanley's 1869–72 expedition to central Africa to find David Livingstone. 1867, founded *New York Evening Telegram*; 1877, scandal in his personal life drove him to France, following which his perspective became increasingly European; 1883, co-founder, Commercial Cable Co., which subsequently laid cable across Atlantic; 1887, established a Paris edition of *New York Herald*. An enormously wealthy and avid sportsman, he sponsored many competitions, notably in the fields of motoring, aeronautics and aviation.

Bettington, Lt. Claude Albemarle, 1875–1912. Born Cape Colony, South Africa. Royal Artillery officer in Boer War, in Ladysmith throughout siege. Subsequently left service, but recommissioned in RFC on formation in 1912. Secured RAeC certificate No.256 on a Bristol biplane, 24 July 1912. Killed 10 Sept. 1912, in the Hotchkiss Bristol military monoplane tragedy.

Billing, Noel Pemberton, 1881–1948. Born Hampstead, son of Birmingham iron founder. A restless early life in Britain and South Africa included periods as a soldier, garage-owner, journalist and yacht dealer, but increasingly drawn to aviation, both as constructor and propagandist. 1909, attempted to establish the 'Colony of British Aerocraft' or 'Essex flying ground' at Fambridge, and with it an 'Imperial Flying Squadron'; 1909–10, founder-editor *Aerocraft*; 1913, founded eponymous flying company, officially registered in June 1914, and re-registered in Nov. 1916 as Supermarine; 1914–16, served as RNAS pilot (retired as Sqn-Cdr); Jan. 1916, contested Mile End as Independent air member, advocating more forceful air policy; Mar. 1916–21, Independent MP for East Herts. Infamous scandalmonger.

Blackburn, Robert, 1885–1955. Born Leeds, educated Leeds Modern School and Leeds University. 1906, elected associate member, Institution of Civil Engineers. Gained practical experience with father's firm (T. Green & Sons, lawnmower manufacturers), and then pursued further training on continent. Here developed interest in aviation. Returned to Leeds late 1908 with plans for his own monoplane. A 35 h.p. (Green engine) Blackburn monoplane appeared following year, but second Antoinette-derived model proved more successful and was marketed together with Blériot monoplanes. Third monoplane design, 'Mercury', exhibited at 1911 Olympia show. Blackburn then joined by B. C. Hucks (*q.v.*), forming limited company in June 1914, on receipt of government BE2 orders.

Blakeney, Brig.-General Robert Byron Drury, 1872–1952. 1891, commissioned 2nd Lt., RE; 1893, joined Balloon Section, Aldershot; 1896–8, served on railway construction, Dongola expedition (Sudan); Sept. 1898, acted as Galloper to Brig.-Gen. Lewis at Omdurman (awarded DSO); 1899, Traffic Manager, Nile Railway; 1900, OC, 3rd Balloon Section, RE, South Africa; 1906, appointed Deputy Gen. Manager, and 1914–18, Deputy Director,

Railway Traffic, Egypt; 1919, appointed Director, with rank of Brig.-Gen.; 1921–3, Gen. Manager, Egyptian Railways.

Blériot, Louis, 1872–1936. Born Cambrai, educated at the French Polytechnique, Paris. Established business as manufacturer and distributor of his own acetylene automobile headlamps. Indefatigable aviation experimenter throughout opening decade of century. Won enormous fame when he crossed the channel in his frail 25 h.p. (Anzani) Blériot XI monoplane on 25 July 1909. Awarded Aero Club de France aviator's certificate No.1 and opened flying school at Pau. By the advent of war Blériot monoplanes were considered outmoded. They were finally struck off the RFC's active list in June 1915, being declared obsolete for all purposes in Sept. 1918: PRO, AIR 1 727/152/6.

Brancker, William Sefton, 1877–1930. 1896, commissioned, RA; 1913, promoted Maj.: GSO 3, Training Directorate, War Office; 18 June 1913, obtained RAeC certificate No.525, Vickers school, Brooklands: July–Aug., CFS: Aug., GSO 2, DMA; Aug. 1914, temp. Lt.-Col., GSO 1, OC Military Aviation, War Office; Aug.–Dec. 1915, OC, 3rd Wing, RFC, France; Dec. 1915, temp. Brig.-Gen.: CO, Northern Brigade, RFC; Feb. 1916, CSO to Gen. Henderson: Mar., Director of Air Organisation, War Office; Feb. 1917, Deputy DGMA: June, temp. Maj.-Gen.: Oct., GOC, RFC, Middle East; Jan. 1918, Controller-Gen. of Equipment, Air Ministry: Aug., Master-Gen. of Personnel, promoted Maj.-Gen.; 1919, retired (Air Vice-Marshal). Joined Airco, in attempt to initiate London-Paris air service. 1922, Director of Civil Aviation. Killed in R101 airship disaster, 5 Oct. 1930.

Brett, Reginald Baliol, 2nd Viscount Esher, 1852–1930. 1895–1902, sec., Office of Works; 1897, superintended Queen Victoria's Diamond Jubilee; 1899, succeeded to title; 1902, superintended coronation of Edward VII: member of Royal Commission on South African War; 1903–4, chairman, War Office Reconstruction Committee. Recommended creation of Army Council and General Staff. 1905, Permanent Member, CID; 1908–9, chairman, CID Aerial Navigation Sub-Committee. Keen early motorist, owning one of first City & Surburban electric broughams.

Brewer, Griffith, 1867–1948. Patent agent, journalist and amateur balloonist. Member of RAeS. 1906–8, participated in Gordon Bennett balloon contests. Flew as Wilbur Wright's passenger, Le Mans, 8 Oct. 1908. (C. S. Rolls, F. H. Butler and B. F. S. Baden-Powell all accompanied Wilbur Wright on same day.) 1909, became Wright brothers' UK patent agent – responsible for introducing Short brothers to the Wrights; 1913, one of the co-founders of Eastchurch-based British Wright Co., which on outbreak of war secured £15,000 from War Office for supposed infringement of Wrights' warping patent. Founder of RAeS Annual Wright Memorial Lecture and guardian of the Wrights' shrine.

Broke-Smith, Brigadier Philip William Lilian, 1882–1963. 1900, commissioned 2nd Lt., RE; Jan. 1902, posted to 5th Balloon Section, Aldershot; 1903–4, conducted ballooning trials, Gibraltar, and Cody man-lifting kite trials, Aldershot (attained record kite ascent of 3,340 ft). Subsequently OC, Experimental Balloon Section, Bengal Sappers. 1910, Instructor, Balloon School; 1911, Adjt, Air Battalion; 1912, Instructor, RFC (MW), CFS. During World War I served as Deputy Asst Director of Aviation, Mesopotamia (DSO). 1919, Asst Director of Works, North West Frontier Force (3rd Afghan War); 1931, Deputy Engineer-in-Chief, India; 1932, Chief Engineer, Eastern Command, India. Retired 1936. During World War II served as Director, Passive Air Defence, Ministry of Supply. Historian of early military aeronautics.

Brooke-Popham, Air Chief Marshal Sir Henry Robert Moore, 1878–1953. 1898, commissioned, OLI; 1910, Staff College; 18 July 1911, awarded RAeC certificate No.108, B&C's Brooklands school; 1912, OC, No.3 Sqn RFC (MW); 1914, SO, BEF HQ; 1915, OC, No.3 Wing, St Omer (DSO); 1916, temp. Brig.-Gen., QMG; 1918, Director of Research, Air Ministry; 1921–6, first Cmdt, RAF Staff College; 1924, Air Vice Marshal; 1928, AOC, Iraq; 1931, Air Marshal: first RAF Cmdt, Imperial Defence College; 1935, Air Chief Marshal: Inspector-Gen., RAF; 1937, Governor of Kenya; 1940–1, C-in-C, Far East.

Brun, General Jean-Jules, 1849–1911. 1867, attended the École Polytechnique and, from there, joined the artillery; 1870, Sub-Lt., School of Instruction, Metz; 1874, Capt. Studied at War College and (1886) appointed Asst Professor of Applied Tactics. 1888, Ordnance Officer to Gen. Ferron, Minister of War; 1897, Professor, War College; 1900, Military Cmdt of the Senate; 1901, Brig.-Gen. and Cmdt, 21st Brigade, Nancy; 1902, Deputy COS, GHQ; 1903, Cmdt, War College; 1906–9, COS, GHQ (1907, attached to GHQ, Russian army, to facilitate military alliance); July 1909–Feb. 1911, Minister of War. Died in office.

Buller, General Sir Redvers Henry, 1839–1908. Member of 'Wolseley ring'. 1879, active service Zululand (VC); 1881, COS to Sir Evelyn Wood, South Africa; 1882, CIS, Egypt; 1884, COS, Khartoum relief expedition; 1890–7, Adjt-Gen.; 1897–9, Aldershot Command; 1899, C-in-C, South Africa; 1900, GOC, Natal Field Force. He took the Chair for the RUSI lecture on military ballooning given by Lt. H. B. Jones in February 1892.

Burgess, William Starling, 1878–1947. Son of successful yacht manufacturer. 1898, saw service with US Navy, Spanish–American War; 1901, graduated in engineering and naval architecture, Harvard; 1904, established W. Starling Burgess Co., yacht and boat manufacturers of Marblehead, Mass.; 1909–10, joined Greely S. Curtis and A. M. Herring in an aeroplane manufacturing enterprise, making first heavier-than-air flights in New England on 30 h.p. pusher-biplane 'Flying Fish'; 1910, Herring left the association, and Burgess Co. & Curtis founded. Burgess met Grahame-White at Boston-Harvard meeting of September 1910, and latter subsequently ordered seven Burgess aeroplanes for import to England. The design, the Burgess Model E, became known as the Grahame-White 'Baby'. 1911, Burgess Co. & Curtis entered into licensing agreement with the Wright (brothers) company to build Wright aeroplanes for the private sports market. Two models were purchased by Thomas Sopwith. September 1913, company acquired the Dunne/Blair Atholl syndicate American patent rights, and developed the type as a seaplane, the US services eventually procuring a number; 1914, Wright licence cancelled (it had been in dispute for some time) and firm reformed as the Burgess Co., to avoid confusion with the (Glenn) Curtiss company; 1916, firm taken over by reconstituted Curtiss Aeroplane & Motor Corp., Burgess becoming an executive director; 1917–19, Lt. Cdr, US Navy Aircraft Planning Board. Retired from aviation in 1919, and returned to yacht and boat building.

Burke, Lt.-Colonel Charles James, 1882–1917. Born Armagh. 1899–1902, served with Royal Irish Militia, South Africa; 1903, commissioned 2nd Lt., Royal Irish Regt; 1904, Lt.; 1905–9, served with West African Frontier Force; 1909, Capt.; 4 Oct. 1910, awarded Ae.C. de France certificate No.260; Nov. 1910–Mar. 1911, attached Balloon School, RE; Apr. 1911–May 1912, attached Air Battalion, RE; May 1912, joined RFC (OC, No.2 Sqn); 1913, brevet Maj.; Nov. 1914, temp. Lt.-Col.: OC, No.2 Wing; 1915, DSO; Feb. 1916, appointed Cmdt, CFS. Killed in action whilst commanding a battalion of the East Lancs Regt, 9 Apr. 1917. Known affectionately as 'Pregnant Percy', due to his rotund size.

Burnaby, Colonel Frederick Gustavus, 1842–85. Famous soldier and adventurer, educated Bedford and Harrow. 1859, gazetted Cornet, 3rd Regt, Household Cavalry; 1866, Capt.; 1879, Maj.; 1880, Lt.-Col.; 1881–5, CO of the Regt. Author of *A ride to Khiva*, London 1876, and *On horseback through Asia Minor*, London 1877. Noted aeronaut and RAeS council member. March 1882, made balloon-crossing of English Channel. Killed in action, Abu Klea, Sudan.

Burney, Commander Sir Charles Dennistoun, 1888–1968. Son of Admiral Sir Cecil Burney. As an RN officer, he supervised a series of Admiralty-sponsored hydroplane trials with B&C between 1911 and 1914. 1914–18, naval war service; 1920–9, Conservative MP for Uxbridge; 1923–30, managing director of the (Vickers) Airship Guarantee Co. Ltd, constructors of the R100 airship; 1929, succeeded to his father's baronetcy.

Burroughes, Hugh, 1883–1985. Educated Manchester School of Technology. 1909–13, tech. asst, Royal Aircraft Factory, Farnborough; 1914–19, managing director, Airco; 1921, director, Gloster Aircraft Co. Subsequently a director of Hawker Siddeley and FRAeS.

Busk, Edward Teshmaker, 1886–1914 (a cousin of Mervyn O'Gorman). Educated Harrow and King's College, Cambridge. Secured 1st Class Hons., Mechanical Sciences Tripos. Engineer with Messrs Halls & Co. of Dartford before joining staff of Royal Aircraft Factory, 10 June 1912. (Asst engineer in charge of physical experimental work.) Main deviser of the RE1 and, from that, BE2c inherently-stable aircraft. Killed when his BE2c caught fire in the air and crashed, 5 Nov. 1914. Posthumously awarded RAeS gold medal.

Busteed, Air Commodore Henry ('Harry') Richard, 1888–1965. Australian test pilot, who came to England in May 1911 with Harry Hawker. 1911–14, test pilot, British & Colonial; 1914–18, RNAS; 1919–30, RAF.

Butler, Frank Hedges, 1855–1928. Wine-merchant, pioneer motorist and prominent aeronaut. Co-founder of the RAeC in 1901: see ch. 2.

Cammell, Lieutenant Reginald Archibald, 1886–1911. Born Trwernen (*sic*). Officer of the Royal Engineers and pioneer army pilot with the Balloon School/Air Battalion. Awarded RAeC certificate No.45, 31 Dec. 1910, taking his tests on a Bristol Boxkite at Larkhill. Killed on a Valkyrie monoplane, Hendon, 17 Sept. 1911: see ch. 6.

Capper, Major-General Sir John Edward, 1861–1955. Officer of the Royal Engineers. 1903–5, OC Balloon Sections RE; 1905–6, OC Balloon Companies RE; 1906–9, CBS and SBF: see chs 4 and 5.

Carden, Colonel Alan Douglas, 1874–1964. 1894, commissioned 2nd Lt., RE; 1900–5, Asst Instructor, SME, Chatham; 1905–6, OC, West Indies Submarine Mining Coy; 1907–10, Asst SBF, Farnborough; 1911–12, Experimental Officer, Air Battalion, RE; Aug. 1912, Sqn-Ldr, RFC, rising to rank of Wing-Cdr (DSO); 1921, appointed to Central Engineering Board, War Office: OC, 1st AA Searchlight Battalion, RE; 1927, retired; 1948–54, attached RAE.

Carew, Lt.-General Sir Reginald Pole-, 1849–1924. 1876–7, private sec., Governor, New South Wales; 1878–9, ADC to Lord Lytton, Viceroy of India; 1879–80, ADC to Roberts, Afghan War, and 1881, South Africa; 1882, ADC to duke of Connaught, Egypt; 1884–90, military sec. to Roberts – consecutively C-in-C Madras and India; 1895–9, CO, 2nd Btn, Coldstream Guards; 1899–1900, commanded consecutively 9th Brigade and 11th Division, South Africa; 1900, KCB; 1903–5, GOC, 8th Division, 3rd Army Corps; 1906, retired with rank of Lt.-Gen.; 1910–16, Conservative MP for Bodmin; 1914, Inspector-Gen. of Territorials.

Chadwick, Roy, 1893–1947. Avro's chief designer, joined A. V. Roe as his personal assistant in September 1911: see ch. 2.

Challenger, George Henry, 1881– . Born Neath, son of Bristol Tramways & Carriage Co.'s general manager, Charles Challenger. 1910, appointed British & Colonial Aeroplane Co.'s chief engineer and works manager: designer of the Bristol Boxkite; 1911, designed Bristol Racing Biplane, Bristol Monoplane (with A. Low) and Bristol Biplane Type T, before being recruited by Capt. H. F. Wood for Vickers's aviation department.

Chanute, Octave, 1832–1910.   Came to prominence as a civil engineer: particularly as a railway/railway bridge constructor. Became interested in aeronautics in 1850s. 1896, inspired by Otto Lilienthal in Europe, established a camp on south shore of Lake Michigan, near Chicago, and with test pilot A. M. Herring began series of multiplane and biplane hang-glider flight trials. His work was a powerful influence on Wright brothers, of whom he became a close confidant. The most important disseminator of aeronautical information – throughout both America and Europe – of the pioneering period.

Cockburn, George Bertram, 1872– .   Born Birkenhead. A Scottish rugby international, awarded RAeC certificate No.5 on 26 Apr. 1910 – having previously learned to fly at the Farman school, Mourmelon, in June 1909 (from where he purchased the first 'Farman III' biplane). Well-known as the only British aviator to have participated in the 1909 Rheims meeting, afterwhich he erected a shed at Larkhill and worked in co-operation with the early army aviators. 1911, instructor to the first naval flying volunteers at Eastchurch. 1914, AID Aeroplane Inspector.

Coanda, Henri, 1885–1972.   Artist, engineer and aircraft designer, son of the Rumanian war minister. Exhibited a prophetic biplane (sesquiplane) employing first attempt at a jet engine at the 1910 Paris Aero Salon. 1911–14, employed as an aircraft designer by British & Colonial: see ch. 3.

Cody, 'Colonel' Samuel Franklin, 1861–1913.   Cowboy, showman and aeronautical pioneer, employed by the War Office 1905–9: see chs 4 and 5. While in service he was granted officer status appertaining to a Colonel, but was not strictly entitled to use the rank.

Coxwell, Henry Tracey, 1819–1900.   The most successful English aeronaut of the 19th century, and influential early advocate of military ballooning: see ch. 4.

Crompton, Rookes Evelyn Bell, 1845–1940.   Born Thirsk, educated Harrow. 1864, ensign, Rifle Brigade, India; 1875, left army, acquiring partnership in Chelmsford engineering firm and becoming supervisor of Stanton Ironworks; 1886, founded Kensington & Knightsbridge Electric Supply Co.; 1890–9, advisor to Indian government on electrical projects; 1897, founder member, Automobile Club; 1899–1900, commanded Volunteer Corps, Electrical Engineers, South Africa. Promoted Col. and retained as War Office consultant. 1914–18, War Office consultant on tank development; 1933, FRS.

Curtiss, Glenn Hammond, 1878–1930.   Born Hammondsport, New York. Left school at fourteen, becoming camera assembler with Eastman Kodak of Rochester and telegraph messenger with Western Union. Took up bicycling (Western New York champion). Opened cycle repair shop and began manufacture of bicycles: subsequently adapted to motor cycles. 1907, established world speed record (136.3 m.p.h.) employing Curtiss engine. 1904–8, developed airship engines in association with 'Capt.' T. S. Baldwin. Attracted notice of Alexander Graham Bell and, 1907, joined latter's Aerial Experiment Association as director of experimental work. 1908, AEA's first aircraft ('Red Wing') produced; 1909, Curtiss left AEA and formed Herring-Curtiss Co. with Augustus Moore Herring (1867–1926); Aug. 1909, won Gordon Bennett Cup at first Rheims meeting. Soon embroiled in litigation with both Herring and Wright brothers. 1910, Herring-Curtiss taken into receivership; 1911, Curtiss (Aeroplane) Motor Co. established. Worked in close association with US Navy.

Daimler, Gottlieb, 1834–1900. Born Schorndorf, near Stuttgart, educated Stuttgart Polytechnic. 1872, joined Gasmotoren-Fabrik Deutz as chief engineer; 1882, with Wilhelm Maybach, established factory in Stuttgart for development of internal combustion engine. Daimler-powered automobile displayed at 1889 Paris Exhibition. 1890, Daimler-Motoren GmbH created.

Dawson, Sir Arthur Trevor, 1866–1931. 1879, entered RN via *Britannia* cadetship; 1885, as acting Sub-Lt., attended Portsmouth and Greenwich RN colleges, and torpedo school, HMS *Vernon*; 1887, Lt., Ordnance Staff, HMS *Excellent*; 1892–6, Experimental Officer, Woolwich Arsenal; 1896, recruited by Vickers as firm's ordnance superintendent. Presently became company's most effective director. Pressed for adoption of both submarines and airships. 1902–14, secured net profit of £1.25m from Admiralty submarine contracts alone. 1909, knighted; 1920, baronet.

de Forest, Arnold Maurice, 1879–1968. Born Bischoffsheim. Adopted son of Austrian-Jewish financier, Baron Hirsch, hereditary baron of the Austrian empire and friend of Edward VII. Educated Eton and Christ Church, Oxford. 1900, became naturalised British subject; 1911, elected Liberal MP, West Ham North. Friend of Churchill and Lloyd George; *bête noire* of George V. He allied a huge fortune with extreme radicalism, and consequently was not generally popular. July 1909, following Blériot's achievement, announced £4,000 prize for longest flight from England to the continent in an all-British aeroplane. 1914, Lt.-Cdr, RNVR. 1932, became a naturalised citizen of and, subsequently, principal diplomatic counsellor to, Liechtenstein.

de Havilland, Sir Geoffrey, 1882–1965. Aviation pioneer, aircraft designer and manufacturer: see ch. 5.

Delagrange, Léon, 1873–1910. Artist and aviation pioneer, born Orleans, son of a wealthy cotton manufacturer. Attended the Ecole des Beaux-Arts, Paris, as did Henri Farman and Gabriel Voisin. Associated with Voisin from 1906. Made first flight at Issy on 14 March 1907 and rapidly became one of the most renown aviators. In 1909 adopted Gnôme-powered Blériot monoplane in preference to the biplane and caused a sensation at Doncaster meeting. Killed at Bordeaux on 4 Jan. 1910, when the wings of his Blériot buckled.

de Rodakowski, E., 1867–1944. Known simply as 'Roda'. Originally a citizen of Austrian Poland, he had served with the 1st Uhlahen (Lancers), becoming one of the Austrian army's top riders. 1889, went to Egypt with his friends, the Locke Kings, and ran latters' Mena House Hotel, in Cairo. Hugh Locke King subsequently established the Brooklands motor-racing circuit on his Weybridge estate, and de Rodakowski was appointed clerk of the course at BARC's first committee meeting, in December 1906. Retained position until 1909. 1911, appointed managing director of United Motor Industries. Subsequently changed name to E. R. Rivers.

de la Vaulx, Comte Henri, 1870–1930. Explorer and sportsman. 1900, made record balloon flight from Paris to Korostyshew, west of Kiev – a distance of 1,925 km – in 35¾ hrs; Oct. 1905, founded FAI: designed 16 h.p. 25,000 cu. ft dirigible capable of being packed and transported for military use; 1907, designed unsuccessful 50 h.p. (Antoinette) Tatin-de la Vaulx pusher-monoplane in association with veteran aeronautical pioneer, Victor Tatin (1843–1913), and Parisian balloon constructor, Maurice Mallet.

Dickson, Captain Bertram, 1873–1913. Officer of the Royal Artillery and successful exhibition aviator. Became a representative of British & Colonial in 1910: see ch. 3.

Ding, William Rowland, 1885–1917. Born Alsager, Cheshire, trained as an engineer. Built successful aircraft models in association with W. F. Sayers, but partnership ran out of funds before full-size Brooklands machine completed. 1913, with Lindsay Bainbridge and others, gained control of Lakes Flying Co. of Bowness-on-Windermere, restyled Northern Aircraft Co.; 27 Apr. 1914, secured RAeC certificate No.774 on a Wright biplane, Beatty school, Hendon. Quickly established reputation as outstanding cross-country and exhibition pilot: nominated as Handley Page's Aerial Derby pilot within days of gaining certificate. 1914, brought Handley Page Type G biplane for Northern Aircraft Co., and intended co-pilot of HP Type L, commissioned for transatlantic attempt. 1916, Northern Aircraft Co. liquidated, Ding becoming freelance test pilot. Killed at Leeds while secretly test flying a revamped BE biplane, 12 May 1917.

Douhet, Major-General Giulio, 1869–1930. Italian artillery officer. 1912–15, Chief of Italy's first Aviation Department; 1912, negotiated creation of the Società Italiana Bristol Aeroplani; 1916, court-martialled for criticising conduct of the war. Sentenced to a year's imprisonment. Feb. 1918, recalled to service and placed at head of Central Aeronautical Bureau. Author of *The command of the air*, Rome 1921. Influential prophet of air power.

Drury, Admiral Sir Charles Carter, 1846–1914. 1900, vice-president, Ordnance Committee; 1903–8, Second Sea Lord; 1907–8, C-in-C, Mediterranean; 1908–11, C-in-C, Nore; 1910–11, issued order accepting Francis McClean's offer of the loan of Short biplanes to the Admiralty, and in overall command of resulting trials.

du Cros, Sir Arthur, 1871–1955.   Born Dublin, third son of William Harvey du Cros (1846–1918), book-keeper and president of Irish cyclist association. Through cycling interest du Cros sen. discerned commercial potential of pneumatic tyre (patented 1888) and in 1889 became chairman, Dunlop Pneumatic Tyre & Cycle Agency Ltd, Coventry. Arthur du Cros joined 1892, becoming general manager and (1896) joint managing director. 1901, founded Dunlop Rubber Co., becoming chairman and, subsequently, president. Responsible for acquisition of Malayan rubber estates. 1908, elected Conservative MP for Hastings; May 1909, founded – and became hon. sec. of – Parliamentary Aerial Defence Committee, Westminster pressure group lobbying for increased military aeronautics expenditure. (Arthur Lee, MP, became chairman: Northcliffe's brother, Cecil Harmsworth – Liberal MP for Droitwich – was vice-chairman.) Du Cros donated £6,000 toward cost of supplying War Office with Clément-Bayard II non-rigid (i.e. lacking internal framework) airship. 1911, provided £10,000 loan for formation of the Grahame-White Aviation Co. 1916, baronet; 1918, Coalition Unionist MP for Clapham: resigned 1922. Renowned art collector and philanthropist.

Dunlop, John Boyd, 1840–1921.   Son of Ayrshire farmer. Trained as vet. 1867, moved to Belfast. Developed pneumatic tyre in 1888: patent granted 1889. Joined with Harvey du Cros in establishing Pneumatic Tyre & Cycle Co. in association with Booth's Cycle Agency of Dublin. 1890, William Thompson's 1845 pneumatic tyre patent discovered; 1892, Dunlop patent invalidated; 1895, Dunlop resigned directorship of company. However, du Cros, aware of public association of tyres with 'Dunlop', kept latter's name for new company, The Dunlop Pneumatic Tyre Co., launched on mainland in 1896.

Dunne, General Sir John Hart, 1835–1924. Saw distinguished active service in Crimea (1854–5) and China (1860). 1881, Maj.-Gen.; 1885–6, GOC, 2nd Infantry Brigade, Aldershot; 1887–90, GOC, Thames District, Chatham; 1893, Gen.; 1894–7, Lt., Tower of London. Col., Wiltshire Regt. Father and active supporter of J. W. Dunne.

Dunne, John William, 1875–1949.   Born Kildare, Ireland. Soldier, aviation pioneer and writer on speculative psychology, employed by the War Office 1906–9: see ch. 5.

Egerton, Maurice, 4th Baron Egerton of Tatton, 1874–1958.   Educated Eton, after which established reputation as explorer-cum-adventurer, big game hunter and pioneer motorist. Took delivery of Short-Wright No.4, 26 Nov. 1909. First flew the aircraft 1 Dec. 1909. Awarded RAeC certificate No.11, 14 June 1910. Crashed another aeroplane shortly afterwards and severely injured. 1913, a co-founder of British Wright Co. Afterwards divided his time between

Tatton Park and his estates in Kenya. Died unmarried and barony became extinct.

Elsdale, Colonel Henry, 1843–1900. Early military balloon pioneer and close friend of Elliott Wood [*q.v.*]. 1864, commissioned, RE; *c.* 1880s, OC, RE balloon field detachment; 1884–5, led balloon detachment, Bechuanaland Expeditionary Force. Retired 1899.

England, Eric Cecil Gordon, 1891– . Born Concordia, Argentina. Acted as pilot for Weiss's early glider trials. 1911–13, test pilot-cum-aircraft designer, B & C; 1913–16, associated with James Radley in seaplane production, Cedric Lee & Co., and consulting engineer/pilot, J. Samuel White & Co.; 1916–18, general manager, F. Sage & Co.'s aviation department; 1930–4, director, Vacuum Oil Co.; 1934–42, managing director, General Aircraft Co.; 1940–4, chairman, engineering industries' association. Responsible for design and development of Hotspur and Hamilcar military gliders of World War II. FRAeS.

Esnault-Pelterie, Robert Albert Charles, 1881–1957. French aviation pioneer. May 1904, built and unsuccessfully tested a Wright-derived glider. Failed to comprehend Wright's warping-cum-rudder control method. Oct. 1904, tested revised version incorporating elevators-cum-ailerons (movable surface fixtures) for flight control – the first use of this device – but still failed to achieve successful gliding flight; 1907, constructed tapered-wing, short-fuselage, steel-framed tractor-monoplane – the REP No.1 – powered by his own 25 h.p. 7-cylinder engine and incorporating primitive wing-warping; Nov.–Dec., achieved flights of up to 600 metres; 1908, REP No.2 (streamlined and incorporating large fin) began making regular flights. Patent subsequently adopted by Vickers. In the inter-war years he became an early theorist of space flight.

Fairey, Sir Charles Richard, 1887–1956. Born Hendon, educated Merchant Taylor's School and Finsbury Technical College. 1902, began work for Jandus Electric Co., Holloway, moving on to Finchley Power Station; 1910, won Crystal Palace model aeroplane competition, but infringed J. W. Dunne's patent; 1911, appointed Blair Atholl Syndicate's works manager; 1913, appointed works manager, Short Brothers Ltd; 1915, founded Fairey Aviation with a contract from Shorts; 1921, firm went into voluntary liquidation and reformed. By 1925 more than half of all British military aircraft were Fairey types. 1942–5, Director-General, British Air Commission, Washington. 1942, knighted.

Farman, Henri (christened *Harry Edgar Mudford Farman*), 1874–1958. His Christian name is sometimes spelt with an 'i', sometimes with a 'y', with Farman himself indiscriminate in the matter, although he tended to use former before and latter after his 1909 split from Voisin. French was his mother tongue, but he remained a British citizen until 1937. Son of an English journalist resident in France, he was initially an art student, who turned to engineering via bicycle and automobile racing, founding his own automobile agency. Began aeroplane construction in association with Gabriel Voisin. Sept. 1907, flew Voisin-Farman 1 pusher-biplane; Jan. 1908, won Grand Prix d'Aviation Deutsch-Archdeacon for flying world's first officially accredited circular kilometre; Mar. 1908, redesignated machine 'Henri Farman No.1–bis', incorporating flap ailerons; Apr. 1909, debut of first entirely Farman-designed machine, the 'Henry Farman III', incorporating a single front-elevator and open biplane tail-unit with rudders. Fitted with 50 h.p. Gnôme rotary engine this became archetypal pre-war pusher-biplane. Subsequently established construction firm with his brother, Maurice.

Fenwick, Robert C., 1884–1912. Born South Shields. 1910, assistant and test pilot to Liverpool patent agent, W. P. Thompson, who founded Planes Ltd aircraft company to develop his own 'pendulum-stability' biplane, which Handley Page had originally been commissioned to build. Awarded RAeC certificate No.35, 29 Nov. 1910: tests taken at Freshfield, Lancashire, on firm's biplane. Retained as test pilot by Handley Page, but crash-landed HP Type D on its maiden flight at Fairlop, 15 July 1911, and immediately sacked. 1911–12, designed Planes Ltd 45 h.p. (Isaacson) Mersey pusher-monoplane, setting up the Mersey Aeroplane Co. at Freshfield with S. T. Swaby, but suffered fatal crash on 13 Aug. 1912, during Larkhill military trials.

Ferber, Captain Ferdinand, 1862–1909. Born Lyon, and attended the École Polytechnique before joining the Corps of Artillery, receiving his Captaincy in 1893. 1898, while an Instructor at the Artillery School of Application, Fontainebleau, began aeronautical experiments. Disciple of German hang-gliding pioneer, Otto Lilienthal, and influenced by reports of Wright brothers' trials. 1904, devised tailplane: seconded to Chalais-Meudon. 1906, joined Antoinette. Killed after losing control of just landed taxi-ing Voisin biplane at ill-fated Boulogne aviation meeting, 22 Sept. 1909.

Fisher, Edward Victor Beauchamp, 1888–1912. Educated Tonbridge School. 1909, assistant to A. V. Roe at Lea Marshes; 1910, worked closely with Howard Flanders at Brooklands on latter's first monoplane. Founding member, Brooklands Aero Club. Took RAeC certificate No.77, 2 May 1911, on a Hanriot monoplane, Hanriot school, Brooklands – where he subsequently became an instructor. 1912, rejoined Howard Flanders. Killed at Brooklands, 13 May 1912, when the 60 h.p. Green engine Flanders monoplane he was piloting (accompanied by American millionaire, Victor Mason) stalled on a turn. Considered potentially one of Britain's finest pilots.

Flanders, Richard Leonard Howard, 1883–1939. Born Italy, of English parents. A graduate of Emmanuel College, Cambridge, he subsequently served a three year apprenticeship with the Bristol-based engineering firm of Brazil, Straker & Co. July–Nov. 1909, assistant to A. V. Roe; Nov. 1909–Aug. 1910, manager, J. V. Neale & Co. Designer of Neale monoplane. From Aug. 1910 experimented on his own account: Feb. 1912, formed limited company based at Richmond and Brooklands. Four Flanders F4 monoplanes procured by War Office, but company subsequently hit by 'monoplane ban'. July 1912, designed Flanders biplane; 1913, met with serious motor cycle accident which incapacitated him until outbreak of war. Company disbanded. 1914–19, employed by Vickers; 1921–7, secretary, Institution of Aeronautical Engineers; 1939, joined Bristol Aeroplane Co.

Fokker, Anthony (Anton Herman Gerard), 1890–1939.   Born Java, Dutch East Indies, where father ran coffee plantation. Grew up in Haarlem. Attended engineering college at Mainz, Germany. 1910–11, constructed 'Spider', inherently-stable monoplane. Established base at Johannisthal airfield, Berlin. 1913, won German military aeroplane competition and established factory and flying school at Schwerin. Constructor of warplanes for German army in First World War, made famous by development of synchronised machine-gun interrupter-device, allowing pilot to shoot directly – through revolving propeller blades. 1922, emigrated to USA. President, Fokker Aircraft Corp.

Folland, Henry Phillip, 1889–1954.   1905–8, apprentice, Lanchester Motor Co.; 1908–12, draughtsman, Daimler Motor Co.; 1912–17, draughtsman-designer, Royal Aircraft Factory, Farnborough (officially designated 'section leader, design office'); 1916, designed SE5 single-seater fighter; 1917–21, chief designer, Nieuport & General Aircraft Co.; 1918, designed Nieuport 'Nighthawk' (fighting scout); 1919, designed Nieuport 'Goshawk' (racing scout); 1921–37, chief engineer-designer and, latterly, director, Gloster Aircraft Co. Designed succession of Gloster aircraft for RAF, including Grebe (1923), Gamecock (1925), Gauntlet (1933) and Gladiator (1935). 1937–51, managing director, Folland Aircraft Co., Hamble. Firm subsequently incorporated within Hawker Siddeley group.

Fox, Captain Alan Geoffrey, 1887–1915.   1908, commissioned 2nd Lt., RE; 1911, Air Battalion: joined RAeS; 30 Jan. 1912, gained RAeC certificate No.176, B&C's Larkhill school;1912, RFC. James McCudden's Flt-Cdr, No.3 Sqn. Died of wounds, 9 May 1915, Cambrin, France.

French, Field Marshal John Denton Pinkstone, 1st earl of Ypres, 1852–1925. 1899–1902, GOC, Cavalry Division, South Africa; 1902–7, C-in-C, Aldershot Command; 1907, Inspector-Gen. of Forces; 1912–14, CIGS; 1913, Field Marshal; Aug. 1914–Dec. 1915, C-in-C, BEF; 1915–18, C-in-C, Home Forces; 1918–21, Lord Lieutenant of Ireland. 1900, knighted; 1916, viscount; 1922, earl.

Friswell, Sir Charles, 1871–1926.   1894, set up one of the first automobile agencies in Britain, specialising in Peugeot cars; 1896, participated in London to Brighton emancipation run. Chosen by the Automobile Mutual Protection Association to lead successful test case against Henry Lawson's British Motor Co.'s monopoly claims. 1905, became sole distributor for Standard Cars; 1907, gained controlling interest in Standard Motor Co. Friend of early aviators, including Sopwith. 1908–9, commissioned construction of abortive A. V. Roe triplane; 1909, knighted; 1911, supplied vehicles for George V's Delhi coronation durbar; 1912, lost controlling interest in Standard Motor Co.

Fullerton, Colonel John Davidson, 1853–1927. Born India, educated Cheltenham and RMA, Woolwich. An officer of the Royal Engineers. 1879–80, active service 2nd Afghan War; 1885–7, active service Burma. Four times mentioned in despatches. Retired 1905. Long-serving Aeronautical Society council member and hon. sec. 1906–8.

Fulton, Major John Duncan Bertie, 1876–1915. 1896, commissioned 2nd Lt., RFA; 1899–1902, promoted Lt.: active service South Africa, including operations for relief of Ladysmith (mentioned in despatches); 1902, Capt.; 1908, 1st Class Interpreter (French); 1909, procured from Grahame-White a 28 h.p. (Anzani) Blériot, which he based at Larkhill; Nov. 1910, first regular officer to gain an aviator's certificate; 1911, seconded to Air Battalion, RE (OC, No.2 [Aeroplane] Coy); 1912, Instructor, CFS (Sqn-Cdr); 1913, promoted Maj.: Chief Instructor, CFS; Dec. 1913, Chief Inspector of Material, RFC; 1914, Chief Inspector (Wing-Cdr), AID; Dec. 1914, temp. Lt.-Col.; Nov. 1915, appointed ADMA. Died 11 Nov.

Gambetta, Léon Michel, 1838–82. Born Cahors, of Genoese-Jewish extraction. 1869, elected Deputy. One of the proclaimers of the Republic on 4 Sept. 1870. Minister of Interior, Government of National Defence. Escaped from Paris by balloon, and for five months acted as dictator of France with aim of continuing war. 1879, President of the Chamber.

Gates, Richard Thomas, 1876–1914. Like J. W. Dunne, served with the Imperial Yeomanry in Boer War. 1911–13, General Manager of the London (Hendon) Aerodrome: awarded RAeC certificate No.225, 4 June 1912; 1913, left to work in the city, remaining a director of Grahame-White Aviation; 1914, commissioned Flt Lt., RNAS. Crash-landed a Farman biplane at Hendon on night of 10 Sept. 1914, on returning from an unofficial RNAS anti-Zeppelin patrol over London, and died from his injuries some four days later.

George, David Lloyd, Earl Lloyd-George of Dwyfor, 1863–1945. 1908–15, Chancellor of Exchequer; 1915–16, Minister of Munitions; July.–Dec. 1916, Secretary of State for War; 1916–22, PM; 1945, created Earl Lloyd George. Displayed a cautious early interest in aviation: Aug. 1908, present at Stuttgart during Zeppelin IV Echterdingen crash, witnessing resulting nationalist fervour; 1908–9, member of the CID Aerial Navigation Sub-Committee; July 1909, attended Sir Benjamin Stone's Commons luncheon in honour of Blériot; Aug. 1909, attended Rheims meeting with Northcliffe.

Gerrard, Air Commodore Eugene Louis, 1881–1963. Born Dublin, educated Warwick School. 1900, commissioned 2nd Lt., RMLI; 1910, served with RN

Airship No.1. One of four naval volunteers selected for flying instruction at Eastchurch in Mar. 1911, replacing George Wildman-Lushington who had fallen sick. 1912, instructor, RFC (NW), CFS; 1914, Sqn Cdr, siege of Antwerp: 22 Sept., led anti-Zeppelin raid on Düsseldorf; Dec. 1914, Cdr. Later Wing Commander, Eastern Mediterranean (DSO 1916); 1919, Group Captain, RAF; 1923, Air Commodore, RAF; 1924–7, AOC, Palestine.

Gibbs, Lieutenant Lancelot Dwarris Louis. Officer of the Duke of Connaught's Own Hampshire and Isle of Wight Royal Garrison Artillery (Militia) Regiment. Pioneer army aviator: see chs 5 and 6. His flying career was prematurely curtailed by a recurrent spinal injury received in a flying accident at Wolverhampton in 1910.

Gilmour, Douglas Graham, 1885–1912. Daring and controversial early aviator. 19 Apr. 1910, awarded Ae.C. de France certificate No.75, after taking tests on a Blériot monoplane at Pau. Operated out of hangar No.7 at Brooklands before joining B&C in Feb. 1911: see ch. 3.

Glaisher, James, 1809–1903. 1833–5, asst, Cambridge University Observatory; 1835–8, asst, Greenwich Observatory; 1838–74, chief of magnetic and meteorological department, Greenwich; 1849, FRS; 1850–72, sec., Royal Meteorological Society; 1862–6, made famous series of balloon ascents with H. Coxwell.

Glazebrook, Sir Richard Tetley, 1854–1935. 1877, Fellow of Trinity College, Cambridge; 1880, appointed demonstrator, Cavendish Laboratory (under Lord Rayleigh); 1882, FRS; 1891, appointed asst director, Cavendish Laboratory; 1898, principal, University College, Liverpool; 1899–1919, first director of the NPL; 1909–20, chairman, Advisory Committee for Aeronautics; 1920–3, Zaharoff Professor of Aviation, Imperial College; 1920–33, chairman, Aeronautical Research Committee.

Grace, Cecil S., –1910. Widely considered to be among the most talented early British aviators: based at Isle of Sheppey, where he flew the Short-Wright No.5 purchased by his brother, Percy. Awarded RAeC certificate No.4, 12 Apr. 1910. Flew Shorts S27 biplane at Wolverhampton and Bournemouth meetings (June and July 1910), and Blériot monoplane at Blackpool and Folkstone meetings (August and September 1910). Lost at sea, 22 Dec. 1910, returning from France on Shorts S29 biplane, after an attempt on the Baron de Forest prize. The first Shorts tractor-biplane had originally been designed for his use.

Graham, Lt.-General Sir Gerald, 1831–99. 1850, commissioned 2nd Lt., RE; 1855–6, active service Crimea (VC); 1860, Third China War (brevet Lt.-Col.); 1877–81, Asst Director of Works, War Office; 1881, promoted Maj.-Gen.; 1882, GOC, 2nd Brigade, 1st Division, Egypt (28 Aug., commanded at Kassassin: 13 Sept., commanded 2nd Brigade, Tel-el-Kebir); 1884–5, GOC,

Eastern Sudan (Suakin) Expeditionary Force; 1890, retired; 1899, Col. Cmdt, RE.

Grahame-White, Claude, 1879–1959.   Prominent pioneer aviator and successful publicist of aviation. Founder of 'The London Aerodrome' at Hendon: see ch. 3.

Green, Charles, 1785–1870.   19th century English aeronaut, son of London fruiterer. July 1821, made debut balloon ascent. He was the first to employ coal-gas. Eventually made 526 ascents, the most celebrated being on 7 Nov. 1836, when he ascended from Vauxhall Gardens and crossed channel, descending next morning at Weilburg, Duchy of Nassau (Germany).

Green, Major Frederick Michael, 1882– .   Jan. 1910, appointed chief engineer, RAeF, after being recruited from Daimler on the recommendation of F. W. Lanchester; Nov. 1910, advised de Havilland to offer his services to RAeF; 1917, chief engineer-designer, Armstrong-Siddeley. Retired 1933, returning to RAE for 1939–45 war. FRAeS.

Green, Gustavus, 1865–1964.   1897, established bicycle repair-cum-construction shop, Bexhill, Sussex; 1904, patented single-cylinder motor-cycle engine and initiated automobile construction; 1908, patented 35 h.p. four-cylinder aero-engine; Oct. 1911, 60/65 h.p. Green engine won Alexander aero-engine competition, Farnborough; 1912, patented 100 h.p. engine; 1913, firm acquired Twickenham works; 1914, 100 h.p. six-cylinder Green engine won £5,000 naval & military aero-engine competition, Farnborough. Company subsequently diversified into marine engines.

Gregory, Commander Reginald, 1883–1922.   Born Westwood Drummond, Inverness. 1905, Lt., RN; 1911, one of four naval volunteers selected for flying instruction at Eastchurch. Had unsuccessfully volunteered for naval aeronautical service as far back as March 1909, at which time he joined the Aero Club (PRO, AIR 1 647/17/122/377). Subsequently a member of Seely's 1912 RFC Technical Sub-Committee. July 1913, Lt.-Cdr; 1914, Sqn Cdr, Great Yarmouth; 1916–17, active service with naval armoured car sqns, including with Roumanian army; 1918, commanded the *Veronica*. Died at Hong Kong, 21 June 1922.

Grey, Charles Grey, 1875–1953.   Influential aviation journalist. Editor of *The Aero* 1909–11, and *The Aeroplane* 1911–39: see ch. 5.

Grosvenor, Hugh Richard Arthur, 2nd duke of Westminster, 1879–1953.   Soldier and aviation philanthropist: see ch. 5, footnote 196.

Grover, Colonel George Edward, 1840–93.   1858, commissioned, RE. Began career in design of coastal defence works (Department of Fortifications). 1862–70, while still occupied with fortification duties, became an active campaigner for adoption of balloon unit within British army; 1870–4, senior member, executive committee, industrial-scientific exhibitions, South Kensington; 1875, Staff College; 1877–80, Intelligence Department, War Office; 1882–5, attached to Admiralty, superintending reconstruction of Portsmouth dockyard; 1885, Deputy Asst Adjt-Gen., Intelligence Department, Eastern Sudan Expeditionary Force (brevet Lt.-Col.); 1887, Asst Inspector-Gen. of Fortifications; 1892–3, Royal Commission representative, Chicago World's Fair.

Groves, James Grimble, 1854–1914. Born Manchester, educated King William's College, Isle of Man, and Owen's College, Manchester. Entered the brewing industry, becoming chairman and managing director of Groves & Whitnall Ltd, and chairman of the Federated Brewers' Association. 1900–6, Conservative MP for South Salford. 1913–14, chairman, A. V. Roe & Co. Ltd. Noted philanthropist.

Grubb, Colonel Alexander Henry Watkins, 1873–1933. Educated Wellington College, and RMA, Woolwich. 1892, commissioned 2nd Lt., RE; 1894, joined Balloon Section, Aldershot; 1895, Lt.; 1899–1900, saw distinguished active service with 1st Balloon Section, South Africa: subsequently joined Mounted Infantry; 1902, Capt., attached to Brig.-Gen. W. Kitchener's column: SO to Chief Engineer, South Africa; 1904, SO to DFW. Subsequently OC, 1st Field Searchlight Section, and OC, 1st Wireless Telegraph Coy. 1912, Maj.; 1914, SO to Chief Engineer, Aldershot; Oct. 1914, posted to 3rd Signal Sqn, Belgium. Served with Signals throughout war, becoming Director of Signals, Salonika Field Force/Army of the Black Sea, Constantinople. 1922, CRE, 1st Division. Retired 1927, as Chief Engineer, Army of the Rhine.

Hadden, Major-General Sir Charles Frederick, 1854–1924. Educated Cheltenham, and RMA, Woolwich. 1893, Chief Inspector, Woolwich; 1901, Ordnance Committee; 1904, Cmdt, Ordnance College; 1904–7, Director of Artillery, GHQ; 1907–13, MGO; 1913–15, President, Ordnance Board.

Haldane, Richard Burdon, Viscount Haldane of Cloan, 1856–1928. Lawyer and politician: 1905–1912, Secretary of State for War; 1912–15, 1924, Lord Chancellor: see ch. 5.

Hamel, Gustav, 1889–1914. Son of German father (well-known London doctor and naturalised Briton). Educated Westminster School and Cambridge. 14 Feb. 1911, took RAeC certificate No.64 on a Grahame-White Blériot monoplane: became chief instructor, Blériot school, Hendon; Mar. 1911, won inaugural Hendon–Brooklands race. Quickly became star aviator. 9 Sept. 1911, carried Britain's first airmail from Hendon to Windsor, to mark coronation of George V. Planned to fly Atlantic in a large Martin-Handasyde monoplane, but on 23 May 1914 vanished during flight from France to England in a Morane-Saulnier monoplane. With onset of war rumours abounded that he had been a German spy, but fishermen had spotted his body in the channel.

Hamilton, Captain Patrick, 1882–1912. 1901, commissioned 2nd Lt., Worcestershire Regt; 1902, Lt.; 1908, Capt.; 1912, attached Air Battalion, joining RFC on formation; 12 Mar. 1912, secured RAeC certificate No.194 on a Deperdussin monoplane at Brooklands, but he had already gained considerable experience. Killed when Deperdussin monoplane he was piloting crashed at Graveley, Herts, 6 Sept. 1912, during army manoeuvres.

Hampton, John, 1799–1851. Served in RN before turning professional aeronaut. Made first ascent in June 1838. Founded Albion Aeronautic Association. First Englishman to make a successful parachute descent, Montpellier Gardens, Cheltenham, 8 Oct. 1838. Early patron of Henry Coxwell.

Handasyde, George Harris, 1877–1958. Educated Edinburgh Royal High School. Trained as a marine engineer, but came to London to work for Charles Friswell. 1908, in association with H. P. Martin, designed and built a 12 h.p. (Humber) monoplane in the disused ballroom of the Old Welsh Harp inn, Hendon; 1910, moved to Brooklands, constructing a JAP-powered Antoinette-derived monoplane, to be flown by Graham Gilmour; 1911, later Gnôme/Antoinette-powered model, 'The Dragonfly', acquired by Thomas Sopwith; 1912, a model ordered by RFC. Martin-Handasyde continued to produce aircraft of a high standard, and were reconstituted a limited company, 'Martinsyde', in 1915. The firm's F4 300 h.p. (Hispano-Suiza) fighter was considered by some the best British single-seat fighter at the time of the Armistice, but it just missed seeing combat. The Martin-Handasyde partnership ended shortly after the war.

Hankey, Maurice Pascal Alers, 1st Baron Hankey, 1877–1963. 1895, commissioned 2nd Lt., RMA; 1902, attached to Naval Intelligence Department, Admiralty; 1908–12, asst sec., CID; 1912–38, secretary to the CID. Also secretary of War Cabinet 1916–19, and secretary of Cabinet 1919–38. 1938, baron; 1939–40, Minister without Portfolio and member of War Cabinet.

Hardwick, Alfred Arkell, 1878–1912. Born London. Aged fourteen, apprenticed Merchant Navy. Aged seventeen, joined Mashonaland (Rhodesia) Police: decorated for service in suppression of Mashonaland revolt of 1896–7. Subsequently employed with Nile Irrigation Works, Egypt. After Boer War spent some three years exploring and hunting in central Africa, experiences recounted in his book, *An ivory trader in North Kenia* [sic], London 1903. Initiated a 'Nigerian Trading Company' and became a special commissioner for *African World*. Afterwards travelled through Morocco and America, before returning to England where, in early 1911, he became asst manager to Handley Page. Killed on 15 Dec. 1912, when a passenger in the Parke-piloted HP Type F which crashed at Wembley.

Hargrave, Lawrence, 1850–1915. Born Greenwich, emigrated to Australia in 1865. 1877, joined Royal Society of New South Wales; 1878, joined Sydney Observatory; 1893, developed the boxkite, constituting a wing-configuration of inherent strength and lifting power, subsequently incorporated into most successful primitive aeroplanes; 1894, conducted man-lifting trials using four tethered kites; 1899, designed a three-cylinder radial rotary engine, again presaging future developments. Refused to patent his inventions and freely disseminated information. 1899, visited England, lecturing before Aeronautical Society.

Harmsworth, Alfred Charles William, 1st Baron Northcliffe, 1865–1922. Born Chapelizod (Dublin), educated Stamford Grammar School and Henley House, Hampstead. Left aged fifteen, and effectively self-educated. 1880, became reporter on *Hampstead & Highgate Express*; 1886, editor, *Bicycling News*. Went on to found *Answers to correspondents* (1888), *Daily Mail* (1896) and *Daily Mirror* (1903). 1905, baron; 1908, acquired the London *Times*; 1917, viscount, chairman of British war mission to USA; 1918, Director of Propaganda in Enemy Countries (Ministry of Information). Enormously influential British advocate and sponsor of aviation throughout the pioneering years.

Harper, Harry, 1880–1960. Aviation journalist and author: see ch. 3, footnote 182.

Hawker, Harry George, 1889–1921. Celebrated Australian pilot, advisor with the Sopwith Aviation Company: see ch. 3.

Hawker, Major Lanoe George, 1890–1916. 1910, joined RAeC, taken up by H. Barber on a Valkyrie monoplane; 1910–11, RMA, Woolwich; July 1911, commissioned 2nd Lt., RE; 1911–13, SME, Chatham; 1911, joined Aeronautical Syndicate's school, Hendon, but organisation closed following Lt. Cammell's death (Sept. 1911); 1912–13, member, Deperdussin school, Hendon (4 Mar. 1913, awarded RAeC certificate No.435); Oct. 1913, promoted Lt., 33rd Fortress Coy; Aug.–Oct. 1914, CFS: subsequently posted to No.6 Sqn, RFC; 25 July 1915, won first VC to be awarded for aerial combat, piloting a Bristol Scout; Sept. 1915, appointed OC, No.24 (DH2) Sqn – Corps' first single-seater fighter sqn. Killed in action 23 Nov. 1916, a celebrated von Richthofen/Albatros victim.

Hearle, Francis Trounson, 1886–1965. Born Penryn, Cornwall. 1907, joined Vanguard Omnibus Co.; 1908, joined Geoffrey de Havilland in aircraft construction venture; Dec. 1910, joined RAeF with de Havilland; 1912, moved to Deperdussin; 1913, joined Airco; 1920, became works manager, de Havilland Aircraft Co. Ltd; 1922, general manager; 1938, managing director; 1950–4, chairman.

Heath, Major-General Gerard Moore, 1863–1929. 1882, commissioned, RE; 1884, joined Telegraph Battalion, accompanying detachment on Warren's Bechuanaland expedition; 1891, Staff College; 1895, OC, Pontoon Section, Chitral Relief Force; 1896–9, OC, Balloon Section, Aldershot; 1899–1900, OC, 2nd Balloon Section, South Africa; 1902–6, Instructor, SME, Chatham;

1910, GSO, Burma Division; 1912, GSO, South Africa; 1915, Chief Engineer, 1st Army; 1917–18, Engineer-in-Chief, British army.

Henderson, Lt.-General Sir David, 1862–1921. Born Glasgow, studied engineering under Lord Kelvin (Glasgow University). 1883, commissioned, Argyll & Sutherland Highlanders; Sept. 1898, ADC to Brig.-Gen. Lyttleton, Omdurman; 1899–1900, active service South Africa. Wounded at Ladysmith and promoted brevet Lt.-Col. (Nov. 1900). Nov. 1900–Sept. 1902, DMI, South Africa. Subsequently wrote official text-book *Field intelligence* (London 1904), to which he added supplementary volume *The art of reconnaissance* (London 1907). Brig.-Gen., SO to Inspector-Gen. of Forces. Aug. 1911, aged forty-nine, learned to fly at B&C's Brooklands school under Howard Pixton (RAeC certificate No.118). 1912, DMT, War Office; 1913–17, DGMA; Aug. 1914, GOC, RFC, France; Oct. 1914, promoted Maj.-Gen., GOC, 1st Division, but disputes within RFC necessitated his return to air service. Remained GOC, RFC, until Oct. 1917. Member of Army Council from Feb. 1916. Jan. 1918, vice-president, Air Council: resigned, Apr. 1918.

Hewlett, Hilda. Pioneering female aviator, wife of novelist, Maurice Hewlett. 1910, learned to fly at Farman school, Mourmelon, where she met Gustave Blondeau. They purchased a Farman biplane and, September 1910, opened the Hewlett & Blondeau Flying School at Brooklands. First British woman to gain RAeC certificate, 29 Aug. 1911. They diversified soon after, with the establishment of the Hewlett & Blondeau Aircraft Manufacturing Company.

Hicks, William Joynson-, 1st Viscount Brentford, 1865–1932. Educated Merchant Taylors School, becoming a solicitor. 1908–10, Conservative MP for Manchester North West; 1911, returned (unopposed), Brentford by-election; 1912–14, chief parliamentary critic of the government's air programme; 1916, chairman, Parliamentary Air Committee; 1918–29, Conservative MP for Twickenham; 1919, baronet; 1923, Postmaster-General and Financial Secretary to the Treasury; 1923–4, Minister of Health; 1924–9, Home Secretary; 1929, viscount.

Hotchkiss, Edward, 1883–1912. Born Shropshire, trained as an engineer. 16 May 1911, awarded RAeC certificate No.87. Joined B&C, becoming chief instructor at the Bristol school, Brooklands. 1912, joined RFC special reserve and killed on 10 Sept., when the Bristol Coanda military monoplane he was piloting on manoeuvres crashed near Port Meadow, Oxford.

Hucks, Bentfield Charles, 1884–1918. Also known as Benjamin or Benny Hucks. Born Stanstead, Essex. Joined Grahame-White as the latter's personal mechanic in early 1910. In an accepted pattern he was part paid in flying lessons. During abortive London–Manchester flights he serviced Grahame-White's Farman aircraft. Subsequently became Grahame-White's leading pilot. 1911, established partnership with Robert Blackburn and tested latter's early monoplanes: 30 May 1911, gained RAeC certificate No.91 at Filey. In the war, after RFC service, he became Airco's chief pilot, testing firm's series of DH machines. Originator of the Hucks non-manual engine starter. Died of pneumonia, November 1918.

Hunter, General Sir Archibald, 1856–1936. 1892–4, Governor, Suakin and Red Sea Littoral; 1894–6, OC, Sudan Frontier Force; 1896–8, Maj.-Gen., Sudan; 1899–1900, consecutively Sir George White's Chief of Staff and GOC, 10th Division, South Africa; 1910–13, Governor of Gibraltar; 1914–17, C-in-C, Aldershot Command; 1918–22, Conservative MP for Lancaster.

Joffre, Marshal Joseph Jacques Césaire, 1852–1931. 1905, presided over committee which sanctioned continued aviation research at Chalais-Meudon; 1914, C-in-C, French north-eastern armies, Western Front; Dec. 1915–Dec. 1916, C-in-C, French armies in the west; 1916, Marshal of France.

Johnson, Claude Goodman, 1864–1926. Educated St Paul's School. 1883, became exhibitions' organiser/clerk, Imperial Institute, South Kensington; 1893, chief clerk; 1897, following planning of 1896 automobile exhibition, appointed secretary of Automobile Club by F. R. Simms; 1900, organised 1,000 miles motor trial; 1901, secretary of (Royal) Aero Club, supervising its establishment in competition with Simms's rival body; 1903, joined C. S. Rolls's automobile agency; 1906, became managing director of newly amalgamated firm of Rolls-Royce. Quickly established company's reputation, but slow to adapt to circumstances of war and, afterwards, firm's attempted American subsidiary failed.

Jones, Colonel Harry Balfour, 1866–1952. 1885, commissioned 2nd Lt., RE; 1890–5, Capt., OC, Balloon Section, RE; 1899–1900, OC, 1st Balloon Section, South Africa (brevet Maj.). Retired from service in 1922.

Jullerot, Major Henri M., 1879–1959. French pilot from Alsace. 2 May 1910, awarded Ae.C. de France certificate No.61; Oct. 1910, joined B&C as pilot-instructor. Saw war service with French air service, RNAS and RAF (AFC), afterwards joining Vickers's aviation department, becoming a special director of Vickers and Supermarine.

Kemp, Ronald Campbell, 1889– . Born Co. Down, Ireland. 9 May 1911, secured RAeC certificate No.80 with Avro Brooklands school and became a pilot-instructor with the firm: also flew the Flanders monoplane. 1912, joined B&C, establishing 'Bristol' school, Halberstadt, Germany; 1914, chief test pilot, RAeF: 23 Feb., crashed FE2a, and his passenger, E. T. Haynes, killed; Aug. 1914–1918, chief test pilot, Short Brothers. Subsequently co-director (with former Avro colleague, F. P. Raynham), Air Survey Co.

Kenworthy, Joseph Montague, 10th Baron Strabolgi, 1886–1953. 1902, entered RN; 1909, persuaded to invest £2,000 in A. V. Roe's fledgling aeroplane enterprise, but his lawyer refused to ratify the agreement; 1914–16, commanded HMS *Bullfinch* and HMS *Commonwealth* (Lt.-Cdr); 1917, Admiralty

War Staff; 1918, Asst COS, Gibraltar; 1919–26, Liberal MP, Central Hull, becoming the only English member to vote against Versailles treaty; 1922–5, president, UK pilots' association; 1926–31, Labour MP, Central Hull; 1934, succeeded to title; 1938–42, opposition Chief Whip, House of Lords.

Kidd, Benjamin, 1858–1916.   Social-Darwinist writer. 1877–94, clerk, Inland Revenue. 1893, published *Social Evolution*, with enormous success. Subsequently a freelance writer. Other works include *The principles of western civilisation*, published in 1902, and *Science of power*, published posthumously in 1918.

Kiggell, Lt.-General Sir Launcelot Edward, 1862–1954.   1909–13, Director of Staff Duties, War Office; 1913–14, Cmdt, Staff College, Camberley; 1914–15, Director of Home Defence, War Office; 1915–18, Chief of General Staff, British armies in France: 1917, Lt.-Gen.

King, Major William Albert de Courcy, 1874–1917.   1894, commissioned 2nd Lt., RE; 1897, Lt.; 1901–2, active service South Africa; 1904, Capt.; Apr. 1906, Chief Instructor and Acting Adjt, Balloon School; June 1916, awarded DSO. Killed in action 27 May 1917.

Lanchester, Frederick William, 1868–1946.   Engineer, automobile manufacturer and pioneering aerodynamicist: see chs 1 and 5.

Langley, Samuel Pierpoint, 1834–1906.   Astronomer by profession. 1887, appointed sec., Smithsonian Institution. Between 1892 and 1894 constructed six steam-driven aeroplane models, perversely designated 'aerodromes', and numbered from No.0 rather than 1. All were failures. 1895, rebuilt No.5 as a tandem-wing monoplane and this remained his standard configuration. May 1896, achieved 3,300 ft unmanned flight; Nov. 1896, model No.6 flew ¾ mile; Dec. 1898, US War Department, then involved in Spanish-American War, commissioned man-carrying prototype. Delays derived from lack of adequate engine. Eventually completed 1903, but untried pilot simply precipitated into the air by catapult-launch. It was a complete failure.

Latham, Hubert, 1883–1912.   Anglo-Frenchman, born Paris and educated Balliol College, Oxford, graduating 1904. Spent much of following year big game hunting, taking up ballooning on return to Europe. 1905, made channel-crossing with balloon constructor, Jacques Faures; 1908, witnessed Wilbur Wright's aeroplane flights at Le Mans. Subsequently became a director of Léon Levavasseur's Antoinette company. Made first flight at Châlons in March 1909, on Antoinette monoplane. Soon became renowned aviator. 5 June 1909, set new French duration record of 1 hr 7 mins 37 secs; 19 July 1909, made unsuccessful attempt to cross English Channel; Aug. 1909, at Rheims, set new world speed record for 100 km, taking 1 hr 28 mins 17 secs. Courting danger, he was killed while big game hunting in 1912.

Laycock, Brig.-General Sir Joseph Frederick, 1867–1952.   Educated Eton and

Oxford: country seat at Wiseton Hall, near Doncaster. As a Captain in the Notts Yeomanry he went to South Africa in 1899, seeing much active service and becoming ADC to General French, Cavalry Division. 1901, elected member of RAC and vice-president, Lincolnshire Automobile Club. Enthusiastic early motorist and sportsman, whose cars were prominent in the inter-city motor-races of the period. 1908, became chairman of the ENV engine company, in which capacity he acted as an aviation philanthropist to the British army. 1914–18, Brevet Lt.-Col., Notts Horse Artillery (Territorial); 1917, CMG; 1919, KCMG. 1940, Zone Commander, Home Guard, Notts.

Lawson, Henry John, 1852–1925. 1876, patented low-balanced bicycle, and 1879, rear-wheel chain-driven bicycle. Became manager of Tangent & Coventry Tricycle Co. and gradually developed reputation as wide-ranging company financier. Through his organisations, like the British Motor Syndicate Ltd (1895), subsequently British Motor Co. Ltd, and the Great Horseless Carriage Co. Ltd (1896), subsequently Motor Manufacturing Co. Ltd, he attempted to gain comprehensive patent acquisitions and monopolise the developing British motor industry. His patents failed to hold, however, and in 1904 he was sentenced to a year's imprisonment for conspiring to defraud investors.

Lebaudy, Paul, 1858–1937. Proprietor of sugar refinery, who with his brother, Pierre, and plant's chief engineer, Henri Julliot, initiated construction of semi-rigid airship in 1899. Nov. 1902, 80,000 cu. ft 'Lebaudy I' airship launched, powered by 40 h.p. Daimler engine.

Lee, Arthur Hamilton, 1st Viscount Lee of Fareham, 1868–1947. Educated Cheltenham and RMA, Woolwich. 1888, commissioned, RA; 1898–9, military attaché, US army (Cuba), and Washington; 1900–18, Conservative MP for Fareham; 1909, chairman, Parliamentary Aerial Defence Committee; 1915, Parliamentary Secretary, Ministry of Munitions; 1916, personal secretary to Lloyd George; 1917–18, Director-General, Food Production; 1919–21, President, Board of Agriculture; 1921–2, First Lord of Admiralty. 1920, presented Chequers to nation as residence for prime minister. 1918, baron; 1922, viscount.

Lee, Lt.-Colonel Henry Pincke, 1842–1911. 1863, commissioned, RE; c. 1880s, OC, RE balloon field detachment; 1882, placed on stand-by for service in Egypt; 1895, retired with rank of Lt.-Col.

Levavasseur, Léon, 1863–1922. Artist turned engineer. 1903, unsuccessfully constructed large, bird-form, twin-propeller monoplane, with curved dihedral on wings. Also designed its engine: first of celebrated Antoinette engines, named after daughter of Levavasseur's financier, Jules Gastambide. (Latter became head of resultant firm, Société Antoinette). 1904, Antoinette engines powered Levavasseur motor boats; 1906, firm's 50 h.p. engine largely perfected. Became mainstay of European aviation until 1909–10. 1908, built Gastambide-Mengin monoplane, leading (same year) to thin, elegant and practical

Antoinette monoplanes IV and V; July 1909, Antoinette IV – piloted by H. Latham – and Blériot XI participated in 'battle of the Channel'. By 1913 Antoinette had gone out of business, superseded by emergence of Gnôme rotary engine.

Lilienthal, Otto, 1848–96. Educated Potsdam technical school and Berlin technical academy. 1871–2, worked for Weber Machine Co., and 1872–80, Hoppe Engineering Works; 1880, founded his own factory specialising in light steam engines and marine signals. Simultaneously developed practical interest in gliding flight (biplane and monoplane). Crashed fatally 9 Aug. 1896, but his writings were widely disseminated.

Lloyd, Major Frederick Lindsay, 1866–1940. Educated Clifton College. 1885, commissioned Lt., RE; 1899–1900, active service South Africa (decorated); 1902, Maj.: Secretary, War Office Committee on Mechanical Transport; 1906, retired from service; 1909–29, clerk of course, Brooklands; 1914, rejoined army with rank of Colonel; 1920, CBE. Vice-chairman, RAC.

Longmore, Air Chief Marshal Sir Arthur Murray, 1885–1970. Born St Leonards, New South Wales. 1904, commissioned, RN; 1911, one of four naval officers selected to undergo flight instruction at Eastchurch; 1912, Instructor, CFS, then OC, Cromarty Air Station and Calshot Experimental Seaplane Station; 1914, OC, No.1 Sqn, RNAS; 1916, transferred to naval duties: present at Jutland (HMS *Tiger*, 1st Battle-Cruiser Sqn); 1917, Wing-Capt., RNAS, attached C-in-C, Mediterranean; 1918, Lt.-Col., RAF, OC, Adriatic Group; 1923, Group-Capt., Air HQ, Iraq; 1925–9, Director of Equipment, Air Ministry; 1929–33, Cmdt, RAF College, Cranwell; 1933–4, AOC, Inland Area; 1934–6, Coastal Area; 1936–8, Cmdt, Imperial Defence College; 1939, AOC-in-C, Training Command; 1940–1, Middle East; 1941–2, Inspector-Gen., RAF: Air Chief Marshal.

Loraine, Robert, 1876–1935. Actor and aviator, born New Brighton, Cheshire. 1900–1, served with Imperial Montgomeryshire Yeomanry, South Africa; 1905, played John Tanner in Shaw's 'Man and Superman' and, 1907, Shaw's 'Don Juan'. A graduate of Blériot's Pau school. As competition pilot participated on Farman biplane at Bournemouth meeting (July 1910) and second Blackpool meeting (Aug. 1910). Jules Védrines, subsequently one of the period's most successful competition pilots, acted as his mechanic. 11 Sept. 1910, made first – extremely hazardous – crossing of Irish Sea, and later same month participated in army manoeuvres on Salisbury Plain; 1911, became manager of Criterion Theatre, London; Aug. 1914, joined RFC. Awarded MC and DSO, and retired from service with rank of Lt.-Col. 1919, returned to the stage as 'Cyrano de Bergerac'; 1927, played Adolf in Strindberg's 'The Father'.

Low, Major Archibald Reith, 1878–1969. Born Aberdeen, son of a clergyman: a first cousin of Lord Reith of the BBC. Educated Watson's School, Edinburgh University and Clare College, Cambridge, studying Latin, Greek and mathematics. Succeeded Handley Page as chief designer with Johnson & Phillips, having previously been assistant designer. 1910, joined B&C as pilot-instructor: awarded RAeC certificate No.34, 22 Nov. 1910. 1911, recruited as chief designer, Vickers aviation department. Designed EFB1 before being dismissed in 1913. Served as Lt.-Cdr, RNAS, during war, including as Senior Technical Officer, Mediterranean Fleet. 1918, Maj., RAF, which unusual rank he retained. 1919, appointed scientific officer/librarian, Air Ministry. FRAeS.

Lowe, Thaddeus Sobieski Coulincourt, 1832–1913. American aeronaut, who organised and led Federal army balloon corps during US Civil War: see ch. 4.

McClean, Sir Francis Kennedy, 1876–1955. Educated Charterhouse and Clifton College. Independent means enabled him to pursue interests in astronomy, ballooning and aviation after retiring from Indian public works department in 1902. Significant aviation philanthropist. Jan. 1909, ordered Shorts biplane No.1; Mar. 1909, ordered Short-Wright No.3; 1910, provided RAeC and Shorts with new flying ground at Eastchurch. Took RAeC certificate No.21, 20 Sept. 1910. 1910–11, loaned Short-Sommer biplanes to Admiralty for instruction of officers, and made similar offer to London Balloon Coy, RE (T); 1911–12, ordered Shorts' first tractor-biplane, S36, and placed it at disposal of naval flying school, Eastchurch. Created a stir in August 1912 by piloting a Short-Sommer (S27 Type) biplane between upper and lower spans of Tower Bridge and then under all subsequent bridges to Westminster. 1913, a co-founder of Eastchurch-based British Wright Co.; Jan.–March 1914, led Nile expedition on 160 h.p. Short pusher float-biplane. Served in RNAS 1914–18 and RAF 1918–19. Chairman of RAeC 1923–4/ 1941–4. Knighted, 1926. FRAeS.

McClellan, Major-General George Brinton, 1826–85. 1848–51, Instructor in Military Engineering, West Point; July 1861, took command of Department of the Potomac; Nov. 1861, appointed Gen.-in-Chief, Union armies. Efficient administrator but hesitant commander. Ordered to take the offensive by President Lincoln, McClellan relinquished supreme command (March 1862), retaining command of Army of the Potomac. Lowe's balloon corps accompanied the army during the Peninsula campaign of March–August 1862. McClellan remained consistently over-cautious. Finally removed from command in November 1862. 1864, Democratic presidential candidate.

Macdonald, Major-General Sir James Ronald Leslie, 1862–1927. 1882, commissioned, RE; 1885, temp. OC, Balloon Equipment Store (covering for Templer, following dispatch of Suakin expedition); 1890, led Kabul River Railway Survey; 1891–4, Chief Engineer, Uganda Railway Survey, and sometime acting commissioner of the Protectorate; 1897–8, led British expedition from Mombasa in attempt to forestall French occupation of Fashoda; 1899–1900, Acting SBF; 1900, temp. OC, 4th Balloon Section, China: Director of Railways, China Expeditionary Force; 1903–4, commanded military escort accompanying

Col. F. Younghusband's controversial political mission to Tibet; 1905, Commander, Presidency Brigade, Calcutta.

Macdonald, Leslie Falconay, 1890–1913. 1910, became first pilot to secure aviator's certificate with either of the B&C flying schools (RAeC No.28, Brooklands, 15 Nov. 1910), becoming a B&C test pilot; Dec. 1910–May 1911, participated in B&C's Australian mission; 1911, recruited by Vickers as chief test pilot; Aug. 1912, Vickers' pilot during Larkhill military trials. Killed 13 Jan. 1913, when new Vickers-Low tractor-biplane he was piloting crashed into the Thames near Dartford.

Macdonogh, Lt.-General Sir George Mark Watson, 1865–1942. An officer of the Royal Engineers. Head of MO5. 1912, member of Flying Corps Committee's think-tank; 1916–18, DMI; 1917, knighted; 1918–22, Adjt-Gen. to the Forces; 1919, Lt.-Gen.; 1924, Col. Cmdt, RE; 1933–4, president, Federation of British Industries.

MacInnes, Brig.-General Duncan Sayre, 1870–1918. Born Hamilton, Ontario. 1891, commissioned 2nd Lt., RE; 1894, Lt.; 1895–6, active service Ashanti; 1899–1902, South Africa (SO to Col. Kekewich, OC, defence of Kimberley); 1902, Capt.; 1902–4, Asst Director of Works, South African Constabulary; 1905–8, successively DAQMG at Halifax, Nova Scotia, and Deputy Asst Adjt-Gen., Canadian Dominion Forces; 1909, Staff College; 1910, gazetted GSO 3, Directorate of Military Training, with responsibility for signal service and aviation; 1911, Maj.; 1912, member of Flying Corps Committee's think-tank; 1913, GSO 2, Staff College; 1914, active service France (retreat from Mons: wounded Nov.); 1915, GSO, War Office; 1916, Brig.-Gen., Director of Air Equipment, DMA. The inability of this department to maintain required procurement levels during the Somme battles led to the breakdown of his health, his removal from office and, some believe, scapegoating – not least by Trenchard himself. 1917, Chief Engineer, 42nd Division; 1918, Inspector of Mines. Killed in action 23 May 1918.

McKenna, Reginald, 1863–1943. 1907–8, President, Board of Education; 1908–11, First Lord of Admiralty; 1911–15, Home Secretary; 1915–16, Chancellor of Exchequer.

Mackenzie, Major Ronald Joseph Henry Louis, 1863–1930. 1882, commissioned, RE; 1885, a member of the balloon detachment led by Templer accompanying the Eastern Sudan (Suakin) Expeditionary Force; 1891, 2nd Miranzai expedition (mentioned in dispatches); 1899–1900, active service South Africa; Dec. 1900, Maj. Retired from service Aug. 1906.

Mackinder, Sir Halford John, 1861–1947. 1887–1905, first reader in geography, Oxford University. Held numerous other academic appointments including director of the LSE. 1919–20, British high commissioner, South Russia; 1926–41, chairman, Imperial Economic Committee. Author of *Britain and the British seas*, London 1902, and *The nations of the modern world*, London 1911.

Expounded Eurasian 'heartland' concept in *Democratic ideals and reality*, London 1919.

MacMunn, Lt.-General Sir George Fletcher, 1869–1952. 1888, commissioned, RA; 1892, active service Burma (wounded, DSO and clasp); 1893, Sima; 1897, Kohat Field Force; 1897–8, Tirah expedition; 1899–1902, South Africa (brevet Majority): acted as an observer with 3rd Balloon Section at Fourteen Streams; 1903, psc; 1914–19, Dardanelles and Mesopotamia: promoted Maj.-Gen.; 1919–20, C-in-C, Mesopotamia; 1920–4, QMG, India; 1925, retired; 1927–39, Col. Cmdt, RA; 1932–38, Comr, Royal Hospital, Chelsea; 1940–42, Home Guard. Prolific author of military and Indian history. 1917, KCB; 1919, KCSI.

Maitland, Air Commodore Edward Maitland, 1880–1921. Educated Haileybury and Trinity College, Cambridge. 1900, commissioned 2nd Lt., Essex Regt; 1902, Lt.; 1911, Capt.; May 1911–May 1912, seconded to Air Battalion, RE (OC, No.1 [Airship] Coy); 19 Sept. 1911, awarded RAeC airship pilot's certificate No.8; 1912–14, RFC; 1915, Maj., Essex Regt; 1915–18, Wing-Cdr, RNAS; Apr. 1918, Brig.-Gen., RAF (subsequently Air Commodore); 1919, senior airship officer, R34 Atlantic-crossing. Killed in R38 airship disaster, 1921.

Martin, Helmuth Paul, 1883– . Born London, educated Wellington College and Central Technical College, London. 1900, built motor engines to be fitted to bicycles; 1903–6, assistant engineer, Libby's Extract Factory (Fray Bentos), Uruguay; 1906, formed Trier & Martin, manufacturing automobile components; 1908, formed aircraft manufacturing partnership with G. H. Handasyde.

Masterman, Charles Frederick Gurney, 1874–1927. 1908, Under-Secretary, Local Govt Board; 1909, Under-Secretary of State, Home Office; 1912, Financial Secretary to Treasury; 1914–15, Chancellor, Duchy of Lancaster; 1914–18, Director, Wellington House (propaganda department). Author and journalist.

Maxim, Sir Hiram Stevens, 1840–1916. Born Sangerville, Maine, USA. 1878, chief engineer, US Electric Lighting Co.; 1881, exhibited inventions at Paris Exhibition and opened workshop in London. Designed automatic gun which fired ten shots a second and won immediate renown. 1884, Maxim Gun Co. formed; 1888, amalgamated with Nordenfeldt Co.; 1896, absorbed into Vickers, Son & Maxim. 1889–94, Maxim tested enormous steam-driven test-rig biplane, with limited success. Became naturalised Briton. 1901, knighted.

Methuen, Field Marshal Paul Sanford, 3rd Baron Methuen, 1845–1932. Member of 'Wolseley ring': served on latter's staff, Egypt (1882). 1884–5, member of Sir Charles Warren's Bechuanaland expedition, commanding mounted corps (Methuen's Horse); 1888–90, Deputy Adjt-Gen., South Africa; 1899–1900, GOC, 1st Division, South Africa; Feb. 1900, demoted by Roberts, and commanded mounted column; 1908–12, C-in-C, South Africa; 1911, Field Marshal.

Money, Major-General John, 1752–1817. Early British advocate of military aeronautics. Began military career in Norfolk Militia. 1770, Capt., 9th Regt of Foot; 1795, Col.; 1798, Maj.-Gen.; 1805, Lt.-Gen.; 1814, Gen. An aeronaut from as early as 1785. 1790, joined rebel party in Austrian Netherlands, receiving commission as Maj.-Gen. Subsequently published a *History of the campaign of 1792*. Published *A short treatise on the use of balloons and field observators in military operations* in 1803.

Montagu, John Douglas-Scott-, 2nd Baron Montagu of Beaulieu, 1866–1929. Noted motor transport advocate. Educated Eton and New College, Oxford. After leaving university gained engineering experience in workshops of London & South Western Railway Co. 1892, elected Conservative MP for New Forest; 1905, succeeded to title; 1909–19, member of the Road Board; 1915–19, advisor on mechanical transport to Indian government; 1916, appointed independent member and deputy chairman, Joint War Air Committee. Chief critic of RAeF/DMA in the Lords and an early advocate of an independent air service.

Montgolfier, Joseph-Michel, 1740–1810, and *Jacques-Etienne Montgolfier*, 1745–99. French aeronautical pioneers: sons of Annonay paper manufacturer. 1782, constructed spherical balloon, lifted by cauldron of paper heating and rarifying air within. In November 1783 a Montgolfier balloon, piloted by Francis Pilatre de Rozier (1757–85), sustained world's first manned flight.

Moore-Brabazon, John Theodore Cuthbert, 1st Baron Brabazon of Tara, 1884–1964. Aviation pioneer and politician: see ch. 2.

Morris, William Richard, 1st Viscount Nuffield, 1877–1963. Born Worcester, educated at the church school, Cowley, Oxford. 1892–1912, flourishing bicycle, motor cycle and automobile tradesman; 1912, formed WRM Motors Ltd and produced Morris-Oxford car; 1919, Morris Motors Ltd incorporated – by 1926, producing 50,000 cars a year; 1927, acquired Wolseley. Noted philanthropist. 1937, founded Nuffield College, Oxford. 1952, merged Morris with Austin Motor Co. to form British Motor Corp. 1929, baronet; 1934, baron; 1938, viscount.

Murray, General Sir Archibald James, 1860–1945. 1907–12, DMT; 1910, Maj.-Gen.; 1912, member of Seely's RFC Technical Sub-Committee; 1912–14, Inspector of Infantry; 1915, Deputy-Chief, then CIGS: Lt.-Gen.; 1916–17, GOC, Egypt; 1917–19, GOC-in-C, Aldershot.

Napier, Montague Stanley, 1870–1931. Born Lambeth, where second generation family business manufactured precision weighing machines. 1899, joined cycle-automobile salesman, S. F. Edge, in establishment of Motor Vehicle Co.; 1903–4, developed first commercially successful 'in-line' six-cylinder engine; 1912, Napier bought Edge out and floated new plc, D. Napier

lor, Cambridge University; 1909, president, Advisory Committee for Aeronautics.

Raynham, Frederick Phillip, 1892–1954. Emigrated from Australia. Secured RAeC certificate No.85, Avro Brooklands school, 9 May 1911. Appointed Avro pilot-instructor. Feb. 1912, recruited as chief instructor, Sopwith school, Brooklands; July 1912, test pilot, Howard Flanders Ltd. Piloted Flanders 100 h.p. ABC tractor-biplane during following month's military trials. Apr. 1913, rejoined Avro as chief test pilot. Also test flew for other manufacturers during war years, including B&C and, most notably, Martinsyde. 1919, unsuccessfully piloted Martinsyde 'Raymor' biplane on transatlantic attempt (crashed on take off); 1921, test pilot, Hawker. Subsequently co-director, Air Survey Co.

Renard, Lt.-Colonel Charles, 1847–1905. Born Damblain, and attended the École Polytechnique before joining the Corps of Engineers in 1868. 1870, active service Franco-Prussian War, awarded Legion of Honour. Secretary, post-war sub-committee on aerostation, which precipitated creation of Central Establishment for Military Aerostation in 1877. With fellow Engineer, Lt. Arthur Krebs, planned construction of navigable balloon. Aug. 1884, 66,000 cu. ft dirigible, 'La France', launched from Chalais-Meudon (powered by electric motor), but practicality limited by need for favourable meteorological conditions. 1898, appointed DMA. Committed suicide Apr. 1905.

Ridge, Theodore John, 1875–1911. Born Enfield. Served as a private, 34th (Middlesex) Imperial Yeomanry, Boer War. Oct. 1909, appointed Asst SBF. At time of his appointment held commission, London Balloon Coy, RE (T). Awarded aviator's certificate day before his death (RAeC No.119, B&C school, Larkhill), becoming with Lt. R. A. Cammell one of only two Englishmen to hold both airship and aviator certificates. Killed 18 Aug. 1911, after crashing the SE1 at Farnborough, having taken it up against advice. Coroner's verdict, death by misadventure.

Roberts, Field Marshal Frederick Sleigh, 1st earl of Kandahar, 1832–1914. Born Cawnpore, India. 1878–80, hero of 2nd Afghan War; 1885–93, C-in-C, India; 1892, baron; 1895, Field Marshal; 1895–9, C-in-C, Ireland; Dec. 1899, appointed Supreme Commander, South Africa, where one of his ADCs was the young duke of Westminster – subsequently an important patron of aviation. 1901–4, C-in-C, British army; 1905, president, National Service League; 1910–14, vice-president, RAeC. An influential supporter of early aviation.

Roe, Sir Alliott Verdon-, 1877–1958. Aviation pioneer, aircraft designer and manufacturer: see ch. 2.

Roe, Humphrey Verdon-, 1878–1949. Brother of A. V. Roe. Army officer turned businessman. Founder of the firm of A. V. Roe & Co.: see ch. 2.

Rolls, The Honourable Charles Stewart, 1877–1910. Celebrated motor car promoter and manufacturer, and pioneer aviator: see chs 2, 5 and 6.

Royce, Sir Frederick Henry, 1863–1933. 1884, founded (with Ernest Clare-mont) F. H. Royce & Co., a Manchester electrical firm; 1894, F. H. Royce & Co. reconstituted as limited company; 1904, produced first car; 1906, com-bined business with that of associated automobile sales agent, C. S. Rolls. Concentrated on celebrated Rolls-Royce 40/50 h.p. 'Silver Ghost'. 1908, works moved to Derby; 1915, Rolls-Royce produced first Eagle aero-engine; 1929, 1931, company engines won Schneider Trophy; 1930, baronet.

Ruck, Major-General Sir Richard Matthews, 1851–1935. 1902–4, Deputy DFW; 1904–8, DFW; 1908, Maj.-Gen., Administration, Eastern Command; 1914, Chief Engineer, Central Force; 1915, GOC, London Defences; 1915–16, Maj.-Gen., Administration, combined Central Force/Eastern Command. 1912–19, chairman of council, RAeS; 1917–19, vice-chairman, air inventions committee.

Sadler, James, 1751–1828. Son of Oxford pastrycook. First English aeronaut: maiden ascent, Oxford, October 1784. 1785, temporarily retired, following alighting accident. Went on to patent improvement in Watts's steam engine, with view to vehicle propulsion. 1812, made unsuccessful attempt to cross Irish Sea by balloon. Father of another noted aeronaut, Windham Sadler.

Samson, Air Commodore Charles Rumney, 1883–1931. Born Cheetham. 1898, midshipman, RN; 1903–4, HMS *Pomone*, Somaliland; 1904, Lt.; 1909–10, HMS *Philomel*, Persian Gulf; 1911, the senior of the four naval officers selected to undergo flying instruction at Eastchurch; Jan. 1912, made flight from platform on HMS *Africa*; Apr. 1912, OC, RFC (NW); Aug. 1914, OC, Naval Air Unit, Ostend; 1915, OC, No.3 Wing, Dardanelles; 1916–17, OC, aircraft carrier HMS *Ben-My-Chree* and escorts, Middle East; 1917–18, Great Yar-mouth Naval Air Station; 1918–19, Felixstowe Aircraft Group; 1919, CSO, HQ Coastal Area; 1920, OC, RAF Mediterranean; 1926, CSO, Middle East Command; 1927, led RAF bomber formation in Cairo to Cape trans-Africa flight. Retired 1929.

Santos-Dumont, Alberto, 1873–1932. Franco-Brazilian aeronaut, son of wealthy Brazilian coffee planter. 1901, awarded Deutsch de la Meurthe airship prize for flight encircling Eiffel Tower; 1906, constructed '14-bis' canard boxkite biplane (so designated because it had been flight-tested beneath S.-D. airship No.14); 13 Sept., at Bagatelle, made first free take-off, covering seven metres. Added new 50 h.p. Antoinette engine and, 30 Oct., hop-flew 60 metres. 12 Nov., flew 220 metres, to be credited with first powered flight in Europe. 1909, constructed successful small monoplane, 'Demoiselle'.

Schneider, Jacques, 1879–1928. Born near Paris, son of Charles-Eugene Schneider, proprietor of the Schneider armaments conglomeration, founded 1836. Trained as mining engineer. Became captivated by aviation following Wright/Voisin/Farman demonstration flights of 1908. Joined Aero Club de France 1910, and obtained certificate following March. Also qualified as balloon pilot, establishing French altitude record of 10,081 metres in 1913. His active aviationary career was curtailed by a 1910 hydroplane accident, which left him with multiple arm fractures; however, a particular interest in fast seaplanes led him to inaugurate £1,000 annual Schneider Trophy contest in Dec. 1912.

Scott-Moncrieff, Major-General Sir George Kenneth, 1855–1924. Educated Edinburgh Academy and RMA, Woolwich. 1878–80, active service 2nd Afghan War; 1893–8, Instructor, SME, Chatham; 1900–1, commanded RE, China Expeditionary Force; 1901, active service Waziristan expedition; 1909–11, Chief Engineer, Aldershot Command; 1911–18, DFW; 1912, member of Seely's RFC Technical Sub-Committee. Retired 1918. Father of C. K. Scott-Moncrieff, celebrated translator of Proust.

Seeley, Sir John Robert, 1834–95. 1863, professor of Latin, University College, London; 1869–95, Regius Professor of Modern History, Cambridge. Author of *The expansion of England*, London 1883.

Seely, John Edward Bernard, 1st Baron Mottistone, 1868–1947. Educated Harrow and Trinity College, Cambridge. Awarded DSO in Boer War: rose to rank of Col. Whilst in South Africa, elected Conservative MP for Isle of Wight. 1904, crossed floor of the House and subsequently re-elected Liberal MP for same constituency. 1908–11, Under-Secretary of State for the Colonies; Mar. 1911–June 1912, Under-Secretary of State at the War Office, serving under Haldane, who had been raised to the peerage. It was Seely's responsibility to answer questions in the Commons on behalf of the Secretary of State. In June 1912 he succeeded Haldane as Secretary of State for War. Forced to resign in March 1914 as a result of the 'Curragh Mutiny'. 1918, Deputy Minister of Munitions; 1919, Under-Secretary of State for Air; 1933, baron.

Short, Horace Leonard, 1872–1917.
Short, Albert Eustace, 1875–1932.
Short, Hugh Oswald, 1883–1969.
Aeronautical engineers and aircraft manufacturers: see ch. 2.

Siddeley, John Davenport, 1st Baron Kenilworth, 1866–1953. Automobile and aero-engine manufacturer: see ch. 1.

Sigrist, Frederick, 1880–1957. Sopwith's permanent assistant, originally engaged as a yacht/boat mechanic. 1912, works manager, Sopwith Aviation Company; 1920, appointed joint managing director of its successor, the Hawker Engineering Co. (subsequently Hawker Aircraft Ltd). Retired 1940.

Sikorsky, Igor Ivanovich, 1889–1972. Born Kiev, educated Kiev Polytechnic. Began experimenting with helicopters as early as 1908, but after witnessing biplane flights at Issy-les-Moulineaux the following year turned to development of aeroplanes. Worked in conjunction with Russian Baltic Railway Car Factory of Petrograd, a general transport-engineering group. Developed huge four-engine biplanes, culminating in 'The Grand' (1913) and 'Ilia Mourometz' (1914). Seventy-five of latter constructed for military purposes before 1917 revolution, which saw the execution of company's chairman. Sikorsky fled. 1919, moved to USA; 1923, founded Sikorsky Aero Engineering Corp; 1928, became naturalised US citizen.

Simms, Frederick Richard, 1863–1944. Automobile pioneer, founder of the Royal Automobile Club and an early advocate of a UK Aero Club: see ch. 2.

Singer, George, 1847–1909. Protégé of James Starley at the Coventry Sewing Machine Co. 1874, founded Singer Cycle Co.; 1891–3, Mayor of Coventry; 1901, acquired rights to Perks & Birch motor-wheel; 1905, launched into automobile production, acquiring licence to build Lea-Francis cars.

Sippe, Sydney Vincent, 1889– . Born London. Awarded RAeC certificate No.172, 9 Jan. 1912, after taking tests on an Avro Type D biplane at Brooklands. Hired by Cdr Schwann to pilot the Type D seaplane prototype at Barrow-in-Furness. Appointed chief pilot, Hanriot Aeroplane Co. Piloted 100 h.p. Gnôme Hanriot monoplane during 1st Aerial Derby (June 1912) and Larkhill military trials (Aug. 1912). During latter, Hanriot monoplane performed with conspicuous success, being denied victory only by judge's idiosyncratic system of assessment. Dec. 1912, appointed test pilot, B&C; Aug. 1914, commissioned Flt-Lt., RNAS; 21 Nov. 1914, participated in celebrated RNAS-Avro 504 raid on Friedrichshafen Zeppelin station. (Awarded DSO.) Subsequently promoted Flt-Cdr. 1919, OBE.

Sommer, Roger, 1877–1965. Son of felt manufacturer from Mouzon, in the Ardennes, educated at the École des Arts et Métiers, Paris. Noted cycling champion and automobile enthusiast. 7 Aug. 1909, established duration record of 2 hrs 27 mins 15 secs in a Farman biplane at Mourmelon and then piloted a Farman during first Rheims meeting. Subsequently initiated design and construction of derived models under his own name. The Sommer biplane was characterised by incorporation of upper-wing extensions (with hanging ailerons) and mono-tailplane section. In 1911 he produced a Blériot-derived monoplane.

Sopwith, Sir Thomas Octave Murdoch, 1888–1989. Pioneering aviator and aircraft manufacturer, becoming hugely successful industrialist: see ch. 3.

Spooner, Stanley, 1857–1940. Early member of the Aeronautical Society (founded 1866) and Aero Club (founded 1901). 1896, established *Automotor*, for which he wrote regular aviation column; 1909, founder-editor *Flight*.

Starley, John Kemp, 1855–1901. Nephew of James Starley, 1830–81, the 'father of the cycle industry'. Born Walthamstow, son of a market gardener. 1872, joined Coventry Machinists; 1878, initiated his own bicycle manufacturing company; 1884–5, produced low-balanced 'Rover' bicycle – the pattern for all modern safety bicycles; 1889, J. K. Starley & Co. converted to limited liability company; 1896, at peak of 'bicycle boom', firm floated as a public company (renamed the Rover Cycle Co.), with Starley remaining managing director. However, slump following boom years induced excessive caution and impeded automobile diversification. Rover only initiated car manufacture in 1904, following Starley's death.

Stewart-Murray, Brig.-General John George, 8th duke of Atholl and marquess of Tullibardine, 1871–1942. 1890–2, Lt., 3rd Btn Black Watch; 1892, Lt., Royal Horse Guards; 1898, SO to Col. Birdwood, Nile expedition; 1899–1902, Lt.-Col., 1st/2nd Scottish Horse; 1907–8, placed Blair Atholl estate at Balloon Factory's disposal for Dunne aircraft trials; 1909, a sponsor of the Blair Atholl Syndicate; 1914–18, Cmdt, Scottish Horse (Brig.-Gen.); 1910–17, Conservative MP for West Perthshire; 1913–20, chairman, and 1921–42, president, RAeC.

Stopes, Marie Charlotte Carmichael, 1880–1958. Educated University College, London. 1904, PhD, Munich; 1905, DSc., London; 1909–20, lecturer in palaeobotany, Manchester and London; 1918, married H. V. Roe and with him founded (1921) mothers' clinic for birth control.

Sueter, Rear-Admiral Sir Murray Fraser, 1872–1960. Born Alverstoke, Gosport. 1886, entered RN via *Britannia* cadetship; 1902–3, pioneered, with Capt. R. H. Bacon, introduction of submarines within the service; 1904, appointed Asst DNO; 1909, Capt.; 1910, OC, RN tender, HMS *Hermione*, and designated Inspecting Capt. of Airships. Branch disbanded Jan. 1912, following 'Mayfly' disaster. 1912–15, Director, Admiralty Air Department. Largely responsible for creation of RNAS. 1914, Commodore 2nd Class; 1915, Commodore 1st Class; 1915–17, Superintendent of Aircraft Construction. Member of Advisory Committee for Aeronautics, 1909–17, and Joint War Air Committee, 1916–17. 1920, retired with rank of Rear-Admiral; 1921–45, Conservative MP for Hertford.

Swann, Air Vice Marshal Sir Oliver, 1878–1948. (He anglicised his surname, originally Schwann, in April 1917.) Born Wimbledon, entered RN via *Britannia* cadetship. Made Cdr 1910, and selected as assistant to M. F. Sueter on construction of RN Airship No.1 (The Mayfly) at Barrow-in-Furness. 1911–12, purchased Avro Type D biplane and formed a syndicate of naval officers to finance seaplane trials; 1912, Asst-Director, Admiralty Air Department. With Sueter, he was instrumental in creation of RNAS. 1914–18, OC, seaplane carrier HMS *Campania*; 1918, transferred to RAF: AOC, Mediterranean; 1922–3, Member for Personnel, Air Council; 1923–6, AOC, Middle East. Retired 1929. 1940–3, Air Liaison Officer, North Midland Region.

Sykes, Major-General Sir Frederick Hugh, 1877–1954. Born Croydon. 1899–1901, served with Imperial Yeomanry Scouts, South Africa; 1901, commissioned, 15th Hussars; 1905, Intelligence Dept, Army HQ, Simla; 1909,

Staff College; 1911, SO, War Office; 20 June 1911, awarded RAeC certificate No.95, B&C's Brooklands school; 1912–14, OC, RFC (MW); 1914–15, COS to GOC, RFC (David Henderson); 1915, OC, RNAS, Eastern Mediterranean; 1916, Asst Adjt-Gen., War Office; 1917, Deputy-Director of Organisation, War Office; 1918–19, Maj.-Gen., Chief of Air Staff; 1919–22, Controller-General of Civil Aviation; 1922–8, Conservative MP for Hallam (Sheffield); 1928–33, Governor of Bombay; 1940–5, Conservative MP for Nottingham (Central).

Templer, Colonel James Lethbridge Brooke, 1846–1924. The dominant figure in British military aeronautics from the 1870s to 1906, when he was replaced as SBF: see ch. 4.

Thomas, George Holt, 1870–1929. Aviation publicist, aircraft manufacturer and post-war air transport pioneer: see ch. 3.

Thompson, Sylvanus Phillips, 1851–1916. Educated Bootham School, York, the Flounders' Institute, Pontefract, and London University (BA/BSc/DSc). 1876, appointed lecturer in physics, Bristol University, becoming professor two years later; 1885–1916, principal, and professor of applied physics and electrical engineering, City and Guilds Technical College, Finsbury: tutor to Handley Page and R. Fairey. 1891, FRS; 1901, president of the Physical Society. Pioneer of applied electricity, prolific author and Quaker minister.

Thurstan, Farnall, –1922. Sir George White's nephew by marriage. 1910–11, led B&C aviation mission to India. Subsequently saw war service with the RNAS, being placed in charge of equipment and supplies in Paris.

Trenchard, Marshal of the Royal Air Force Hugh Montague, 1st Viscount Trenchard, 1873–1956. 1893, commissioned, Royal Scots Fusiliers; 31 July 1912, obtained RAeC certificate No.270, Sopwith school, Brooklands: joined staff of CFS; 1915–18, GOC, RFC (France), and latterly Cdr, Independent Bombing Force. Chief of Air Staff, Jan.–Apr. 1918 and 1919–29. 1927, first Marshal of RAF; 1931–5, Chief Commissioner, Metropolitan Police. 1930, baron; 1936, viscount.

Trollope, Lt.-Colonel Francis Charles, 1858–1913. 1877, commissioned 2nd Lt., Grenadier Guards; from 1883, attached to Royal Engineers' Balloon Factory; 1884–5, accompanied Balloon detachment, Bechuanaland; 1889, Capt.; 1894, Maj.; 1900, Acting-Superintendent, Balloon Factory (Sept. 1900, British delegate, Paris International Aeronautical Congress); 1901, a founder member, RAeC; 1902, Lt.-Col.: joined RAeS, serving on its council for many years and becoming a vice-president.

Turner, Rowley, 1840–1917. Paris agent of the Coventry Sewing Machine Company: see ch. 1.

Voisin, Gabriel, 1880–1973. Architectural student turned engineer from

Lyons. Worked in association with his brother, Charles Voisin (1882–1912). 1905, designed and built in co-operation with, respectively, Archdeacon and Blériot, two Wright-cum-'Hargrave-boxkite'-derived float-gliders, each incorporating a forward-elevator and stabilising boxkite tailplane, but no method of flight control. Both aircraft were towed off river Seine by Antoinette-powered motor boat and the Voisin-Archdeacon (piloted by Gabriel Voisin) flew 300 metres on 18 June 1905. Later same year Voisin established aircraft factory at Billancourt. The Voisin-Archdeacon became the prototype European biplane of the primitive period. 1907, Voisin standardised stable pusher-biplane configuration, building a 50 h.p. Antoinette Voisin-biplane in association with Léon Delagrange and, more successfully, H. Farman. Through latter's modifications there emerged a practical aeroplane. The original Voisin-biplane was outmoded by 1910.

Volkert, George Rudolph, 1892–1978. 1912, recruited straight from Northampton Institute as chief designer, Handley Page Ltd. Closely involved in design and subsequent production of Handley Page wartime bomber series (Types 0/100, 0/400 and V/1500). Acted as chief designer from 1912–21, 1924–31 and 1935–48, responsible for, amongst other aircraft, the HP42 and the Halifax bomber of World War II.

Ward, Colonel Bernard Rowland, 1863–1933. Educated Winchester and RMA, Woolwich. Early service career curtailed by ill-health. In 1887 he matriculated at Balliol College, Oxford, with a view to a teaching career, but rejoined corps. 1889, OC, Balloon detachment, during the period leading up to establishment of permanent section in 1890; 1890–5, Asst Chief Engineer, Madras District, India; 1895, OC, Balloon Section, RE (compiling first official instruction manual); 1897–1903, Instructor, RMA; 1904–7, attached, Royal Canadian Engineers; 1906, Lt.-Col.; 1907–11, OC, Depot Battalion, Chatham; 1911, Col.; 1912–17, OC, RE Records; 1917, Chief Engineer, London District. Author and historian. Retired 1919.

Warren, General Sir Charles, 1840–1927. 1876–7/1879–80, Special Commissioner, Griqualand West (South Africa); 1877–8, OC, Diamond Fields Horse, 9th Kaffir War; 1880–4, Chief Surveying Instructor, SME, Chatham; 1882, Special Service Officer attached to Admiralty, Egypt; 1884–5, GOC, Bechuanaland Expeditionary Force; 1886–8, Chief Commissioner, Metropolitan Police; 1889–94, GOC, Singapore; 1899–1900, GOC, 5th Division, South Africa; 1905, Col. Cmdt, RE.

Watson, Colonel Sir Charles Moore, 1844–1916. Born Dublin, educated Trinity College, Dublin, and RMA, Woolwich. 1866, commissioned, RE; 1873, replaced Frederick Beaumont on RE Committee, Balloon Sub-Committee; 1878, instrumental in creation of Balloon Equipment Store, Woolwich: ADC to Sir L. Simmons, War Office; 1880, Staff College; 1882, Intelligence Service, Egypt (brevet Majority); 1883, Surveyor-General, Egypt; 1886, Governor,

Suakin and Red Sea Littoral; 1888–9, OC, balloon field detachment, RE; 1896, Col. Retired 1902.

Weir, William Douglas, 1st Viscount Weir of Eastwood, 1877–1959. Born Glasgow, eldest child of James Weir, engineer, descendant of poet Robert Burns. Educated Glasgow High School (left at sixteen). Joined father's firm, becoming managing director (1902) and chairman (1912). July 1915, became Director of Munitions, Scotland; Feb. 1917, made Controller of Aeronautical Supplies, Ministry of Munitions, and member of the Air Board; Dec. 1917, Director General of Aircraft Production; Apr. 1918, succeeded Rothermere as Secretary of State for Air. Advocate of air power and advisor to government on rearmament in the 1930s. 1939, Director General of Explosives, Ministry of Supply, and deputy chairman of Supply Council. Director of ICI, Shell and International Nickel. 1918, baron; 1938, viscount.

Weiss, José, 1859–1919. Artist and gliding-flight pioneer. Early associate of Handley Page: see ch. 3.

Wells, Herbert George, 1866–1946. Son of impecunious tradesman from Bromley, Kent. Apprenticed to drapery shop. Obtained studentship, Normal School of Science, South Kensington. Taught at Henley House, Kilburn (Northcliffe's old school). 1890, tutor, University Correspondence College; 1895, published *The time machine* and soon developed into prolific and popular novelist, with scientific-socialistic prophetic bent.

Westland, Lt.-Colonel Francis Campbell, 1884–1941. 1903, commissioned 2nd Lt., RE. From SME joined Balloon Coy/School, Aldershot. 1907–8, participated in Dunne Blair Atholl trials; 1908, 36th Coy, Sierra Leone; 1912, 9th Field Coy; 1916, OC, 218th Field Coy, 32nd Division; 1924, OC, 55th Field Coy; 1925, Asst. CRE, Northumbria; 1926, DCRE, Salisbury; 1929, CRE, Lahore; 1930–3, SORE, Poona.

White, Sir George, 1854–1916. Transport promoter, tramway entrepreneur and aeroplane manufacturer. Founder of the British & Colonial/Bristol Aeroplane Company: see ch. 3.

Wildman-Lushington, Captain George ('Gilbert') Verron, 1887–1913. An officer of the Royal Marine Artillery. One of the original four naval volunteers selected for flying instruction at Eastchurch in March 1911. Dropped out because of ill-health and rejoined the group at a later date. Awarded RAeC certificate No.290, CFS, 17 Sept. 1912. The First Lord's (Churchill's) favourite flying instructor, he was killed in a landing accident on a Short biplane on 2 Dec. 1913, the day after instructing Churchill on the same machine.

Wilson, Field Marshal Sir Henry Hughes, 1864–1922. 1903–4, member of War Office Committee on Military Ballooning; 1910–14, DMO; 1915, Chief Liaison Officer, French HQ; 1917, C-in-C, Eastern Command; 1918–22, CIGS; 1919, Field Marshal.

Wilson, Walter Gordon, 1874–1957. Born Blackrock, Co. Dublin. After naval cadetship, entered King's College, Cambridge (1894). 1896, elected to honorary exhibition. Made acquaintance of C. S. Rolls. 1897, co-founder, Wilson & Pilcher Ltd motor company; 1901, first Wilson-Pilcher car produced, incorporating epicyclic gears; 1904, company acquired by Armstrong, Whitworth & Co., Wilson becoming firm's automobile designer; 1908–14, worked for J. & E. Hall of Dartford, designing Hallford lorry; 1914, joined RNAS armoured car section. Soon embroiled in creation of offensive armoured vehicles, attached for this purpose to RN Armd Car Sqn No.20. With (Sir) William Tritton, played crucial role in the tank's design and development. 1916, transferred to army, becoming Maj., Heavy Branch, Machine-Gun Corps (renamed Tank Corps, 1917). Chief of design, mech. warfare dept, War Office, until war's conclusion. Subsequently invented the Wilson self-changing gearbox and founded the Self-Changing Gears Co., of Coventry.

Windham, Sir Walter George, 1868–1942. 1884, joined HM's Indian Marine; 1901–9, King's Messenger (foreign service). Early motoring enthusiast, manufacturing the Windham detachable body from Clapham factory, and initiating London motor cab service. 1901, Maj., Motor Volunteer Corps; 1908, founder and president, Aeroplane Club of Great Britain and Ireland – a reaction to the prevailing ethos of the official Aero Club; 1909, helped organise 1st Rheims meeting and controller, Doncaster aviation meeting; 1911, inaugurated trial aeroplane mail services, United Provinces Exhibition, Allahabad (Feb.), and Hendon-Windsor (Sept.); 1914–18, served with RN; 1918, OC, Buncrana (Londonderry), supervising entry of US troops into Europe; 1923, knighted.

Wolseley, Frederick York, 1837–99. Born Co. Dublin, younger brother of Garnet Wolseley. Emigrated from Ireland to Australia as a young man and in 1877 founded the Wolseley Sheep Shearing Machine Company. 1889, founded English sub-division, subsequently recruiting Herbert Austin to run it. Resigned 1894.

Wolseley, Field Marshal Garnet Joseph, 1st Viscount Wolseley, 1833–1913. Born Co. Dublin. 1873, GOC, Ashanti expedition, famous for assemblage of 'Wolseley ring' of selected officers; 1879, sent to retrieve situation in Zululand: High Commissioner, Transvaal; 1880, QMG, War Office; 1882, Adjt-Gen.: crushed rebellion of Egyptian army, Tel-el-Kebir; 1884–5, GOC, Khartoum relief expedition; 1890–5, C-in-C, Ireland; 1895–9, C-in-C, British army. Celebrated for his reforms.

Wood, Major-General Sir Elliott, 1844–1931. 1864, commissioned, RE; 1878, Maj.; 1882, active service Egypt, including Tel-el-Kebir; 1884–5, active service Sudan; 1888, Col.; 1889–94, AAG, RE Headquarters; 1894–99, CRE, Malta; 1899, CRE, Aldershot; 1899–1902, Engineer-in-Chief, South Africa: promoted Maj.-Gen.; 1902–5, GOC RE, 1st Army Corps. Retired 1906.

Wood, Field Marshal Sir Henry Evelyn, 1838–1919. 1873, member of Wolseley's 'Ashanti ring'; 1879, KCB for services in Zulu War, in which he commanded a column; 1881, Royal Commissioner for settlement of Transvaal; 1882, accompanied Wolseley to Egypt (first British Sirdar, Egyptian army); 1886, appointed to Eastern Command (UK); 1889, Aldershot Command: supported claims of balloon detachment, leading to establishment of permanent Balloon Section in 1890; 1893, QMG, War Office; 1897, Adjt-Gen.; 1903, Field Marshal.

Wood, Major Herbert Frederick, 1883–1919. Born Rawal Pindi, India, and commissioned with the 9th Lancers in 1901 (not the 12th Lancers, as is generally reported). Retired from the service in 1911, becoming manager of Vickers's aviation department: see ch. 3.

Wright, Howard Theophilus, 1867–1944. Born Dudley, served engineering apprenticeship with Joseph Wright & Co. (his father's firm), which made boilers and pit-head gear. 1889, boiler division acquired by Hiram Maxim; 1905, joined his brothers (Warwick and Walter) in formation of Howard T. Wright Brothers Ltd, to develop gas turbines and retail automobiles; 1906, they founded Warwick Wright Ltd to further automobile interests; 1908–12, Howard Wright formed partnership with William Oke Manning (b. 1877: educated St Paul's School and via Callenders Cables Co.). Initially Manning led design process, Wright construction, but skills increasingly synthesised. 1910, produced Howard Wright biplane; 1911, firm acquired by Coventry Ordnance Works. Howard Wright and W. O. Manning designed company's 1912 military trials biplanes. Nov. 1912, Wright left to become manager and chief designer of J. Samuel White & Co.'s aviation department.

Wright, Wilbur, 1867–1912, and *Orville Wright*, 1871–1948. Self-taught inventors of the aeroplane. Made first controlled heavier-than-air powered flight on a 12 h.p. biplane at Kitty Hawk, North Carolina, 17 Dec. 1903. Petrol engine drove two oppositely rotated pusher-propellers. Flight controlled by means of rudder and wing-warping mechanism, enabling airframe's centre of gravity to remain constant. Sept. 1904, flew first circular flight; Oct. 1905, flew twenty-four mile circular flight; 1908, Wilbur travelled to France; Aug. 1908, initiated public demonstration flights, Le Mans; Jan. 1909, moved to Pau, to teach first pupils – Comte Charles de Lambert, Paul Tissandier, Capt. Lucas de Girardville – re. French Wright Co. Brothers then became enmeshed in patent litigation. Wilbur died of typhoid fever, May 1912.

Wyness-Stuart, Lieutenant Athole, 1882–1912. 1909, commissioned, RFA reserve; Aug. 1912, appointed to RFC: killed on 6 Sept., when acting as passenger on board the Capt. Hamilton-piloted Deperdussin monoplane which crashed on manoeuvres.

Yorke, Charles Alexander, 8th earl of Hardwicke and Viscount Royston, 1869–1936. Mining engineer. 1901, Lt., Motor Volunteer Corps; 1906, joined

Oxford, Nuffield College
Mottistone papers (J. E. B. Seely)

Weybridge, Brooklands Museum
Hilda Hewlett, unpublished memoir
miscellaneous papers

Yeovilton, Fleet Air Arm Museum
miscellaneous papers

Government papers

Public Record Office, Kew

Air Ministry Records
AIR I

Admiralty Records
ADM 116

Board of Trade Records
BT 31

Cabinet Office Records
CAB 2
CAB 16
CAB 21
CAB 38

Foreign Office Records
FO 368
FO 412

War Office Records
WO 32
WO 105
WO 108
WO 163

Serial publications

The Aero, ed. C. G. Grey, 1909–11
Aerocraft, ed. N. Pemberton Billing, 1909–10
Aeronautics
The Aeroplane, ed. C. G. Grey, 1911–39
Automobile Club Journal

Automotor Journal
Flight
Journal of the (Royal) Aeronautical Society
Journal of the Royal Artillery
Royal Engineers Journal

Contemporary accounts, articles and memoirs

'The Aeroplane' directory: who's who in British aviation, London 1949

Ardagh, Lady S., *The life of Sir John Ardagh*, London 1909

The army list, London 1860–1914

Atholl, Katharine, duchess of, *Working partnership*, London 1958

Bacon, R., *From 1900 onward*, London 1940

Baden-Powell, B. F. S., 'Military ballooning', *Journal of the Royal United Services Institution* xxvii (1884), 735–56

——— 'Kites: their theory and practice', *Journal of the Society of Arts* xlvi (1898), 359–70

——— 'War kites', *Journal of the Aeronautical Society* iii (1899), 1–6

——— 'The war balloon in South Africa', *Journal of the Aeronautical Society* vi (1902), 14–15

Bannerman, Sir A., 'The difficulty of aerial attack', *Journal of the Royal United Services Institution* liii (1909), 638–45

——— 'Some problems of aviation in war', *Army Review* i (1911), 123–4

——— 'Aeronautics and the army', *Army Review* i (1911), 332–8

——— 'Aircraft for use in war', *Journal of the Royal Artillery* xxxix (1912), 179–85

Beaumont, Capt. F., 'On balloon reconnaissances as practised by the American army', *Professional Papers of the Royal Engineers* xii (1863), 94–103

Billing, N. P., *P-B, the story of his life*, London 1917

Brabazon of Tara, Lord, *The Brabazon story*, London 1956

Brett, M. V. (ed.), *Journals and letters of Reginald Viscount Esher*, London 1934–8

Brewer, G., 'Wilbur Wright', *Journal of the Aeronautical Society* xvi (1912), 148–53

——— 'The genesis of the flying industry: reminiscences of Short Brothers' (1938), RAeS archive

——— *Fifty years of flying*, London 1946

——— and P. Y. Alexander, *Aeronautics: an abridgement of aeronautical specifications filed at the Patent Office 1815–91*, London 1891

Brock, M. and E. Brock (eds), *H. H. Asquith: letters to Venetia Stanley*, Oxford 1982

Burnaby, F. G., *A ride across the channel and other adventures in the air*, London 1882

Burney, Sir C. D., *The world, the air and the future*, London 1929

Busk, M., *E. T. Busk: a pioneer in flight*, London 1925

Butler, F. Hedges, *5,000 miles in a balloon*, London 1907
────── *Fifty years of travel by land, water and air*, London 1920
Cammell, Lt. R. A., 'Aeroplanes with cavalry', *Cavalry Journal* vi (1911), 197–9
Churchill, R. S. (ed.), *W. S. Churchill: companion* vol. ii, London 1969
Churchill, W. S., *London to Ladysmith via Pretoria*, London 1900
────── *Ian Hamilton's march*, London 1900
Clark, A. (ed.), *A good innings: the private papers of Viscount Lee of Fareham*, London 1974
Cormack, A. (ed.), 'No.1 Balloon Section, Royal Engineers, in the Boer War', contemporary diary, repr. *Journal of the Society for Army Historical Research* lxviii, lxix (1990–1)
Coventry up-to-date (trade review), Coventry 1896
Coxwell, H., *Balloons for warfare: a dialogue between an aeronaut and a general*, Tottenham 1854
────── *My life and balloon experiences: with a supplementary chapter on military ballooning*, London 1887
de Havilland, G., *Sky fever: the autobiography of Sir Geoffrey de Havilland*, London 1961
du Cros, Sir A., *Wheels of fortune*, London 1938
Dunlop, J. B., *The history of the pneumatic tyre*, Dublin 1924
Edwards, R. F. (ed.), *Roll of officers of the Corps of Royal Engineers: 1660–1898*, Chatham 1898
Egerton, M., 'A pioneer aviator from Cheshire: the logbook of Maurice Egerton', *Cheshire History* xxv–xviii (1990–1)
Elsdale, Maj. H., 'Military ballooning', *Minutes of Proceedings of the Royal Artillery Institution* xvii (1890), 41–57
Farman, D. and H. Farman, *The aviator's companion*, London 1910
Fisher, H. A. L., *An unfinished autobiography*, London 1940
Fokker, A. and B. Gould, *The flying Dutchman: the life of Anthony Fokker*, London 1931
Fox, Capt. A. G., 'A few notes on the employment of aeroplanes in warfare', *Royal Engineers Journal* xvii (1913), 333–6
Glaisher, J., *Travels in the air*, London 1871
Green, F. M., 'The first ten years', *Journal of the Royal Aeronautical Society* lxx (1966), 346
Grover, Lt. G. E., 'On the uses of balloons in military operations', *Professional Papers of the Royal Engineers* xii (1863), 71–86
Haldane, R. B., *An autobiography*, London 1928
Hamel, G. and C. Turner, *Flying: some practical experiences*, London 1914
Hamilton, E., *Captain Patrick Hamilton: soldier and aviator*, London 1912
Hankey, Lord, *The supreme command 1914–1918*, 2 vols, London 1961
Harmsworth, Alfred C. and others, *Motors and motor-driving*, London 1902
Hart's army list, London 1900–14

Holls, G. F. W., *The Peace Conference at the Hague and its bearings on international law*, New York 1900

Hubbard, T. O'B. and J. H. Ledeboer (eds), *The aeronautical classics*, London 1910–11

Jones, H. B., 'Military ballooning', *Journal of the Royal United Services Institution* xxxvi (1892), 261–79

—————— 'Note on a performance of the "Bristol" war balloon during the South African campaign', *Journal of the Aeronautical Society* vi (1902), 65

Jourdain, P. R., *Aviation in France in 1908* (annual report, Smithsonian Institution), Washington, DC 1909

Joynson-Hicks, W., *The command of the air or prophecies fulfilled: being speeches delivered in the House of Commons*, London 1916

Kane's list of officers of the Royal Regiment of Artillery, Sheffield 1914

King, Capt. W. A. de C., 'Aerial reconnaissance: its possible effect on strategy and tactics', *Royal Engineers Journal* xviii (1913), 147–54

Lanchester, F. W., *Aircraft in warfare: the dawn of the fourth arm*, London 1916

Longmore, Sir A., *From sea to sky, 1910–1945*, London 1946

Macmillan, N. (ed.), *Sir Sefton Brancker*, London 1935

Masterman, C., *The condition of England*, London 1909

Maxim, Sir H., *Artificial and natural flight*, London 1908

—————— *My life*, London 1915

Macdonald, D., *How we kept the flag flying: the story of the siege of Ladysmith*, London 1900

McFarland, M. W. (ed.), *The papers of Wilbur and Orville Wright*, New York 1953

McInnes, I. and J. V. Webb, *A contemptible little Flying Corps: being a definitive roll of those warrant officers, NCOs and airmen who served in the RFC prior to the outbreak of the First World War*, London 1991

Merriam, F. Warren, *First through the clouds: the autobiography of a boxkite pioneer*, London 1954

Möedebeck, Maj. W. L., 'The development of aerial navigation in Germany', *Journal of the Aeronautical Society* vi (1902), 24–8

Montagu of Beaulieu, Lord, 'Aerial machines and war', Aldershot 1910

—————— and B. F. S. Baden-Powell, *A short history of balloons and flying machines*, Edinburgh 1907

Nevinson, H. W., *Ladysmith: the diary of a siege*, London 1900

O'Gorman, M., *Problems relating to aircraft*, London 1911

Roe, A. V., 'Trials, troubles and triplanes', *The Aero* vi (1912), 95–8

—————— *The world of wings and things*, London 1939

Roskill, S. (ed.), *Documents relating to the Naval Air Service*, London 1969

Samson, C. R., *Fights and flights*, London 1930

Scott, J. B. (ed.), *The Hague conventions and declarations of 1899 and 1907*, New York 1915

Seely, J. E. B., *Adventure*, London 1930

Short Brothers, *Maps and rate book entries for the Battersea railway arches rented by Short Brothers, 1906–11*, RAeS archive n.d.

Sikorsky, I., *The winged S: an autobiography*, New York 1938

Sopwith, Sir T. O. M., 'My first ten years in aviation', *Journal of the Royal Aeronautical Society* lxv (1961), 236–51

Stone, Col. F. G., 'The Rheims aviation week and its value from a military point of view', *Journal of the Royal Artillery* xxxvi (1909), 353–64

Sueter, Rear-Admiral Sir M. F., *Airmen or Noahs?: fair play for our airmen; the great 'Neon' air myth exposed*, London 1928

Sykes, Sir F., *From many angles: an autobiography*, London 1942

Templer, J. L. B., 'Military balloons', *Journal of the Royal United Services Institution* xxiii (1879), 173–84

———— 'British war balloon operations in South Africa', *Journal of the Aeronautical Society* v (1901), 54–6

Unwin, N. H. F., *Facsimile of Geoffrey de Havilland's log book: Royal Aircraft Factory 1911–12*, RAE 1971

Vetch, Col. R. H., *Life, letters and diaries of Lieutenant-General Sir Gerald Graham*, Edinburgh 1901

Voisin, G., *Men, women and 10,000 kites*, London 1963

War Office, *Military aeroplane competition 1912: report of judges committee*, London 1912

———— *Report of departmental committee on accidents to monoplanes*, London 1912

Watson, Col. C. M., 'Military ballooning in the British army', *Professional Papers of the Royal Engineers* xxviii (1902), 39–59

Weiss, J., *Gliding and soaring flight*, London 1922

Wells, H. G., *Experiment in autobiography*, London 1934

Where to buy at Coventry: an illustrated trades' review, Coventry 1889

Who was who, London 1901–1991

Williams, E. E., *The foreigner in the farmyard*, London 1897

Windham, Sir W. G., *Waves, wheels, wings: an autobiography*, London 1943

Wood, Field Marshal Sir E., *From midshipman to field marshal*, London 1906

Official histories

Brown, Brig.-Gen. W. Baker, *History of the Corps of Royal Engineers*, iv, Chatham 1952

Colville, H. E., *History of the Sudan campaign: compiled in the Intelligence Division of the War Office*, 2 vols, London 1889

Girouard, Lt.-Col. Sir P., *History of the railways during the war in South Africa*, Chatham 1904

Headlam, Sir J., *History of the Royal Artillery 1860–1914*, 3 vols, Woolwich 1931–40

History of the Ministry of Munitions, 12 vols, London 1920–4

Maurice, Sir F., *History of the war in South Africa*, i–ii, London 1906–7

Porter, Maj.-Gen. W., *History of the Corps of Royal Engineers*, ii, London 1889

Raleigh, Sir W., *The war in the air: being the story of the part played in the Great War by the Royal Air Force*, i, Oxford 1922

Sandes, E. W. C., *The Royal Engineers in Egypt and the Sudan*, Chatham 1937

Victoria county history, VIII: *Warwickshire/Coventry*, London 1969

Waller, Col. S., 'History of the Royal Engineer operations in South Africa 1899–1902', unpubl. manuscript, 1904

War in South Africa: prepared in the historical section of the General Staff, Berlin, trans. Col. W. H. H. Waters, Berlin 1904

Watson, Sir C. M., *History of the Corps of Royal Engineers*, iii, Chatham 1915

Secondary sources

Aero Club of America, *Navigating the air*, New York 1907

Aldcroft, D. H., 'The entrepreneur and the British economy, 1870–1914', *Economic History Review* xvii (1964), 113–34

——— (ed.), *The development of British industry and foreign competition, 1875–1914*, London 1968

——— and H. W. Richardson, *The British economy 1870–1939*, London 1969

Althuser, J., *Pierre Michaux et ses fils*, trans. D. Roberts, Kenilworth 1988

Amery, L. S. (ed.), *The Times history of the war in South Africa*, 7 vols, London 1900–9

Andrew, C., *Her Majesty's Secret Service: the making of the British intelligence community*, New York 1986

Andrews, C. F. and E. B. Morgan, *Vickers aircraft since 1908*, London 1988

Andrews, P. and E. Brunner, *The life of Lord Nuffield: a study in enterprise and benevolence*, Oxford 1955

Armstrong, A. C., *Bouverie Street to Bowling Green Lane: fifty-five years of specialised publishing*, London 1946

Atkinson, K., *The Singer story*, Godmanston 1996

Bailey, P., *Leisure and class in Victorian England*, London 1978

Barker, R., *The Schneider Trophy races*, London 1971

——— *The Royal Flying Corps in France: from Mons to the Somme*, London 1994

Barnes, C. H., *Handley Page aircraft since 1907*, London 1976

——— *Bristol aircraft since 1910*, London 1988

——— *Shorts aircraft since 1900*, London 1989

Bartleet, H. W., *Bartleet's bicycle book*, London 1931

Best, G., *Humanity in warfare: the modern history of the international law of armed conflicts*, London 1983

Boddy, W., *The story of Brooklands: the world's first motor course*, 3 vols, London 1948–50

Bond, B. (ed.), *Victorian military campaigns*, London 1967

Bowyer, C., *Handley Page bombers of the First World War*, Bourne End 1992

Brabazon of Tara, Lord, *Forty years of flight*, Oxford 1949

——— *The internal combustion engine and its effects*, London 1962

Bramson, A., *Pure luck: the authorised biography of Sir Thomas Sopwith, 1888–1989*, Wellingborough 1990

Brett, R. D., *The history of British aviation 1908–1914*, London 1933

Broke-Smith, P. W. L., 'The history of early British military aeronautics', *Royal Engineers Journal* lxvi, lxvii (1952)

Broomfield, G. A., *Pioneer of the air: the life and times of Colonel S. F. Cody*, Aldershot 1953

Bruce, E. S., 'The balloon work of the late Mr Henry Coxwell', *Journal of the Aeronautical Society* iv (1900), 118–20

Bruce, G., 'Shorts, origins and growth: the sixteenth Short brothers commemorative lecture' (1977), RAeS archive

———— 'Shorts aircraft: some new evidence on the early years to 1912' (1977), RAeS archive

———— *C. S. Rolls: pioneer aviator*, Monmouth 1978

———— *Charlie Rolls: pioneer aviator*, Derby 1990

Bruce, J. M., 'A history of Martinsyde aircraft', Journal of the Royal Aeronautical Society lxxii (1968), 755–770

———— *The aeroplanes of the Royal Flying Corps (Military Wing)*, London 1982

Bullen, R. J., H. Pogge von Strandmann and A. B. Polonsky (eds), *Ideas into politics: aspects of European history 1880–1950 (essays in honour of James Joll)*, London 1984

Bulman, Maj. G. P., 'Captain F. S. Barnwell: the 1st Frank Barnwell memorial lecture', *Journal of the Royal Aeronautical Society* lviii (1954), 382–95

Casey, L. S., *Curtiss: the Hammondsport era 1907–1915*, New York 1981

Caunter, C. F., *The history and development of cycles as illustrated by the collection in the Science Museum*, London 1955

———— *Motor cycles: an historical survey*, London 1982

Chadeau, E., *L'Industrie aeronautique en France 1900–1955*, Paris 1987

Chilton, E., 'Rear Admiral Sir Murray Sueter, CB', *Cross and Cockade Journal* xv (1984), 49–55

Christienne, C. and P. Lissarrague, *A history of French military aviation*, Washington, DC 1986

Churchill, R. S., *Winston S. Churchill: young statesman 1901–1914*, London 1967

Clare, P., 'The Edwardians and the constitution', in D. Read (ed.), *Edwardian England*, London 1972.

Clark, C. S., *The Lanchester legacy, a trilogy of Lanchester works: volume one, 1895–1931*, Coventry 1995

Clew, J., *JAP: the vintage years*, Yeovil 1985

Coleman, D. C., 'Gentlemen and players', *Economic History Review* xxvi (1973), 92–116

Combs, H. B., *Kill Devil Hill: the epic of the Wright brothers*, London 1980

Cooper, M., *The birth of independent air power: British air policy in the First World War*, London 1986

Corn, J. J., *The winged gospel: America's romance with aviation 1900–1950*, New York 1983

Cross, Capt. J. R., 'British military ballooning', *Army Air Corps Journal* (1980), 27–34

Crouch, T. D., *A dream of wings: Americans and the airplane 1875–1905*, New York 1981

—— *The eagle aloft: two centuries of the balloon in America*, Washington, DC 1983

—— *The bishop's boys: a life of Wilbur and Orville Wright*, New York 1989

Cullingham, G., *Patrick Y. Alexander: patron and pioneer of aeronautics*, Bath 1984

Dictionary of Business Biography, London 1984–6

Dictionary of National Biography, London–Oxford 1892–1990

Dictionnaire de Biographie Française, Paris 1929–1995

Donne, M., *Pioneers of the skies: a history of Short Brothers plc*, Belfast 1987

Edgerton, D., *England and the aeroplane: an essay on a militant and technological nation*, Basingstoke 1991

Facon, P., 'L'Armée Française et l'aviation 1891–1914', *Revue Historique Des Armées* clxiv (1986), 77–88

—— and others, *French military aviation: a bibliographical guide*, New York 1989

Fearon, P., 'The formative years of the British aircraft industry, 1913–1924', *Business History Review* xliii (1969), 476–95

—— 'The vicissitudes of a British aircraft company: Handley Page Ltd between the wars', *Business History* xx (1978), 63–86

Fischer, F., *Germany's aims in the First World War*, London 1967

Fletcher, C. R. L. and R. Kipling, *A history of England*, London 1911

Fletcher, J. (ed.), *The Lanchester legacy, a trilogy of Lanchester works: volume three, a celebration of genius*, Coventry 1996

Floud, R., *The British machine tool industry, 1850–1914*, Cambridge 1976

Folland, E. H. S., 'The life and work of H. P. Folland', *Aerospace* i (1974), 12–20

Foreman-Peck, J., 'Diversification and the growth of the firm: the Rover company to 1914', *Business History* xxv (1983), 179–92

Fraser, D., 'The Edwardian city', in D. Read (ed.), *Edwardian England*, London 1972

Fraser, P., *Lord Esher: a political biography*, London 1973

Freedman, L., *The evolution of nuclear strategy*, London 1981

French, D., *British economic and strategic planning, 1905–1915*, London 1982

Fuller, J. F. C., *The conduct of war 1789–1961*, London 1961

Gamble, C. F. Snowden, *The air weapon: being some account of the growth of British military aeronautics*, London 1931

Gardner, C. (ed.), *Fifty years of Brooklands*, London 1956

George, A. D., 'Aviation and the state: the Grahame-White Aviation Company, 1912–23', *Journal of Transport History* ix (1988), 209–13

Gibbs-Smith, C. H., *A directory and nomenclature of the first aeroplanes, 1809 to 1909*, London 1966

—— *The invention of the aeroplane, 1809–1909*, London 1966

Powers, B. D., *Strategy without slide-rule: British air strategy 1914–1939*, London 1976

Pritchard, J. Laurence, 'Sir Francis Kennedy McClean, AFC', *Journal of the Royal Aeronautical Society* lix (1955), 721–6

—————— 'Major B. F. S. Baden-Powell: an appreciation', *Journal of the Royal Aeronautical Society* lx (1956), 9–24

—————— 'Sir Alliott Verdon-Roe, 1877–1958', *Journal of the Royal Aeronautical Society* lxii (1958), 231–8

—————— and others, 'Mervyn O'Gorman, 1871–1958', *Journal of the Royal Aeronautical Society* lxii (1958), 469–75

—————— ' "H.P." Sir Frederick Handley Page, C.B.E.', *Journal of the Royal Aeronautical Society* lxvi (1962), 737–42

Pudney, J. S., *Laboratory of the air: an account of the Royal Aircraft Establishment of the Ministry of Supply Farnborough*, London 1948

Rae, J. B., 'The engineer-entrepreneur in the American automobile industry', *Explorations in Entrepreneurial History* viii (1955), 1–11

—————— 'Financial problems of the American aircraft industry 1906–1940', *Business History Review* xxxix (1965), 99–114

Read, D. (ed.), *Edwardian England*, London 1972

Reader, W. J., *Architect of air power: the life of the first Viscount Weir*, London 1968

Richardson, K., *The British motor industry 1896–1939*, London 1977

Roberts, D., *The invention of the safety bicycle*, Mitcham 1990

—————— *Cycling history: myths and queries*, Birmingham 1991

Robertson, B., *Sopwith: the man and his aircraft*, Letchworth 1969

Robinson, B. R., *Aviation in Manchester: a short history*, Manchester 1977

Robinson, D. H., *Giants in the sky: a history of the rigid airship*, Henley-on-Thames 1973

Robson, B., 'Mounting an expedition: Sir Gerald Graham's 1885 expedition to Suakin', *Small Wars and Insurgencies* ii (1991), 232–9

—————— *Fuzzy-wuzzy: the campaigns in the Eastern Sudan 1884–85*, Tunbridge Wells 1993

Robson, G., *The Rover story*, Cambridge 1977

Roe, A. V., 'Trials, troubles and triplanes', *The Aero* vi (1912), 95–8

Roe, G. V., 'Was Roe first?: the grounds for appeal', *Flypast* (January 1989), 18–21

Rolt, L. T. C., *The Dowty story*, London 1962

—————— *Victorian engineering*, London 1974

—————— *The aeronauts: a history of ballooning 1783–1903*, Gloucester 1985

Roseberry, C. R., *Glenn Curtiss: pioneer of flight*, Syracuse 1991

Roskill, S., *Hankey, man of secrets: I, 1877–1918*, London 1970

Rubinstein, D., 'Cycling in the 1890s', *Victorian Studies* xxi (1977), 47–71

Russell, G. C. D., 'The pioneers, God bless 'em', *Journal of the Royal Aeronautical Society* lxx (1966), 140–2

Sanderson, M., 'The University of London and industrial progress, 1880–1914', *Journal of Contemporary History* vii (1972), 243–62

——— *The universities and British industry 1850–1970*, London 1972

Saul, S. B., 'The American impact on British industry', *Business History* iii (1960), 19–38

——— 'The motor industry in Britain to 1914', *Business History* v (1962), 22–44

Scott, J. D., *Vickers: a history*, London 1962

Sears, S. W., *To the gates of Richmond: the Peninsula campaign*, New York 1992

Sharp, C. M., *DH: a history of de Havilland*, London 1960

Sharp, G., *The siege of Ladysmith*, London 1976

Shaw, W. H. and O. Ruhen, *Lawrence Hargrave: explorer, inventor and aviation experimenter*, New South Wales 1977

Short Brothers Ltd, *Seventy-five years of powered dynamic flight*, Belfast 1978

Simms Motor Units Ltd, *The Simms century: F. R. Simms 1863–1944*, privately published 1963

Smith, C. B., *Testing time: a study of man and machine in the test-flying era*, London 1961

Sommer, D., *Haldane of Cloan: his life and times 1856–1928*, London 1960

Spaight, J. M., *The beginnings of organised air power*, London 1927

Spiers, E. M., *Haldane: an army reformer*, Edinburgh 1980

Spratt Bowring, F. T. N., 'The work of the RE in the China or "Boxer" War of 1900–1901', *Royal Engineers Journal* xiii (1911), 169–88

Squires, J. D., 'Aeronautics in the Civil War', *American Historical Review* xlii (1936), 652–69

Stansbury Haydon, F., *Aeronautics in the Union and Confederate Armies*, Baltimore 1941

Starley, W., 'The evolution of the cycle', *Journal of the Society of Arts* xlvi (1898), 601–16

——— *The life and inventions of James Starley*, Coventry 1902

Studer, C., *Sky storming Yankee: the life of Glenn Curtiss*, New York 1937, repr. 1972

Tagg, A. E., *Power for the pioneers: the Green and ENV aero engines*, Newport 1990

Tapper, O., *Armstrong Whitworth aircraft since 1913*, London 1973

Taylor, A. J. P., *English history 1914–1945*, Oxford 1965

Taylor, H. A., *Jix, Viscount Brentford: being the authoritative biography of the Rt Hon. William Joynson-Hicks*, London 1933

Taylor, H. A., *Fairey aircraft since 1915*, London 1974

Taylor, J. W. R., *CFS: birthplace of air power*, London 1987

Taylor, M. J. H., *Shorts: the planemakers*, London 1984

Thetford, O., *British naval aircraft since 1912*, London 1988

Thompson, R., *The Royal Flying Corps*, London 1968

Thoms, D. and T. Donnelly, *The motor car industry in Coventry since the 1890s*, London 1985

Thurstan, J., 'Charles Grey and his pungent pen: personal recollections of a great aeronautical journalist', *Journal of the Royal Aeronautical Society* lxxiii (1969), 839–52

Till, G., *Air power and the Royal Navy 1914–1945*, London 1979

Tobin, G. A., 'The bicycle boom of the 1890s', *Journal of Popular Culture* vii (1974), 838–49

Trebilcock, C., 'British armaments and European industrialisation 1890–1914', *Economic History Review* xxvi (1973), 254–72

────── *The Vickers brothers: armaments and enterprise 1854–1914*, London 1977

Tredrey, F. D., *Pioneer pilot: the great Smith Barry who taught the world to fly*, London 1976

Tritton, P., *John Montagu of Beaulieu: motoring pioneer and prophet*, London 1985

Troubridge, Lady and A. Marshall, *John Lord Montagu of Beaulieu: a memoir*, London 1930

Tuchman, B. W., *The proud tower: a portrait of the world before the war, 1890–1914*, London 1966

Turner, Maj. C. C., *The old flying days*, London 1927

Venables, R., 'The history of the Blackburn Company', *Motor Sport* xxiv (1948), 69–71

Villard, H. S., *Contact!: the story of the early birds*, Washington, DC 1987

────── *Blue ribbon of the air: the Gordon Bennett races*, Washington, DC 1987

Walker, P. B., *Early aviation at Farnborough: the history of the Royal Aircraft Establishment*, 2 vols, London 1971–4

Wallace, G., *Flying witness: Harry Harper and the golden age of aviation*, London 1958

────── *Claude Grahame-White: a biography*, London 1960

Waller, P. J., *Town, city and nation: England 1850–1914*, Oxford 1983

Ward, D., 'The public schools and industry in Britain after 1870', *Journal of Contemporary History* ii (1967), 37–52

Waters, Maj. R. S., 'Ballooning in the French army during the revolutionary wars', *Army Quarterly* xxiii (1932), 327–40

Watt, D. C., 'Restraints on war in the air before 1945', in M. Howard (ed.), *Restraints on war: studies in the limitation of armed conflict*, Oxford 1979

White, G., *Tramlines to the stars: George White of Bristol*, Bristol 1995

Wiener, M. J., *English culture and the decline of the industrial spirit 1850–1980*, Cambridge 1981

Williams, W. W., *The life of General Sir Charles Warren*, Oxford 1941

Williamson, G., *Wheels within wheels: the story of the Starleys of Coventry*, London 1966

Wilson, C. H. and W. J. Reader, *Men and machines: a history of D. Napier & Son (Engineers) Ltd 1808–1958*, London 1958

Wilson, H. W., *With the flag to Pretoria: a history of the Boer War of 1899–1900*, 2 vols, London 1900–1

Wright, P., *The Royal Flying Corps 1912–1918 in Oxfordshire*, Oxford 1985

Unpublished works

Gibbs-Smith, C., 'Shorts connection with the Wright brothers, 1908–1909', memo, RAeS archive n.d.

Munro, R. L., 'Flying shadow: Captain Bertram Dickson', unpubl. manuscript 1991, Museum of Army Flying

Harrison, A. E., 'Growth, entrepreneurship and capital formation in the UK cycle and related industries, 1870–1914', unpubl. PhD diss., York 1977

Scrope, H. E., 'Golden wings: 50 years of aviation by the Vickers group of companies', unpubl. manuscript 1960

Waller, Col. S., 'History of the Royal Engineer operations in South Africa 1899–1902', unpubl. manuscript 1904, RE Library

Index